Prematurity in Scientific Discovery

Prematurity in Scientific Discovery

On Resistance and Neglect

EDITED BY

Ernest B. Hook

UNIVERSITY OF CALIFORNIA PRESS

Berkeley   Los Angeles   London

University of California Press
Berkeley and Los Angeles, California

University of California Press, Ltd.
London, England

© 2002 by the Regents of the University of California

Library of Congress Cataloging-in-Publication Data

Hook, Ernest B.
　Prematurity in scientific discovery : on resistance and neglect. Edited by
Ernest B. Hook.
　　p.　　cm.
　Includes bibliographical references and index.
　ISBN 0–520–23106–6 (Cloth : alk. paper).
　1. Discoveries in science—Congresses.　2. Science—Philosophy.
3. Stent, Gunther Siegmund, 1924–.　I. Title.

Q180.55.D57 H66　2002
DeweyNumber—dc21
　　　　　　　　　　　　　　　　　　　　　2001004245

Manufactured in the United States of America
10　09　08　07　06　05　04　03
10　9　8　7　6　5　4　3　2　1

The paper used in this publication is both acid-free and totally chlorine-
free (TCF). It meets the minimum requirements of ANSI/NISO
Z39.48–1992 (R 1997) (*Permanence of Paper*).

For Gunther Stent, of course

CONTENTS

FIGURES AND TABLES

FIGURES

TABLES

PREFACE

This volume originated from a symposium by the same title held at the University of California at Berkeley, 5–7 December 1997. Gunther S. Stent had outlined his concept of prematurity in two articles in the early 1970s. The better-known but condensed version appeared in the December 1972 *Scientific American*. An expanded version appeared the same year in *Advances in the Biosciences*, a series not widely circulated and now discontinued. These articles, both titled "Prematurity and Uniqueness in Scientific Discovery," addressed, in addition to prematurity, the potential aesthetic content and uniqueness of scientific work. Material from the expanded version that is pertinent to prematurity appears as chapter 2 of this volume.

The value of a conference on this topic became clear after a presentation on 26 January 1994 by Gunther Stent titled "Can Scientific Discovery Be Premature? The Case of DNA, 1944–1953" at the fifth of an ongoing interdisciplinary seminar series, Patterns of Discovery in the Sciences, at the University of California at Berkeley. His thesis on prematurity then appeared relatively unknown, although it had been published almost twenty years earlier. I found it to be of great heuristic attraction, as did many working scientists with whom I have since discussed it. In chapter 1 of this volume, I note in particular why Stent's formulation may appear more attractive than Thomas Kuhn's (on differences in paradigms) in explaining why some scientific work is ignored, and how, in contrast to Kuhn's, Stent's has actual utility for science policy, both social and personal. This chapter, as well as chapters by Lawrence Stern and others, addresses possible reasons for the relative neglect of Stent's thesis by historians and philosophers of science and its appeal in particular to scientists among those relatively few who have been aware of it.

In seeking contributors to the conference and this volume, which emanated from it, I circulated copies of both of the Stent essays and a draft of my introductory chapter. I asked potential participants—scientists, philosophers, historians, sociologists, and one political scientist—to develop their thoughts stimulated by these discussions of prematurity in science. Many declined, but fortunately a sufficient number agreed to attend and present a paper, enabling a productive interdisciplinary interchange over a very full two days and more. The wide range of views, comments, and responses herein from individuals in vastly different fields illustrate some marked differences in interpretation of what Stent was getting at. This variation in focus and understanding of his formulation—which indeed had some ambiguities, as discussed in particular in the last chapter of the volume and by Lawrence Stern in chapter 18—reflects in an interesting way the different perspectives of creative scientists and perceptive metascientists upon what to many clearly is a slippery notion. Yet each contributor considers factors associated in some manner with "hang-ups" in science, as one author termed them.

The chapters by the physical scientists Glenn Seaborg and Charles Townes, while not invoking Stent's concept of prematurity explicitly, touch on several fascinating historical exemplars of the notion. Their accounts, as well as that of the geneticist Norton Zinder, illustrate in a striking manner how three eminent researchers in basic science, toward the end of lengthy, productive careers, view their own outstanding work and that of others in the light of Stent's original papers.

Many authors modified their papers as a consequence of discussion at the conference. Martin Jones and Elihu Gerson wrote greatly expanded versions of their comments. Lawrence Stern's scientometric analysis of Stent's *Scientific American* paper and my own summary of Ida Noddack's "premature" suggestion of nuclear fission were written after the meeting specifically for the volume.

As Stern's analysis was not presented at the conference and was completed sometime afterward, the other contributors could not comment explicitly upon his results or interpretations. Fortunately, his quantitative evaluation confirms the rough subjective impressions of the differential impact of Stent's papers that I note in chapter 1, to which others allude elsewhere in their own chapters, so I have not asked them to modify their essays as a consequence.

In the last part of the volume especially, one will find criticisms of Stent's notion and attempts to alter or extend it, some more sympathetic than others. Fortunately, in his comments in chapter 24, Gunther Stent engages some of these. I thought it important to conclude the volume with some further defense of his core notion from at least some critiques, implicit or explicit, that he had not addressed, and to reemphasize its potential utility. To synthesize the variety of opinions expressed and take them into account while fully developing the notion of prematurity, or even to respond in detail to the different viewpoints, would require another volume.

Whatever the differences among the contributors, or whatever one's views of prematurity as Stent defined it or as modified by some herein, at the very least Stent's concept has provoked an extensive interdisciplinary exchange and reflection on scientific discovery processes. I hope this volume will be useful not only because it presents that variety of views but because it clarifies some of the multiple ways in which scientific knowledge expands.

Ernest B. Hook
Berkeley, California

ACKNOWLEDGMENTS

Kenneth Carpenter, Ed Hackett, Roger Hahn, Horace Judson, Elizabeth Lloyd, and Gunther Stent, among many others, provided helpful advice on the planning of or arrangements for the conference from which this volume emerged. Patricia Podzorski made valued and valuable efforts to assist in securing support from the National Science Foundation that assured funds at the planning stage of the conference. She, Johanna Weber, and Jessica Madarasz assisted ably in the preplanning steps, and the latter two supervised the handling of the mind-numbing logistical details before and at the symposium itself. Noreen Killorin Hook donated large amounts of time and energy, which contributed to its success. Wayne Wu and Buen Ortiz helpfully assisted with editing chores and provided comments and suggestions that improved the final product.

Most of all, of course, I am indebted to Gunther Stent not just for starting the hare of prematurity almost three decades ago but also for his affable, good-humored, and somewhat bemused cooperation throughout the long process that stretched over a number of years and resulted in this volume.

The material in chapter 2, "Prematurity in Scientific Discovery" by Gunther Stent, originally appeared in "Prematurity and Uniqueness in Scientific Discovery," *Advances in the Biosciences* 8 (1972): 433–40, and paragraphs 3 and 4 on p. 446, paragraphs 1–4 on p. 447, and paragraph 1 on p. 448, copyright 1972, and is reprinted by permission from Elsevier Science.

Chapter 6, "Scotoma: Forgetting and Neglect in Science" by Oliver Sacks, is an edited and shortened version of an article by the same title that appeared in *Hidden Histories of Science*, ed. Robert B. Silvers (New York: New York Review of Books, 1995), pp. 141–90, copyright 1995, and appears by permission of Oliver Sacks.

Figures 1 and 2 of Glenn Seaborg's chapter were kindly provided by, and are used with permission of, the Lawrence Berkeley National Laboratory.

Kenneth J. Carpenter, Professor Emeritus, Department of Nutritional Sciences, University of California, Berkeley

Nathaniel C. Comfort, Deputy Director, Center for the History of Recent Science, and Associate Professor, Department of History, George Washington University, Washington, D.C.

Elihu M. Gerson, Director, Tremont Research Institute, San Francisco, California

Michael T. Ghiselin, Chair, Center for the History and Philosophy of Science, California Academy of Sciences, San Francisco

William Glen, Editor-at-Large, Stanford University Press, and Visiting Scientist-Historian, U.S. Geological Survey, Menlo Park, California

Norriss S. Hetherington, Research Associate, Office for the History of Science and Technology, University of California, Berkeley

Frederic L. Holmes, Professor, Department of the History of Medicine, Yale University, New Haven, Connecticut

Ernest B. Hook, Professor, School of Public Health, University of California, Berkeley, and Department of Pediatrics, University of California, San Francisco

David L. Hull, Professor, Department of Philosophy, Northwestern University, Evanston, Illinois

Martin Jones, Associate Professor, Department of Philosophy, Oberlin College, Oberlin, Ohio

Ilana Löwy, Historian of Biomedical Science, INSERM-SERES, Paris, France

Arno G. Motulsky, Professor Emeritus, Departments of Medicine and Genetics, University of Washington, Seattle

Gonzalo Munévar, Professor of Humanities and Social Sciences, Lawrence Technological University, Southfield, Michigan

Mary Jo Nye, Horning Professor of the Humanities and Professor of History, Oregon State University, Corvallis

Michael Ruse, Lucyle T. Werkmeister Professor of Philosophy, Department of Philosophy, Florida State University, Tallahassee

Oliver Sacks, neurologist and author, New York, New York

Glenn T. Seaborg, Professor Emeritus, Department of Chemistry, and Associate Director, Lawrence Hall of Science, University of California at Berkeley (deceased)

Gunther S. Stent, Professor Emeritus, Department of Molecular and Cell Biology, University of California, Berkeley

Lawrence H. Stern, Professor, Department of Sociology, Collin County Community College, Plano, Texas

Charles H. Townes, University Professor Emeritus, Department of Physics, University of California, Berkeley

George Von der Muhll, Professor Emeritus, Department of Politics, University of California, Santa Cruz

Norton D. Zinder, John D. Rockefeller II Professor Emeritus, Laboratory of Genetics, Rockefeller University, New York, New York

PART ONE

Introduction

CHAPTER ONE

A Background to Prematurity and Resistance to "Discovery"

Ernest B. Hook

Scientists and historians can cite many cases of scientific and technological claims, hypotheses, and proposals that, viewed in retrospect, have apparently taken an unaccountably long time to be recognized, endorsed, or integrated into accepted knowledge and practice.[1] Indeed, some have had to await independent formulation. While some frequently cited cases, such as the particulate theory of heredity attributed to Gregor Mendel, are, on closer examination, somewhat problematic exemplars of the thesis, there are, I contend, many clear examples of what may be termed, for want of a better term, "delay."[2]

Such delay, of course, inhibits and even may deny contemporary or posthumous recognition of individual achievement. More important, considerable social loss

---

1. Barber (1961) enumerates some of these and discusses some underlying factors. Some earlier references and commentary on related issues not cited by Barber appear in Stern 1941. See also Stern 1927.

2. Most geneticists believe Mendel enunciated a particulate theory of heredity. This includes Gunther Stent, whose classification of it as premature I discuss below. But the implied historical context of his efforts has been challenged by, among others, Olby 1979 and Brannigan 1981. In their view Mendel was retrospectively recognized for a position he did not enunciate or accept himself, and he regarded his own work, as did others of the time, as simply continuing in the tradition of nineteenth-century hybridists. Olby's suggestion revolves mainly around Mendel's denotation of—what we would today term—homozygotes with a single symbol, i.e., as "a" or "A" rather than "aa" or "AA," as done currently to designate two alleles. Ernst Mayr grants that Mendel did not have a clear picture of pairs of alleles that separate during gamete formation. But he maintains that Mendel's discovery of segregation (at least of different alleles), constant ratios, and independent assortment of characters well warrants the general view of him. This would support the view that Mendel's work is emblematic of Stent's concept of prematurity. See Mayr 1982, pp. 710, 726. Sandler and Sandler 1985 emphasize the importance of Mendel's recognition that "events of transmission could be detached from development and studied separately." For further discussion, see Sandler 2000; and Holmes, chapter 12, and Stent, chapter 24, in this volume.

may result as a consequence.[3] Among the most poignant examples are those from the medical field: for instance, Ignaz Semmelweis's report on the virtues of adoption of hand washing by physicians to prevent puerperal fever, sepsis, and death among women delivered of newborns; John Snow's association of a cholera epidemic with a water source; and Humphrey Davy's report on the pain-ameliorating effects of nitrous oxide and suggestion that the gas might be useful for pain relief during surgery.[4] The first two of these are now regarded as classics in public health. None had immediate impact when first proposed.

Identification and dissection of the factors that contribute to "delay" are not only of interest to scientists, historians, philosophers, and sociologists. Their recognition may also lead to useful scientific and personal practices and be of value to those making science and technology policy.

RESISTANCE AND REJECTION

One may classify at least five grounds on which scientific claims or hypotheses—even those later achieving widespread recognition or endorsement—may be rejected at first offering. In addition to prematurity (defined further below), investigators may reject or choose not to follow up on a scientific report or hypothesis because (1) they are unaware of it, (2) having reviewed it, they judge it to be of no immediate relevance to their current work and thereafter ignore it, (3) they harbor inappropriate prejudice against some aspect of the claim or its proponent, or (4) it appears to clash directly with their observation or experience—for instance, it is based on an experimental finding they cannot replicate.

---

3. In this volume, Seaborg (in chap. 3) and I (in chap. 10) discuss an episode that illustrates one possible counterexample to this contention. Had Noddack's correct interpretation of nuclear fission as an explanation of the findings interpreted by Fermi's group as evidence of transuranium elements been taken seriously by those in a position to confirm it, the Nazis might well have had an atomic bomb early in World War 2 (Noddack 1934; Fermi 1934). See also Hull, chapter 22 in this volume.

4. For traditional views on Semmelweis see Ackerknecht, 1982 (pp. 187–88). The summary by Carter 1983 (pp. 3–58) suggests that initial local resistance to Semmelweis's views was to some extent exacerbated by a power struggle within the medical school at Vienna to which he was affiliated (pp. 22, 42). Furthermore, and more important, Semmelweis insisted strongly that every recognized case of puerperal fever was due to decaying organic matter. This obviously did not apply to the disease as then defined (more broadly than today) (p. 43). As I read the history of this episode, his monolithic attitude appears to have diverted attention from, and/or led to resistance to, the adoption of hand washing by physicians before examination of each patient. There might have been a more congenial response had he argued that this practice would affect only some fraction of mortality. On the other hand, a more tactful suggestion in the early 1840s, by that most congenial of physicians, Oliver Wendell Holmes, was not followed up either (Holmes 1842–43). For further discussion and references, see Mettler 1947, pp. 964–72, 975–77. On Snow's cholera study, see Stern 1941, pp. 205–6. See also Stern 1927, pp. 11–96. According to one view, Snow's work was attacked because his theory of waterborne disease undermined the notion that certain offensive trades caused cholera. On Humphrey Davy's report on nitrous oxide, see Bergman 1998, p. 277. See also Hook, chapter 25 in this volume.

Lack of awareness of a claim or hypothesis may derive from at least four causes, all ameliorable, or avoidable by the proponent. It may stem from the limitation of publication to a language other than the lingua franca of the period.[5] It may stem from presentation of the work in difficult-to-comprehend and/or atypical, unfamiliar prose or terminology.[6] And it may stem from publication in an article or monograph focused primarily on other events and/or in an obscure outlet read by an audience highly unlikely to appreciate the work.[7] One may readily suggest to individual investigators, as well as those influencing the policies of the scientific community, strategies to overcome these barriers. Strategies pertinent to science policy, some of which have been partially adopted in the United States, include providing readily available translations in the lingua franca and indexing abstracts and key words of as many publications as practical, even those that appear in obscure sources, and so on. On the investigator's part, it may not be sufficient merely to exposit a claim, hypothesis, or proposal in a well-recognized professional outlet and in a clear or cogent style bereft of hyperbole likely to encourage either incomprehension or suspicion.[8] The investigator may well have to follow up on it and publish further, not only

5. Tobias 1996, for instance, cites the case of *Grundzuge einer Theorie von Phylogeteischen Systematik* by Willi Hennig, published in 1950. He reports that it had hardly any influence until translated in 1966 as *Phylogenetic Systematics*. Thereafter it had a marked impact and, in Tobias's view, provided for many researchers "virtually the only acceptable way to construct phylogenies." Tobias provides a somewhat different conceptual approach to resistance or delay of discovery, with a number of interesting examples, especially from paleoanthropology.

6. A possible example in microbial genetics is the work of C. H. Browning (1908), about which Zuckerman and Lederberg note, "In principle the investigation of [of bacterial recombination] was technically feasible by 1908 as demonstrated by Browning's use of drug resistance as a selective marker"; however, not only did Browning report a negative finding, but he "used terminology not readily transferable to the case of bacterial recombination" (Zuckerman and Lederberg 1986). The latter was discovered by Lederberg and Tatum in 1946. In mathematics Evariste Galois's insistence upon a constrained and somewhat idiosyncratic style delayed recognition of his significant achievements. Joseph Liouville (in *Journal de mathematiques pures et appliquées* 1846, cited by Singh 1997, p. 249) commented on Galois's "exaggerated desire for conciseness" in mathematical exposition.

7. That William Charles Wells's anticipation in 1813 of Darwin's and Wallace's theory of natural selection had little if any impact may be in part because it was mentioned in an article devoted primarily to other issues and its title gave no clue to the hypothesis within. See Shryock 1966; and Ghiselin, chapter 16 in this volume. Another example is the paper by Grosse (1935) about which Edoardo Amaldi, a collaborator of Enrico Fermi, says, "Unfortunately this paper was not read in Rome neither, I believe in Berlin, Paris nor any other place where the experiments on the "transuranic elements" [those later recognized as pseudo elements] were carried out during the 30's. We learned about the existence of this paper only years later" (Amaldi 1989, p. 17). Grosse pointed out that elements 93 and 94 could either belong to a second group of rare elements that Bohr had proposed earlier, as is now known to be correct, or (as Grosse did say "seems more probable") be homologues of manganese and iron in the main body of the table, as the rest of the scientific community, including apparently Bohr himself, presumed without question. It was the latter belief that kept Emilio Segrè (1939) from recognizing he had truly been the first to isolate the real element 93 and not a fission product. (As he wrote later, "I had it [element 93] in my hands and did not recognize it" [1993, p. 153].) See also Seaborg, chapter 3, and Hook, chapter 10, in this volume.

8. As this implies, the rhetoric and method of presentation of a discovery may affect its reception by those aware of it and to whom it is presumably pertinent. George M. Gould (1901) cites a number

to call his or her work to the attention of those who overlooked or ignored it but also to convince the scientific community that in fact he or she has been able to replicate the work or has found further grounds to regard a hypothesis as important.

Less readily overcome obstruction may stem from strong social forces—religious, ideological, political, and economic—that lead to challenge, rejection, or suppression.[9] In *practice,* the only remedy may be to seek expression and circulation of the unrecognized, inhibited, or suppressed ideas, proposals, and inventions in areas and social climates where the prohibitive factors do not reign. But in principle, in an enlightened society one may suggest some goals, some general social solutions to overcome the barriers. As obvious as these may be, I believe it worthwhile to list some of them: limitation of economic suppression of new inventions or useful technology, encouragement of ideological tolerance, opposition to implacable doctrinaire social forces, and most important tactically, attempts to disconnect the apparent implications of scientific discoveries from the feared ideological consequences.

Factors related to but distinct from the more global social forces concern resistance at the *individual* level. New scientific and technical discoveries may threaten not one's economic welfare or ideological persuasion but rather the "psychic capital" invested in current scientific views—some involving one's own work—challenged implicitly or explicitly by a new report.[10] Of course, the longer one has held views and invested energy in them, the more reluctant one may be to alter them.

---

of alleged historical medical and scientific examples of resistance—e.g., Semmelweis and puerperal sepsis—and provides a discussion of general predisposing factors. But the tenor of Gould's subsequent commentary illustrates why many may have objected to scientific and medical doctrines later widely subsequently adopted. After citing accepted historical examples, he attempts to put his own work in the same context of "unaccepted truth." He goes into extensive hyperbolic detail and denunciation of those who object to his views about the use of glasses to cure the ills of eyestrain and abrogate its consequent debilitating effects on the body. He concludes that "the self assumed [*sic*] leaders who oppose new truth are criminals[;] . . . when [they] . . . do this it is plain homicide, murder at least in the second degree," conflating the alleged consequences of untreated or unrecognized eyestrain with preventable deaths from puerperal sepsis. He may have intended his remarks to frighten readers into accepting his views, but succeeded in repelling many who found his comments merely the rhetoric of a crank to be ignored.

9. One may cite, among many obvious examples of intrusion of ideology, the resistance in the former Soviet Union to certain scientific theories because of their apparent discordance with Marxist-Leninist dialectical materialism. Many view the rejection of Mendelian genetics in favor of Lysenko's doctrines as such an example. Louis Althusser (1977, p. 13) writes, for instance, that "the official [Soviet] version of dialectical materialism *guaranteed* Lysenko's theories, while these theories in turn served to '*verify*' this official version and to strengthen it" (emphasis in the original). Close historical reexamination, however, suggests that this episode is more complex. Many local factors also contributed to the state's reaction (Graham 1972; Krementsov 1997). A better, i.e., "purer," example of rejection on ideological grounds is the Soviet condemnation of Linus Pauling's resonance theory of the chemical bond because it was *viewed* (and feared) as idealistic and in conflict with a determinist Marxist-Leninist ideology (Graham 1972, pp. 297–323, 530–37; Goertzel and Goertzel 1995, pp. 118–19). For an example of claimed suppression for economic reasons in communications technology, see Wiener 1993.

10. For a forceful example of the consequence of obstruction by a single powerful authority whose psychic capital was apparently jeopardized by a new proposal, see Kameshar Wali (1991), who describes how in Subrahmanyan Chandrasekhar's view, Arthur Eddington delayed developments in theoretical

This inevitably results in conceptual inertia that some have associated with aging.[11] And ranker reasons than those produced by hardening of cerebral arteries or of scientific beliefs may arise from the prejudices of culture, nation, gender, ethnicity, or race.[12]

All these sources of resistance to discovery originate in what some have termed the "externalist" factors influencing science.[13] And for all the above factors, one may, in principle, suggest some types of science policies to address them. For instance, the review of work by referees without knowledge of its authors, as currently practiced by some journals, clearly diminishes effects of some types of prejudices that inappropriately inhibit publication.[14] Editors' close scrutiny of reviewers' judgments may enable them to distinguish opinions based on wounded psychic capital from legitimate methodological objections.

---

astronomy for two generations by his objections to Chandrasekhar's theory and essentially forced him to abandon a line of investigation. (Some of this is summarized in Sacks 1995, p. 184.) Threats to psychic capital may in part explain some of the objections raised to reports by Adrian M. Wenner and his colleagues that honey bees communicate about the location of food through mechanisms other than, or in addition to, the "dance" language proposed by Karl Von Frisch, work for which the latter shared a Nobel prize. (See Wenner and Wells 1990, esp. p. 399.) Wenner and Wells, admittedly partisan, analyze this interesting ongoing controversy in the context of reviewing general resistance to unpopular work. For an overview sympathetic to Wenner, see Veldink 1989. See also Wenner et al. 1991.

11. Thomas Nickles cites Einstein's poignant comments in his memorial to Paul Ehrenfast following the latter's suicide in 1933, in which he stated that, added to Ehrenfast's personal and professional conflicts "was the difficulty of adaptation to new thoughts which always confronts the man past fifty. I do not know how many readers of these lines will be capable of fully grasping that tragedy" (in Albert Einstein, *Out of My Later Years*, 1956, cited in Nickles 1980, p. 42). According to Nickles, "This last, of course, was exactly what Werner Heisenberg and Max Born wrote of the older Einstein himself to explain Einstein's resistance to the ideas of the new quantum theory" (p. 42).

12. I have found it difficult to locate a blatant example in which such prejudice was *exclusively* responsible for rejection. One may find many episodes in which retrospectively those who found a work uncongenial or implausible on other grounds have pejoratively associated it with its disliked origin, leading to stronger psychic grounds for rejection. For instance, I suspect those Nazi physicists who dismissed relativity as "Jewish science" would almost certainly have dismissed it even if proposed by a Prussian Junker, albeit more respectfully. For another example like this, see Hook, chapter 10, this volume. Certainly there are numerous cases in which *attribution* of a particular well-accepted discovery has been affected inappropriately by chauvinistic concerns, perhaps most notoriously the dispute between partisans of Isaac Newton and Gottfried Leibniz as to the discovery of calculus.

13. For those not familiar with the term, it refers to factors extrinsic to the putative value-free application of the scientific method. Economic and/or social factors influencing scientific inquiry are externalist. This is opposed to an "internalist approach," which focuses on those aspects of scientific inquiry seen traditionally as free of values except for the search for truth. The image most scientists have of the ideal working of science is of course the latter. Concern with issues of acceptance of a theory based on replication, falsification, and so on may be regarded as primarily internalist, and concern with those of class and economic factors as primarily externalist. But as has been pointed out on many occasions, it is really not possible to separate these absolutely. See, for example, Nagel 1950, esp. p. 22.

14. I am aware of two that have adopted this policy: *Epidemiology* and the *American Journal of Public Health*. The report by Peters and Ceci (1982) provides some objective evidence of the value of blind review.

For factors intrinsic to the scientific process, however, one finds it more difficult to suggest appropriate strategies. For instance, an "internalist" ground such as failure to replicate a reported observation (see n. 13) constitutes an understandable reason for rejecting the claim and any theories based on it. Yet there are fascinating accounts of reports at first rejected because of widespread failure to replicate results, but which were later confirmed, after the realization that some taken-for-granted component had not been duplicated in the earlier attempts.[15] It is somewhat difficult to generalize from such cases, except to note the obvious need for attention to detail, as in animal physiology, where apparently "unimportant" factors such as the precise strains, age, and care of animals, and so on, may be critical.

PREMATURITY AS A CAUSE OF DELAY OR REJECTION

There remains, nevertheless, one residual factor that delays recognition of scientific discovery and (by extension) technological innovation, but for which one may argue there is *no* obvious solution, social or otherwise. The polymath Gunther Stent has denoted this as scientific prematurity.[16] A discovery is premature if it cannot be connected by a series of simple logical steps to canonical knowledge of the time. He suggests it is appropriate that the scientific community ignore (if not actually reject) work that is premature, until it can be so connected. In this view, delayed recognition of Mendel's laws, for instance, is the necessary price that society and scientists pay at the time to prevent being overwhelmed by useless cacophony.

A terminological issue arises here. One must distinguish claims not yet established from retroactively recognized discoveries.[17] Stent designates a claimed discovery not connected to canonical knowledge as an example of "here-and-now prematurity." One might argue that an announced finding or theory not yet accepted should be classified, for these purposes, not as a discovery but as a claim, hypothesis, proposal, potential discovery, or alleged discovery, or so forth, and retroactively, if found incorrect, as a pseudo discovery, failed claim, or false hypothesis. This perspective con-

15. Ludwik Gross's reports of the viral transmission of leukemia, published in 1951 and 1952, were rejected for a long period, and this led to questions of his probity. Then Jacob Furth repeated his experiments precisely, with the same strain, the same type of leukemia, and "absolutely" newborn mice. Furth's report of confirmation, and the medical community's widespread regard for him, as well as, one must suspect, appreciation of how previous attempts to replicate had failed, finally led to acceptance of Gross's work and of the broader concept of possible viral transmission of leukemia. Furth's reputation in the field led to a quicker reversal of opinion than would have occurred had the work been replicated by an unknown researcher. Nevertheless, the key point, which would have resulted in change sooner or later, was the recognition and report that replication required the precise age and strain of mice and type of leukemia studied by Gross. On the reception of Gross's work, see Kevles 1995, pp. 81–85; Klein 1990, p. 127 (cited in Kevles 1995, p. 111); and Bessis 1976 (cited in Kevles 1995, p. 110).

16. Stent 1972b. A significantly expanded but less well-known exposition appears in Stent 1972a. For the extracts of the latter, bearing on prematurity, see Stent, chapter 2 in this volume.

17. For more extensive comment on the need for this distinction in defending Stent's notion, see the next section and Hook, chapter 25 in this volume.

cedes that ambiguity may arise from usage of the term "discovery" in the context of Stent's discussion of prematurity, but it insists that any such ambiguity present on this account is irrelevant to the analysis of fundamental interest here and is readily avoidable with some simple terminological reexpression.

WHIGGISM AND OTHER POTENTIAL PROBLEMS

Stent's concept has been received more positively by scientists than historians or philosophers of science.[18] The responses of the latter may be explained in part by the views of one philosopher of science who wrote me in declining to participate in the conference from which this volume emanated. This individual found no value in Stent's formulation because

> the concept seems heuristic only with hindsight (because you have to know what came later to label something premature with respect to it). But hindsight is exactly what historians label "Whiggish." Although we might, from our own points of view, be interested in our precursors, these points of view cannot be used to interpret how the past unfolded (on pain of appeal to pernicious teleology). The same is true of evolutionary explanations.

As contributions elsewhere in this volume indicate, the fear of Whiggism—perhaps one might term it "Whiggophobia"—has concerned many who have considered prematurity. It appears worthwhile to consider this notion here in some detail.

Whiggism, or more appropriately, "the Whig interpretation of history," after the title of Herbert Butterfield's book that gave rise to this notion, is the tendency to write of the past from the perspective of those who "won." In England from 1688 on, in many senses the Whig position triumphed, hence the name. In Butterfield's original formulation Whiggism, or the "Whig interpretation," is the tendency "to praise revolutions provided they have been successful, to emphasize certain principles of progress in the past and to produce a story which is the ratification if not the glorification of the present."[19] Whiggism is understood as something to be deprecated, although even Butterfield indulged in it in a book about English history he wrote in the dark days of World War 2.[20] In the history of science and technology, for example, a tendency to ignore the extent and consequences of alchemy or astrology would be clearly Whiggish.

"Whiggism" is often used as a synonym for, and surprisingly often confused with, "presentism." The latter term refers simply to an approach to history from the perspective of the present, one that does not necessarily entail the affirmation

---

18. See Stern, chapter 18 in this volume, for quantitative evidence.

19. Butterfield [1931] 1965. See in particular p. v.

20. Butterfield 1944. But here he had a *political* goal: to rally his countrymen by emphasizing their "glorious" political heritage at risk from the Nazis, and not, I suspect, a genuine historical one. The volume was based on radio programs for the public.

of current views implied by Whiggism. A focus on past episodes whose examination may give some guidance to present concerns is presentist but need not be Whiggish. As Whiggism is understandably a pejorative term, and yet the best-known form of presentism—albeit a crude one—understandably its negative connotations have tended to tarnish any presentist approach or indeed any attempts to examine the past consciously in the light of present knowledge, such as those undertaken in some chapters that follow.[21]

These considerations explain my perplexity with those who, like the philosopher of science cited above, can write: "Hindsight is exactly what historians label 'Whiggish.'" Even should *any* historian do this, justification of such a label would depend critically on the nature of the "hindsight" and how it was applied.

Certainly there are other objections one might raise to Stent's concept of prematurity. Many are related to the notion of causality in history.[22] And one raised by an anonymous sympathetic reviewer of a proposal for this volume questioned whether the notion of prematurity is self-contradictory in the sense that, if someone or some group did discover something at some time, then it was possible to do so, and hence it was not premature. In this view, as actuality proves possibility, no discovery can ever be premature. Elsewhere in this volume, one will find expressed virtually the opposite view, that every "true" discovery is premature.[23] For reasons elaborated upon in chapters 24 and 25 of this volume, these extreme viewpoints almost by fiat prevent the notion of prematurity from doing any useful "work" for us in distinguishing one category of resistance to scientific claims, hypotheses, or proposals.

Some of these questions, fascinating as they are, revolve about implicit ambiguities in deep concepts such as "discovery" or "cause." But in any event, I suggest that, with some slight terminological reformulation of the term "prematurity" such as that discussed in the section above, these and related objections do not jeopardize logically, heuristically, or practically the value of Stent's formulation.

STENT'S EXAMPLES OF PREMATURITY

Stent's exposition cited five examples of prematurity: Mendel's laws; the implications for genetics of the report on DNA as the mediator of bacterial transformation, written by Oswald Avery and colleagues in 1944; Michael Polanyi's theory (1914–16) of gaseous adsorption of solids; and, as examples in particular of here-

---

21. I emphasize "consciously" because in one sense one could claim that *all* history must be presentist at the time written. What one emphasizes, what one ignores, etc., must be influenced by one's time and place. The only important distinctions are whether one is a conscious or unconscious presentist, and if the former, the extent to which one recognizes and attempts to diminish the effects of this bias.

22. See Rigby 1995 on this theme.

23. On this point, see Zinder, chapter 5, and the reply by Stent, chapter 24 in this volume.

and-now prematurity, claims for extrasensory perception, and claims in the 1960s of the transfer of memory from animal to animal by nucleic acid extracts.[24] Close historical scrutiny of the three now-recognized discoveries associated with these episodes suggests that, as examples of the phenomenon defined by Stent, they are not as straightforward as they initially appear. Indeed, shortly after Stent published his article, some individuals challenged his classification of the work of Avery and his colleagues as premature.[25] Problems with the others as classic exemplars—such as the particulate nature of heredity attributed to Mendel's work—were already in the literature or have emerged subsequently.[26]

Despite the objections raised about specific cases, Stent's concept—with some modest refinements regarding terminology —still holds up, I believe. And it provides, for reasons discussed below, a rich, useful, and provocative perspective on the phenomenon of scientific discovery that leads to insights for the working scientist, for the historian, philosopher, or sociologist of science, and for those making science policy.

SPECIFIC ANTICIPATIONS AND EXTENSIONS

René Taton anticipated Stent in 1957 with an alternative but more expansive definition of premature scientific discovery. According to Taton, it occurred when "the level [of] science as a whole fails to lead to its satisfactory explanation *or* to the derivation of useful conclusions from it" (emphasis added).[27] The first criterion is similar to but not identical to Stent's. An example cited by Taton is the observation "by Abbé Picard in 1675 of luminous spots in the 'empty space' of a mercury barometer when transported at night. For this phenomenon, connected with electric discharge in rarefied gases, to be understood and interpreted correctly, a prior knowledge of electrical theories and the structure of gases was essential. Thus the discovery, although it had fruitful repercussions in many experiments to which it gave rise, did not effectively become part of scientific knowledge until the second half of the nineteenth century."[28]

24. On ESP, see my chapter 25 in this volume. On Avery, see Holmes (chap. 12) and Zinder (chap. 5); on Mendel, see Holmes; on Polanyi, see Nye (chap. 11)—all in this volume.

25. Carlson 1973; and reply by Stent 1973.

26. For references on Mendel, see n. 2 above. For explicit objections to Stent's view of Avery's work, see Carlson 1973 and Lederberg 1995 (and for the implicit objection, see Lederberg 1986 and 1990). For discussion of pertinence that does not consider prematurity explicitly, see the fascinating discussion by Stadler (1997) on studies of induced mutation by ultraviolet light and its perceived pertinence to the chemical nature of the gene *before* Avery et al. 1944. On Polanyi's theory of surface adsorption of gases, see Polanyi 1963, reprinted in Polanyi 1969. See also Mulkay 1972; and Nye, chapter 11 in this volume.

27. Taton 1962. Stent, as I read him, would not classify a claim connected to canonical knowledge as premature in his sense merely because one could derive no "useful conclusions" from it.

28. Ibid., p. 160. As described, this example does not appear to be an example of prematurity in Stent's sense.

Taton's formulation, unlike Stent's, does not enlighten scientific disputes. The potential use of Stent's approach to explain some disputes past and present, and consequently to suggest guideposts to current science policy, I regard as a particularly attractive feature that I do not find in Taton's. (See further below.)[29]

Thomas N. Tarrant, a patent attorney, extended Stent's definition in an interesting manner to define a "premature invention." An invention is premature if it depends on undeveloped technology in order to provide any substantial contribution in terms of utility. He cited two specific patents that illustrated this concept.[30]

Harriet Zuckerman and Joshua Lederberg extended Stent's notion conceptually in another direction by defining, analogously, what they termed "postmature" discoveries: those made far later than they might readily have been, as judged retrospectively by contemporary knowledge and available technology.[31] This postulates a different type of delay with some greater methodological difficulties of analysis. Prematurity, they suggest, "is a matter of actual historical observation," while postmaturity "is a matter of retroactive conjecture." (Of course they are here implicitly excluding Stent's cases of current here-and-now prematurity, which differs even more from postmature discovery in that the classification depends on current, not historical, judgment or observation.)

---

29. J. R. Ravetz (1971) has discussed an apparently analogous concept that he designates by a similar term. He defines scientific "immaturity" as the state of an entire field defined "in terms of absence of 'facts,' a condition caused by the absence of criteria of adequacy appropriate for the detection and avoidance of pitfalls in research." In essence, Ravetz regards an immature field as an ineffective one. He cites as one example medicine up to the end of the nineteenth century (pp. 364–66). But here Ravetz means by "ineffective" something different from what the term usually implies. He states that, even if a society develops a "body of genuine craft skills" and "successful special 'arts,' " this does not constitute evidence of maturity. For a field to be mature and effective in his view, such arts should be derived from the "application of a solidly established body of fact" (p. 373), by which, apparently, he means also to include a solidly established body of theory sufficient to enable such derivation. It is not enough for practices to *work* for the field to be effective in his view, as I read the implications of his formulation. One must derive the utility of practices—or they must be derivable—from some other facts and principles. In analogy to Stent, one might state that they must be "connectable" to some existing body of canonical knowledge, but only one that includes a good deal of theory. In this interpretation an alleged discovery in such a field without much theory would necessarily be immature—to extend Ravetz's term from an entire field to claims within it—as well as here-and-now premature in Stent's sense. Irrespective of this issue, Stent's notion of prematurity applies, quite usefully, to alleged findings in what Ravetz would view as a mature field. It has important implications for science policy (see discussion below) that I do not find in Ravetz's formulation.

30. Tarrant 1972. Tarrant cited in his communication two examples of a premature invention. The first was sound on film, invented by Charles E. Fritts, who filed his patent application in 1890 (U.S. Patent 1,203,190). The audio amplifier was unknown at the time, and sound movies appeared commercially only in the late 1920s. The second was the field effect transistor, for which J. E. Lilienfeld filed a patent in 1926 (U.S. Patent 1,745,175). It did not become commercially practical until after the introduction of silicon planar technology. As far as I know, Tarrant's proposed definition of a premature invention was not previously published. Tarrant's letter extending Stent's concept to premature inventions appears to have been submitted as a letter to the editor of *Scientific American*, but it did not appear there.

31. Zuckerman and Lederberg 1986.

Particular examples of postmaturity that they offer may provoke even greater difficulties in analysis than claims of historical observations of prematurity of the type alluded to above. Zuckerman and Lederberg suggest as possible exemplars both bacterial recombination and Linus Pauling's proposal of the protein alpha-helix, which, in Pauling's view, he could have formulated a decade earlier than he did.[32] I suggest as another example the discovery of the correct human chromosome number in 1956—and the subsequent discovery of karyotype abnormalities in those with Down's or other syndromes in 1959—with methods that could have been employed in the 1920s or earlier.[33]

Zuckerman and Lederberg suggest that solutions to problems having "no socially and cognitively defined disciplinary home" are, among others, especially likely to be postmature. Solution of significant problems of this type would, according to this interpretation, benefit from attempts by science-policy analysts to establish a "socially defined" discipline to facilitate their investigation.

THE PREMATURE AND THE NONPARADIGMATIC

Stent published his concept of prematurity in 1972. It has received serious but only modest attention in the literature of the history, philosophy, or sociology of science. Indeed, it appears to have engendered more interest and comment in scientific journals and related outlets than in those of the metasciences. I attribute this in part to its appearance, in its better-known version, in a semipopular scientific journal and, in an expanded version, in an obscure scientific outlet—rather than in a more traditional journal of history, philosophy, or sociology of science.[34]

---

32. As I interpret their concluding discussion, Zuckerman and Lederberg believe that, while discovery of bacterial sex was technologically feasible by 1908 (see above), only in the 1930s or later could one have reasonably regarded its discovery as postmature. See also Zinder in this volume. On Pauling and the protein alpha-helix, see Pauling 1974.

33. The discovery depended in essence on the introduction of the very simple hypotonic treatment, to which there were no apparent cognitive or technological barriers even before the 1920s. There were simply few geneticists, in particular cytogeneticists, interested in and focusing upon medical problems or working with human or even mammalian material. Many, of course, worked with immediately fruitful topics in lower organisms. Few physicians then had sufficient knowledge of and background in the techniques of cytogenetics. The very few medical or clinical geneticists of the time focused on disorders with Mendelian or potentially Mendelian segregation patterns, in pursuit of the explanatory power in humans provided by the identification of mutants in lower organisms. There were no cognitive barriers to recognition of chromosomal abnormalities in humans after the general acceptance of Morgan's chromosomal theory of inheritance and Bridges's observation of nondisjunction. See Kottler 1974 and Hsu 1979, pp. 15–29.

34. Zuckerman is a sociologist of science, yet her paper on postmaturity, coauthored with Joshua Lederberg, appeared in a science journal, *Nature.* One exception in the metasciences is Lamb and Easton 1984, which explicitly presents Stent's formulation (pp. 178–84 and 227–29). Among additional examples cited by Lamb and Easton is Charles McMunn's report of what was later termed cytochrome in 1886, which was ignored until David Keilin's work 38 years later. Lamb and Easton provide the only citation to Taton's earlier alternative formulation of prematurity of which I am aware. See also Stern, chapter 18 in this volume. For the two versions of Stent's article, see Stent 1972a, 1972b.

In addition, many individuals have not understood fully or recognized the distinctiveness of Stent's formulation. Some have commingled it with other factors responsible for scientific and technological resistance discussed above.[35] And some may have ignored it because they conflated it with the implications of one of Thomas Kuhn's conceptions of a scientific paradigm.[36] In discussion with scientists, I have found that some view a premature discovery, as defined by Stent, as nothing more than one that is inconsistent with an existing paradigm in Kuhn's sense. They see nothing usefully new in Stent's formulation.[37]

Stent's notion and the views of Kuhn, as the latter are widely perceived, can be seen as related in some senses. But I see three important differences, one of which has some practical implications for investigators.

In a prevalent interpretation of Kuhn's views, competing paradigms cannot coexist, indeed are not even commensurable—and a major shift, sometimes a scientific revolution, follows when a new paradigm is adopted. One could attempt to embed Stent's formulation within Kuhn's by viewing a premature claim or proposal as part of a different paradigm than the reigning one. For the claim or proposal to be accepted, then, there must be a major change. Yet changes in paradigms, from one "incommensurable" position to another, occur rarely. Canonical knowledge grows every day. Clearly, there are many premature discoveries, in Stent's sense, whose acceptance need not await a whole scientific revolution or even a significant paradigm change. There may be a single or only a few "missing" logical steps by which, for instance, a claim, hypothesis, or proposal could be connected to canonical knowledge—without requiring a so-called Kuhnian paradigm shift—and yet still be premature in Stent's classification.

Another difference is that the concept of prematurity implies direct, actual scientific and technological progress. The concept of paradigm shift as used by Kuhn—for instance, in analogy with a gestalt shift, which he discusses explicitly—does not necessarily do so. This may explain why, in private discussion with scientists after it is explained, I find some are particularly attracted to Stent's, rather than

35. Tobias 1996 adopts Stent's terminology and applies it to several examples. But in my reading, his discussion, which frequently invokes the term "paradigm," appears to imply, perhaps unwittingly, that Stent's is little more than a reexpression of Kuhn's formulation (Kuhn 1970), and that lack of connection to canonical knowledge can be rectified only by some major change such as a revolution.

36. Kuhn 1970.

37. I do not suggest that this is necessarily a correct interpretation of Kuhn's original exposition, but rather that it is simply a widely held view. But one colleague insisted to me quite the opposite, that Kuhn's and Stent's notions are completely distinct and have no relevance to each other. Some of the differences in interpretation of Kuhn stem from ambiguities in his original exposition in *The Structure of Scientific Revolutions,* ambiguities that Kuhn partially conceded subsequently in the postscript to the second, 1970 edition (1970, pp. 174–210). See also his "Second Thoughts on Paradigms" (Kuhn 1977a, pp. 293–319, or Kuhn 1977b), as well as the preface to Kuhn 1977a, especially pp. xix–xx, in which he elaborates on this point. For instance, he notes that there were two separate ways in which he had used the term "paradigm": to denote an "exemplar" and to denote a "disciplinary matrix." He expressed some regret for introducing the term "paradigm," because as a consequence of his varying usage "the

Kuhn's, formulation as a preferable heuristic in conceptualizing one reason why some work, in particular some of their own, may not be widely accepted. There appears to be some psychic comfort in the realization that regarding their results as premature implies these results may well be correct but it is work for which the world is not yet ready, so to speak. Kuhn's formulation provides less assurance.[38]

There also appear to be some important differences in practical implications of these perspectives. Certainly, irrespective of the precise distinctions between the concepts of Stent and Kuhn, one may concede that for social and science policy there are no obvious solutions to the loss engendered either by the nonacceptance of contemporary (but eventually accepted) premature claims or by proposals rejected because they are outside a current paradigm. (Again note that novel and unusual ideas and suggestions are not always or necessarily premature in Stent's sense. Some may be readily connected with canonical knowledge but unaccepted for other reasons.)

Yet to investigators whose claims, hypotheses, or proposals are premature—that is, manifest in here-and-now prematurity, as Stent terms it—and who recognize the fact, one may suggest several useful strategies beyond the rather vacuous advice to seek or await a paradigm shift in a field. If the premature reported claim, hypothesis, or proposal originates in experiment, then the investigator will likely find it useless to try to convince the scientific community by repeating the same type of experiment. Two other strategies are more likely to succeed. He or she should undertake work that *can* link the findings with canonical knowledge and, in fact, by linking it, perhaps alter a canon in some respect. Alternatively, the investigator should attempt to harness the alleged findings for some practical, even commercial use. Success of the latter type may not provide a "connection" to canonical knowledge but will ensure that the scientific and technological community pays serious attention to the work, and that others will join the search for the connection. Until either of these strategies produces results, one may defend the scientific community for ignoring or even rejecting such claims, hypotheses, or proposals. It would otherwise be overwhelmed by attention to false and useless leads.

UNEXPLAINED SUCCESSFUL TECHNOLOGY

Practical or commercial use of a discovery indicates a potential problem in Stent's formulation, for a discovery that actually "works" practically may or may not be

---

inevitable result has been confusion" (p. xx). For exchanges on these and other difficulties in his formulation, see Suppe 1977, pp. 500–17. Many with whom I have discussed both Stent's and Kuhn's views, especially those outside the field of the philosophy of science, are not aware of Kuhn's subsequent altered position.

38. J. B. Rhine, the well-known proponent of extrasensory perception, wrote Stent a laudatory letter of gratitude about his *Scientific American* article shortly after it was first published, even though in essence Stent had argued forcefully that the work of Rhine and others in this field should be ignored because it was an example of here-and-now prematurity (Rhine 1973).

linked to existing canonical knowledge. Cold fusion could not be linked to canonical knowledge, but had it been clearly replicable and had it worked, it would have been accepted even though premature in Stent's sense. Henri Giffard's heat injector worked for 50 years before Henri Poincaré provided a theoretical explanation. Prior to that it was an object of great astonishment because it appeared to be an example of perpetual motion, doing work without power. It was premature in Stent's sense but still widely employed.[39]

Medicine abounds with such examples. Among the most dramatic is the discovery of the effects of sulfuric ether for gaseous anesthesia in the nineteenth century.[40] Another striking example is the discovery of x rays by Wilhelm Röntgen in 1895. It was accidental, with no connection to previous canonical knowledge.[41] Yet its obvious success and ready confirmation led to quick acceptance and application. Of course, Antoine-Henri Becquerel's discovery of radioactivity in 1896 shortly thereafter made the finding appear more coherent to physicists. But with or without that later discovery by Becquerel, one can defend the view that the ready replicability of the effects of x rays and their obvious applications would have led to their acceptance quickly. Radioactive theory or no theory, one could now literally see the image of an internal fracture or a hidden bullet lodged in a wound and confirm its location at surgery.

One might argue that the discovery of the very fact that, say, x rays revealed the location of hidden objects was an expansion of canonical knowledge, so once that was accepted, the discovery itself was not premature. That is to say, the link between the finding and canonical knowledge is the trivial fact that it *is* now regarded as part of canonical knowledge. But should the generally recognized observation that it works, without the understanding of why, be sufficient to make such knowledge canonical, although there is other knowledge regarded as canonical that entails a deeper level of fundamental understanding? I believe yes. But because of variation in the quality or depth of canonical knowledge, we must qualify the notion of prematurity accordingly. (See also chapter 25 herein.)

One pertinent issue clearly relates to replication or replicability. James Lind's discovery in the eighteenth century of the use of citrus fruit to treat scurvy illustrates this. This knowledge was lost in the nineteenth century, at least to authorities of the British navy, because use of canned citrus juice did not prevent scurvy among men on Arctic expeditions, most notably the Franklin expedition, and because even the consumption of some fresh limes did not appear to work in other circumstances. It was not the later discovery of vitamin C but rather the conse-

---

39. Kranakis 1982.

40. See Bergman 1998. In the case of nitrous oxide, another agent introduced at that time, there had been some prior experimental evidence in lower animals and the consequent suggestion that it be applied to humans, both of which had been ignored for over 40 years.

41. Lamb and Easton (1984, p. 174) report that A. W. Godspeed and W. J. Jennings in 1890 had taken what only six years later was recognized to be an "X-ray photograph.".

quences of the earlier discovery of an animal model for scurvy, in the guinea pig, that eventually led to the British navy's reacceptance of citrus fruit or extracts as antiscorbutic or preventative.[42] This led to methods of assay of the antiscorbutic content of foods and, in particular, to recognition that certain processing and storage methods had inactivated that substance in canned juices provided on expeditions, and that certain atypical limes had very little of the antiscorbutic substance. In a sense, one may argue that, by expanding canonical knowledge, the animal model enabled a link to previously rejected claims. The practical value of Lind's work was clear in the eighteenth century. But its loss—to the British navy, at any rate—was associated with insufficient understanding of its mechanism. If one maintains that Lind's discovery expanded canonical knowledge, in Stent's formulation, then the canon contracted after the Franklin expedition and later reexpanded.

Clearly, one should not ignore a finding such as Lind's that works, even if it cannot be connected to canonical knowledge at the time. But this example also illustrates something less obvious but important: observing and using something that works will not necessarily ensure that it will continue to work, especially if its mechanism is not understood. It gives a forceful example of the great utility of seeking underlying explanations for empirical knowledge. This provides an important rationale for science policy that supports basic research in the quest for explanations of incompletely understood but important technology or other phenomena.

An episode recounted by Charles H. Townes, the inventor of the maser and laser, illustrates a related issue. Before publication of his report, Townes described his mechanism for the laser in private conversation with John von Neumann, perhaps the most outstanding mathematical physicist of the twentieth century. Von Neumann told him it would not work, to which Townes replied, "Well, yes [it does work], and we have it." Von Neumann walked away, thought about it for a while, and then came back and said to Townes, "You are right."[43] The fact that it *did* work—at least Townes's authority for this claim, as perceived by von Neumann, persuaded him. And this led him to reflect and then recognize a misunderstanding of the implication of (his) canonical knowledge.

THE UTILITY OF STENT'S FORMULATION

The considerations discussed above suggest two possibly fruitful, albeit somewhat controversial, approaches to reexamining historical episodes involving scientific and technological work initially overlooked, ignored, or rejected but later accepted.

42. Among the mammals, only guinea pigs, primates, and humans appear to lack the capacity to synthesize vitamin C. Thus rats, sheep, goats, dogs, and cats could thrive while sailors suffered scurvy on board ship. See Carpenter 1986, and chapter 7 in this volume, for a fascinating exposition.

43. See Townes, chapter 4 in this volume.

These are to inquire, first, what simple alterations in scientific policies, or in social factors affecting such policies at the time, if any, would have diminished the delay in acceptance, and, second, what else an investigator (or sympathetic contemporary investigator) could reasonably have done at the time, if anything, to *overcome* the sources of resistance.

I recognize that these queries come close to being counterfactual conditionals, which many historians regard as anathema.[44] But if an understanding of the historical processes that have affected scientific development is to enlighten current scientific policy, effective *use* of that understanding can be promoted by careful posing of precisely these questions. The answers to the first query may well indicate whether the discovery constitutes a genuine case of prematurity in Stent's sense, and in any event will provide useful perspectives to those making current policy. The answers to the second query may provide guidance to a present-day Semmelweis whose insights would otherwise be lost to the world until independently rediscovered.

BIBLIOGRAPHY

Ackerknecht, E. H. 1982. *A Short History of Medicine.* Rev. ed. Baltimore: Johns Hopkins Press.

Althusser, L. 1977. "Introduction: Unfinished History." Translated by G. Lock. In *Proletarian Science? The Case of Lysenko,* by D. Lecourt, pp. 7–16. Translated by B. Brewster. London: NLB.

Amaldi, E. 1989. "The Prelude to Fission." In *Fifty Years with Nuclear Fission,* ed. J. W. Behrens and A. D. Carlson, 1:10–19. La Grange Park, Ill.: American Nuclear Society.

Barber, B. 1961. "Resistance by Scientists to Scientific Discovery." *Science* 134:596–601.

Bergman, N. A. 1998. *The Genesis of Surgical Anesthesia.* Park Ridge, Ill.: Wood Library–Museum of Anesthesiology.

Bessis, M. 1976. "How the Mouse Leukemia Virus Was Discovered: A Talk with Ludwik Gross." *Nouvelle Revue Française d'Hematologie* 16:296.

Brannigan, A. 1981. "The Law Valid for *Pisum* and the Reification of Mendel." In *The Social Basis of Scientific Discoveries,* pp. 89–142. Cambridge: Cambridge University Press.

Brock, T. D. 1990. *The Emergence of Bacterial Genetics.* Cold Spring Harbor, N.Y.: Cold Spring Harbor Laboratory Press.

Browning, C. H. 1908. "Chemo-Therapy in Trypanosome-Infections: An Experimental Study." *J. Path. Bact.* 12:166–190.

Butterfield, H. [1931] 1965. *The Whig Interpretation of History.* New York: Norton.

———. 1944. *The Englishman and His History.* Cambridge: Cambridge University Press.

Carlson, E. A. 1973. Letter to the editor. *Scientific American* 228 (January): 8.

44. Fischer 1970, pp. 15–21. (Gunther Stent informs me that, in Berlin, counterfactuals are referred to metonymously as " 'If my grandmother had had wheels, then she would have been a bus' propositions.") The major objection to counterfactual questions appears to be the fact that, once one starts asking fictional questions and posing contrary-to-fact history, this invites one to begin a process that may not only appear limitless but imposes no clear constraints on the nature of admissible evidence pertinent to the original query.

Carpenter, K. J. 1986. *The History of Scurvy and Vitamin C.* New York: Cambridge University Press.

Carter, K. C. 1983. Translator's introduction. In *The Etiology, Concept, and Prophylaxis of Childbed Fever,* by I. Semmelweis. Translated and edited by K. C. Carter, pp. 3–58. Madison: University of Wisconsin.

Fermi, E. 1934. "Possible Production of Elements of Atomic Number Higher Than 92." *Nature* 133:898–99.

Fischer, D. H. 1970. *Historians' Fallacies: Toward a Logic of Historical Thought.* New York: Harper and Row.

Goertzel, T., and B. Goertzel. 1995. *Linus Pauling: A Life in Science and Politics.* New York: Basic Books.

Gould, G. M. 1901. "The Reception of Medical Discoveries." *Arch. Ophthal.* 13:715–49.

Graham, L. R. 1972. *Science and Philosophy in the Soviet Union.* New York: Alfred A. Knopf.

Grosse, A. V. 1935. "The Identity of Fermi's Reactions of Element 93 with Element 91." *J. Am. Chem. Soc.* 57:438–39.

Holmes, O. W. 1842–43. "The Contagiousness of Puerperal Fever." *New Eng. Quart. J. Med. Surg.* 1:503–30.

Hsu, T. C. 1979. *Human and Mammalian Cytogenetics: An Historical Perspective.* New York: Springer-Verlag.

Kevles, D. J. 1995. "Pursuing the Unpopular: A History of Courage, Viruses, and Cancer." In *Hidden Histories of Science,* ed. R. B. Silvers. New York: New York Review of Books.

Klein, G. 1990. *The Atheist and the Holy City: Encounters and Reflections.* Trans. T. Friedman and I. Friedman. Cambridge: MIT Press.

Kottler, M. J. 1974. "From 48 to 46: Cytological Technique, Preconception, and the Counting of Human Chromosomes." *Bull. Hist. Med.* 48:465–502.

Kranakis, E. F. 1982. "The French Connection: Giffard's Injector and the Nature of Heat." *Technology and Culture* 23:3–38.

Krementsov, N. 1997. *Stalinist Science.* Princeton: Princeton University Press.

Kuhn, T. S. 1970. *The Structure of Scientific Revolutions.* 2d ed. Chicago: University of Chicago Press.

———. 1977a. *The Essential Tension.* Chicago: University of Chicago Press.

———. 1977b. "Second Thoughts on Paradigms." In *The Structure of Scientific Theories,* ed. F. Suppe, pp. 459–82. 2d ed. Urbana: University of Illinois.

Lamb, D., and S. M. Easton. 1984. *Multiple Discovery: The Pattern of Scientific Progress.* N.p.: Avebury Publishers.

Lederberg, J. 1986. "Forty Years of Genetic Recombination in Bacteria: A Fortieth Anniversary Reminiscence." *Nature* 324:627–28.

———. 1990. "Introduction: Reflections on Scientific Biography." In *The Excitement and Fascination of Science: Reflections by Eminent Scientists,* ed. J. Lederberg, pp. xvii–xxiv. Vol. 3, pt. 1. Palo Alto: Annual Reviews.

———. 1995. "Greetings." In *DNA: The Double Helix, Perspective, and Prospective at Forty Years,* pp. 176–79. New York: New York Academy of Sciences. Also printed in *Ann. NY Acad. Sci.* 758: 176–79.

Mayr, E. 1982. *The Growth of Biological Thought: Diversity, Evolution, and Inheritance.* Cambridge: Harvard University Press, Belknap Press.

Mettler, C. C. 1947. *History of Medicine.* Ed. F. A. Mettler. Philadelphia: Blakiston.

Mulkay, M. J. 1972. "Conformity and Innovation in Science." *Sociological Review Monograph* 18:7.

Nagel, E. 1950. "The Methods of Science: What Are They? Can They Be Taught?" *Scientific Monthly* 70 (January): 19–23.

Nickles, T. 1980. "Introductory Essay: Scientific Discovery and the Future of the Philosophy of Science." In *Scientific Discovery, Logic, and Rationality.* Dordrecht: D. Reidel.

Noddack, I. 1934. "Über das element 93." *Z. Agnew. Chem.* 47:653–55.

Olby, R. 1979. "Mendel No Mendelian?" *Hist. Sci.* 17:53–72.

Pauling, L. 1974. "Molecular Basis of Biological Specificity." *Nature* 248:769–71.

Peters, D. P., and S. J. Ceci. 1982. "Peer-Review Practices of Psychological Journals: The Fate of Published Articles Submitted Again." *Behavioral and Brain Sciences* 5:187–200.

Polanyi, M. 1963. "The Potential Theory of Adsorption." *Science* 141:1010–13.

———. 1969. *Knowing and Being.* Ed. M. Greene. London: Routledge and Kegan Paul.

Ravetz, J. R. 1971. "Immature and Ineffective Fields of Inquiry." In *Scientific Knowledge and Its Social Problems,* pp. 364–402. Oxford: Oxford University Press.

Rhine, J. B 1973. Letter to Gunther Stent, 4 January 1973. Gunther Stent Papers, Bancroft Library, University of California at Berkeley.

Rigby, S. H. 1995. "Historical Causation: Is One Thing More Important Than Another?" *History* 80:227–42.

Sacks, O. 1995. "Scotoma: Forgetting and Neglect in Science." In *Hidden Histories of Science,* ed. R. B. Silvers. New York: New York Review of Books.

Sandler, I. 2000. "Development: Mendel's Legacy to Genetics." *Genetics* 154:7–11.

Sandler, I., and L. Sandler. 1985. "A Conceptual Ambiguity That Contributed to the Neglect of Mendel's Paper." *Pubbl. Stn. Zool. Napoli II* 7:3–70.

Segrè, E. 1939. "An Unsuccessful Search for Transuranium Elements." *Phys. Rev.* 55:1104–5.

———. 1993. *A Mind Always in Motion: The Autobiography of Emilio Segrè.* Berkeley and Los Angeles: University of California Press.

Shryock, R. H. 1966. "The Strange Case of Well's Theory of Natural Selection, 1813: Some Comments on the Dissemination of Scientific Ideas." In *Medicine in America: Historical Essays,* pp. 259–72. Baltimore: Johns Hopkins Press.

Singh, S. 1997. *Fermat's Last Theorem.* London: Fourth Estate.

Stent, G. S. 1972a. "Prematurity and Uniqueness in Scientific Discovery." *Advances in the Biosciences* 8:433–49.

———. 1972b. "Prematurity and Uniqueness in Scientific Discovery." *Scientific American* 227 (December): 84–93.

———. 1973. Letters. *Scientific American* 228 (January): 8.

Stern, B. J. 1927. *Social Factors in Medical Progress.* New York: Columbia University Press.

———. 1941. "Resistance to Medical Change." In *Society and Medical Progress,* pp. 175–213. Princeton: Princeton University Press.

Suppe, F., ed. 1977. *The Structure of Scientific Theories,* pp. 507–17. 2d ed. Urbana: University of Illinois Press.

Tarrant, T. N. 1972. Letter to Gunther Stent, 13 December 1972. Gunther Stent Papers, Bancroft Library, University of California at Berkeley.

Taton, R. 1962. *Reason and Chance in Scientific Discovery.* Trans. A. J. Pomerans. New York: Science editions.

Tobias, P. V. 1996. "Premature Discoveries in Science, with Special Reference to *Australopithecus* and *Homo habilis.*" *Proc. Amer. Phil. Soc.* 140:49–64.

Veldink, C. 1989. "The Honey-Bee Language Controversy." *Interdisciplinary Science Reviews* 14:166–75.

Wali, K. 1991. *Chandra: A Biography of S. Chandrasekhar.* Chicago: University of Chicago Press.

Wenner, A. M., D. E. Meade, and L. J. Friesen. 1991. "Recruitment, Search Behavior, and Flight Ranges of Honey Bees." *Amer. Zool.* 31:768–82.

Wenner, A. M., and P. H. Wells. 1990. *Anatomy of a Controversy: The Question of a "Language" among Bees.* New York: Columbia University Press.

Wiener, N. 1993. *Invention: The Care and Feeding of Ideas.* Cambridge: MIT Press.

Youngston, A. J. 1979. *The Scientific Revolution in Victorian Medicine.* London: Croom Helm.

Zuckerman, H. A., and J. Lederberg. 1986. "Postmature Scientific Discovery?" *Nature* 324:629–31.

CHAPTER TWO

Prematurity in Scientific Discovery

Gunther S. Stent

One of the depressing by-products of the fantastically rapid progress that was made in molecular genetics in the past twenty-five years is that now merely middle-aged participants in its early development are obliged to look back upon their early work from a depth of historical perspective that, in the case of biological specialties that came into flower in earlier times, had opened up only after all the witnesses of the first blossoming were long dead.[1] I have been trying to make virtue out of necessity and actually exploit this singular position for fathoming the evolution of a scientific field. Thus, in looking back on the history of molecular genetics from the viewpoint of my own experiences, I have found that one of its most famous incidents, Oswald T. Avery's identification of DNA as the active principle in bacterial transformation and, hence, as genetic material, illuminates a general problem of cultural history.[2] The case of Avery brings, I think, insights into the question of whether it is meaningful, or merely tautological, to allege that a discovery is "ahead of its time," or premature.

PREMATURITY

In 1968, I published a brief retrospective essay on molecular biology with particular emphasis on its origins.[3] In this historical account, I mentioned neither Avery's

Editor's note: The text included here is a slightly modified version of material in Stent 1972a pertaining to prematurity and is reprinted with permission of Elsevier Scientific. All discussion of uniqueness has been excluded.

1. Editor's note: This progress refers to work in the period 1945–1970.
2. Avery et al. 1944.
3. Stent 1968a.

name nor DNA-mediated bacterial transformation. My essay brought forth a letter to the editor by Carl Lamanna, who complained that "it is a sad and surprising omission that Stent makes no mention of the definitive proof of DNA as the basic hereditary substance by O. T. Avery, C. M. MacLeod and M. McCarty.[4] The growth of [molecular genetics] rests upon this experimental proof.... I am old enough to remember the excitement and enthusiasm induced by the publication of the paper by Avery, MacLeod and McCarty. Avery, an effective bacteriologist, was a quiet, self-effacing, non-disputatious gentleman. These characteristics of personality should not prevent the general scientific public represented by the audience of *Science* to let his name go unrecognized."[5]

I was taken aback by Lamanna's letter and replied that I agreed that I should have really mentioned in my essay Avery's proof in 1944 that DNA is the hereditary substance.[6] But, I went on to say, in my opinion it is not true that the growth of molecular genetics rests upon Avery's proof. For many years that proof had actually made a surprisingly small impact on geneticists, both molecular and classical, and it was only the Hershey-Chase experiment of 1952 which caused those people to focus on DNA.[7] The reason for this delay was neither that Avery's work was unknown to or mistrusted by geneticists nor that the Hershey-Chase experiment was technically superior. Instead, Avery's discovery, so I declared, had been merely "premature." And in the last two sentences of my reply to Lamanna, I sketched out the argument about prematurity that I shall try to develop here in somewhat greater detail.

My prima facie reason for considering Avery's discovery premature is that it was not appreciated in its day. But is it, in fact, *true* that Avery's discovery was not appreciated? Lamanna, for example, mentions his own excitement and enthusiasm induced by the publication of Avery's paper, and several participants in the 1946 Cold Spring Harbor Symposium on Heredity and Variation in Microorganisms have told me that Avery's discovery formed the subject of intense discussion at that symposium. So how can I say that it was not appreciated? By lack of appreciation I do not mean that Avery's discovery went unnoticed, or even that it was not considered important. What I do mean is that no one seemed to be able to do much with it, or build upon it, except for the students of the transformation phenomenon. That is to say, Avery's discovery had virtually no effect on general genetic discourse.[8]

By way of support of this allegation, I invite examination of the 1946 Cold Spring Harbor Symposium volume. It contains a paper by M. McCarty, Harriet

4. Avery et al. 1944.

5. Lamanna 1968.

6. Stent 1968b.

7. Hershey and Chase 1952.

8. Editor's note: Stent has asked me to note that what he meant here was that lack of appreciation of or inability to build upon the discovery of Avery et al. only made it a candidate for *consideration* as premature in the sense he defines later in his paper.

Taylor, and Avery, whose main concern is not the meaning of the discovery for genetics but the elucidation of the role of serum in the DNA-mediated transformation phenomenon. Although many of the other papers of the volume are followed by discussants' remarks, no discussant of the McCarty, Taylor, and Avery paper is on record. Only five of the other 26 symposium papers refer to Avery's discovery.

Three phage workers, T. F Anderson, A. D. Hershey, and S. E. Luria, venture the opinion that the phenomenon is probably of wide biological importance. L. Dienes concludes that since DNA is "a substance without apparent organization," Avery's discovery means that "bacteria possess a mechanism for the exchange of hereditary characteristics, [that is] different from the usual sexual processes," and S. Spiegelman is under the impression that Avery discovered "the induction of a particular enzyme with a nucleoprotein *[sic]* component."[9] Neither Max Delbrück nor J. Lederberg and E. L. Tatum mention Avery at all in their now famous 1946 symposium papers.

An even more convincing demonstration of the lack of appreciation of Avery's discovery is provided by the 1950 Golden Jubilee of Genetics symposium "Genetics in the 20th Century."[10] Here some of the most eminent geneticists of that time presented essays that surveyed the progress of the first 50 years of genetics and assessed its present status. Only one of the 26 essayists saw fit to make more than a passing reference to Avery's discovery, then six years in the past, namely A. E. Mirsky, who still expressed some doubts that the active transforming principle is really pure DNA. H. J. Muller's 1950 symposium essay on the nature of the gene contains no mention of Avery or DNA.

So, *why* was Avery's discovery not appreciated in its day? Because it was "premature." But is this really an explanation or is it merely an empty tautology? In other words, is there a way of providing a criterion of the prematurity of a discovery other than its failure to make an impact? Yes, there is such a criterion: *A discovery is premature if its implications cannot be connected by a series of simple logical steps to contemporary canonical [or generally accepted] knowledge.*[11] This criterion is not to be confused with that of an *unexpected* discovery, which *can* be connected with the canonical ideas of its day but might overthrow one or more of them. For instance, the finding of a "reverse transcriptase" would fall into the category of unexpected discoveries—provided, of course, that the function attributed to that enzyme of catalyzing the assembly of a DNA replica from an RNA template can eventually be shown to occur in vivo.[12] Although prior to that finding, it had been generally assumed by molecular geneticists that there is no reverse flow of "information" from RNA to DNA, there is no difficulty

9. Dienes 1946, p. 58; Spiegelman 1946, p. 269.

10. Dunn 1951.

11. Editor's note: The words "generally accepted" were inserted in Stent 1972b but do *not* appear in the published version of Stent 1972a, reproduced here.

12. The discovery of reverse transcriptase is described in Baltimore 1970; and Temin and Mizutani 1970. Editor's note: Shortly after publication of Stent's article, the function was shown to occur in vivo.

at all in understanding such a process from the viewpoint of the previous current ideas of polynucleotide synthesis.

Why could Avery's discovery not be connected with canonical knowledge? By 1944, DNA had long been suspected of exerting *some* function in hereditary processes, particularly after R. Feulgen [with H. Rossenbeck] had shown in 1924 that DNA is a major component of the chromosomes.[13] But the then current view of the molecular nature of DNA made it well nigh inconceivable that DNA *could* be the carrier of hereditary information. First of all, until well into the 1930s DNA was generally thought to be merely a *tetranucleotide* composed of one residue each of adenylic, guanylic, thymidylic, and cytidylic acid. Secondly, even when it was finally realized by the early 1940s that the molecular weight of DNA is actually much higher than that demanded by the tetranucleotide theory, it was still widely believed that the tetranucleotide is the basic repeating unit of the large DNA polymer in which the four purine and pyrimidine bases recur in regular sequence. DNA was therefore viewed as a monotonously uniform macromolecule which, like other monotonous polymers such as starch or cellulose, is always the same no matter what its biological source. The ubiquitous presence of DNA in the chromosomes was, therefore, generally explained in purely physiological or structural terms. Instead, it was usually to the chromosomal protein that the informational role of the genes had been assigned since the great differences in the specificity of structure that exist between heterologous proteins in the same organism, or between homologous proteins in different organisms, had been appreciated since the beginning of this century. The conceptual difficulty of assigning the genetic role to DNA had by no means escaped Avery, for in the conclusion of his paper he states that "if the results of the present study of the transforming principle are confirmed[,] then nucleic acids must be regarded as possessing biological specificity the chemical basis of which is as yet undetermined."

However, by 1950, the tetranucleotide theory had been overthrown, thanks largely to the work of Erwin Chargaff who showed that, contrary to the demands of that theory, the four nucleotide bases are not necessarily present in DNA in equal proportions.[14] Chargaff found, furthermore, that the exact base composition of DNA differs according to its biological source, suggesting that DNA may not be a monotonous polymer after all. So when, two years later, Hershey and Chase showed that upon infection of the host bacterium at least 80% of the phage DNA enters the cell whereas at least 80% of the phage protein remains outside, it was now possible to connect *their* conclusion that DNA is the genetic material with canonical knowledge.[15] For Avery's "as yet undetermined" chemical basis of the biological specificity of nucleic acids could now be envisaged as the precise sequence of the four nucleotide bases along the polynucleotide chain. The general impact of the Hershey-

13. Feulgen and Rossenbeck 1924.
14. Chargaff 1950.
15. Hershey and Chase 1952.

Chase experiment was immediate and dramatic. DNA was suddenly in and protein was out, as far as thinking about the nature of the gene was concerned. Within a few months, there arose the first speculations about the genetic code, and Watson and Crick were inspired to set out to discover the structure of DNA.

Naturally, the case of Avery is only one of many premature discoveries in the history of science. I have presented it here for consideration mainly because of my own failure to appreciate it when I joined Delbrück's phage group and took the Cold Spring Harbor phage course in 1948. Since then, I have often wondered what my later fate would have been if only I had been intelligent enough to appreciate Avery's discovery and infer from it four years before the Hershey-Chase experiment that DNA must also be the genetic material of the phage.

Probably the most famous case of prematurity in the history of biology is that of Gregor Mendel, whose discovery of the particulate nature of heredity in 1865 had to await 35 years before it was "rediscovered" at the turn of the century.[16] Mendel's discovery made no immediate impact, so it can be argued, because the concept of discrete hereditary units could not be connected with the (mid 19th century) canonical knowledge of anatomy and physiology. Furthermore, the statistical methodology by means of which Mendel interpreted his data was wholly foreign to the way of thinking of his contemporary biologists. By the end of the 19th century, however, chromosomes, mitosis, and meiosis had been discovered, and Mendel's results could now be accounted for in terms of microscopically visible structures and processes. Furthermore, by then the application of statistics to biology had become commonplace. In some respects, however, Avery's case is a more dramatic example of prematurity than Mendel's. Whereas Mendel's discovery seems to have been hardly mentioned by anyone until its rediscovery, Avery's discovery *was* widely discussed, and yet could not be appreciated for eight years.

A striking example of delayed appreciation of a discovery in the physical sciences, as well as an explanation of that delay in terms of the concept to which I refer here as prematurity, has been provided by Michael Polanyi.[17] In the years 1914–1916, Polanyi published a theory of the adsorption of gases on solids which assumed that the force attracting a gas molecule to a solid surface depends only on the position of that molecule, but not on the presence of other molecules, in the force field. Despite the fact that Polanyi was able to provide strong experimental evidence in favor of his theory, it was generally rejected. Not only was the theory rejected, but it was considered so ridiculous by the leading authorities of the time that Polanyi believes continued defense of his theory would have ended his professional career had he not managed to publish work on other more palatable ideas. The reason for the general rejection of Polanyi's adsorption theory was that, at the very time he put it forward, the role of electrical forces in the architecture of matter had just been discovered. And hence, there seemed to be no doubt that

16. Mendel 1866.
17. Polanyi 1963.

gaseous adsorption must also involve electrical attraction between gas molecules and solid surfaces. That point of view, however, was irreconcilable with Polanyi's basic assumption of the mutual independence of individual gas molecules in the adsorption process. Instead of Polanyi's theory, the theory of I. Langmuir, which did envisage a mutual interaction of the gas molecules of the kind expected from electrical forces, found general acceptance. It was only in the 1930s after F. London developed his new theory of cohesive molecular forces based on quantum mechanical resonance rather than electrostatic attraction, that it became conceivable that gas molecules *could* behave in the way in which Polanyi's experiments indicated they are actually behaving. Meanwhile, Langmuir's theory had become so well-established, and Polanyi's had been consigned so authoritatively to the ash can of crackpot ideas, that Polanyi's theory was rediscovered only in the 1950s.[18]

We may now consider whether the notion of prematurity is actually a useful historical concept. First of all, is prematurity the only possible explanation for the lack of contemporary appreciation of a discovery? No, evidently not. For instance, Lamanna suggested the "quiet, self-effacing, non-disputatious" personality of Avery as the cause for the failure of general recognition of his discovery. And Chargaff is another believer in the idea that personal modesty and reticence for self-advertisement accounts for lack of contemporary appreciation.[19] For instance, Chargaff has attributed the 75-year hiatus between F. Miescher's discovery of DNA in 1869 and the appreciation of its importance to Miescher being "one of the quiet in the land," who lived when "the giant publicity machines, which today accompany even the smallest move on the chess-board of nature, with enormous fanfares were not yet in place." Indeed, the 35-year hiatus in the appreciation of Mendel's discovery is often attributed to Mendel having been a modest monk living in an out of the way Moravian monastery. Hence, the notion of prematurity provides an alternative to the—in my opinion, for the cases mentioned here, false—invocation of lack of publicity as an explanation for delayed appreciation.

But, more importantly, does the prematurity concept pertain only to retrospective judgments made with the wisdom of hindsight? No, I think it can be used also to judge the present. For some discoveries have been made recently that are still premature at this very time. One example of here-and-now prematurity is the alleged finding that sensory information received by an animal can be stored in RNA or other macromolecules.

In the early 1960s, there began to appear reports by experimental psychologists purporting to have shown that the memory trace, or engram, of a task learned by a trained donor animal can be transferred to a naive recipient animal by injecting or feeding the recipient with an extract made from the tissues of the donor.[20] At that

18. Editor's note: For further discussion of this case and additional references, see Nye, chapter 11 in this volume.

19. Chargaff 1971.

20. For a critical summary, see Quarton 1967.

time, the central message of molecular genetics that nucleic acids and proteins are "informational macromolecules" had just gained wide currency, and the facile equation of sensory information with genetic information soon led to the proposal that macromolecules—DNA, RNA, or protein—store memory. As it happens, the experiments on which the macromolecular theory of memory is based have been very difficult to repeat, and the results claimed for them may indeed not be true at all. But it is significant that few neurophysiologists have even bothered to check these experiments, despite everybody having heard about them and being aware that the possibility of chemical memory transfer would constitute a fact of capital importance. The lack of interest of neurophysiologists in the macromolecular theory of memory can be accounted for by recognizing that this theory, whether true or false, is clearly premature: there is no chain of reasonable inferences by means of which our present, albeit very imperfect, view of the functional organization of the brain can be reconciled with the possibility of its acquisition, storage, and retrieval of experiential information by encoding such information in nucleic acid or protein molecules. Thus for the community of neurobiologists there is no point in devoting its time to checking on experiments whose results, even if they were true as alleged, could not be connected with canonical knowledge.

The concept of here-and-now prematurity can be applied also to the troublesome subject of extrasensory perception, or ESP. During the summer of 1948, while taking the Cold Spring Harbor phage course, I happened to witness a heated argument between two future mandarins of molecular biology, S. E. Luria and R. E. Roberts. Roberts was then interested in ESP and felt it had not been given fair consideration by the scientific community. As far as I remember, he thought one might be able to set up some experiments with molecular beams which could provide more definitive data on the possibility of mind-induced departures from random distributions than J. B. Rhine's then much discussed card-guessing procedures.[21] Luria declared that not only was he not interested in Roberts proposed experiments, but that in his opinion it was unworthy of anyone claiming to be a scientist even to discuss such rubbish. How could an intelligent fellow like Roberts entertain the possibility of phenomena totally irreconcilable with the most elementary physical laws? Moreover, a phenomenon which is manifest only to specially endowed subjects, as claimed by parapsychologists to be the case for ESP, is outside the proper realm of science, which must deal with phenomena accessible to every observer. Roberts replied that far from his being unscientific, it was Luria whose bigoted attitude toward the unknown is unworthy of a true scientist. The fact that not everybody has ESP only means that it is an elusive phenomenon, such as musical genius. And just because a phenomenon cannot be reconciled with what we now know, we need not shut our eyes to it. On the contrary, it is the duty of the scientist to try to devise experiments designed to probe its truth or falsity.

21. Rhine 1948.

It seemed to me then that *both* Luria and Roberts were right, and, in the intervening years, I often thought about this puzzling disagreement, unable to resolve it in my own mind. Finally, I read C. W. Churchman's review of a book on ESP,[22] and I began to see my way toward a resolution. Churchman set forth that there are three different possible scientific approaches to ESP. The first of these is that the truth or falsity of ESP, like that of the existence of God or the immortality of the soul, is totally independent of either the methods or findings of empirical science. And hence, an adherent of the tenets of logical positivism would relegate ESP to the class of meaningless propositions. Thus the problem of ESP is defined out of existence. I imagine that this was more or less Luria's position.

Churchman's second approach is to reformulate the ESP phenomenon in terms of currently acceptable scientific notions, such as *unconscious* perception or conscious fraud. This procedure is not as arbitrary as it might seem on first sight, because the "extra" in extrasensory perception is a conceptually fuzzy negative property anyhow. Thus, rather than defining ESP out of existence, it is trivialized. The second approach probably would have been acceptable to Luria too, but not to Roberts.

Finally, the third approach is to take the proposition of ESP literally and to attempt to examine in all seriousness the evidence for its validity. That was, more or less, Roberts' position. But, as Churchman points out, this approach is not likely to lead to satisfactory results. Parapsychologists can maintain with some justice that the existence of ESP has already been proven to the hilt, since no other set of hypotheses of psychology has received the degree of critical scrutiny that has been given to ESP experiments. And many other phenomena have been accepted on much less statistical evidence than that which has been offered for ESP. The reason that Churchman advances for the futility of a strictly evidential approach to ESP is that, in the absence of a hypothesis of how ESP *could* work, it is not possible to decide whether any set of relevant observations can be accounted for *only* by ESP, to the exclusion of alternative explanations. Churchman thus applies to the problem of ESP the principles of Karl Popper's "hypothetic deductive" theory of scientific discovery, according to which facts gain scientific meaning only within the framework of preconceived hypotheses.

After reading Churchman's review, I realized that Roberts would have been ill-advised to proceed with his ESP experiments, not because, as Luria claimed, they would not be "science," but because any positive evidence he might have found in favor of ESP would have been, and would still be, premature. That is, until it is possible to connect a phenomenon like telepathy with canonical knowledge of, say, electromagnetic radiations and neurophysiology, no demonstration of its occurrence can be appreciated.

Is the lack of appreciation of premature discoveries merely attributable to the intellectual shortcoming of scientists, who, if they were only more perceptive, would

22. Churchman 1966.

give immediate recognition to any well-documented scientific proposition? Polanyi is not of that opinion. Upon reflecting on the cruel fate of his theory half a century after first advancing it, he declares that . . . "this miscarriage of the scientific method could not have been avoided. . . . There must be at all times a predominantly accepted scientific view of the nature of things, in the light of which research is jointly conducted by members of the community of scientists. A strong presumption that any evidence which contradicts this view is invalid must prevail. Such evidence has to be disregarded, even if it cannot be accounted for, in the hope that it will eventually turn out to be false or irrelevant." [23]

This is a view of the operation of science rather different from that commonly held, under which acceptance of authority is seen as something that must be avoided at all costs. The good scientist is seen as an unprejudiced man with an open mind who is ready to embrace any new idea supported by the facts. As the history of science shows, its practitioners do not appear to act according to that popular view. . . . [24]

STRUCTURALISM

It is only since about mid-century, more or less contemporaneously with the growth of molecular biology, that a resolution of the age-old epistemological conflict of materialism versus idealism emerged in the form of what has come to be known as structuralism. [25] This development is another example of R. K. Merton's multiple discovery concept, since structuralism emerged simultaneously, independently, and in different guises in several diverse fields of study, for example in psychology, linguistics, anthropology, and biology. [26]

Both materialism and idealism take it for granted that all the information gathered by our senses actually reaches our mind; materialism envisages that thanks to this information reality is *mirrored* in the mind whereas idealism envisages that thanks to this information reality is *constructed* by the mind. But structuralism has provided the insight that knowledge about the world enters the mind not as raw data but in an already highly abstracted form, namely as *structures*. And in the preconscious process of converting step-by-step the primary data of our experience into structures, information is necessarily lost, for the creation of structures, or the recognition of patterns, is nothing else than the selective destruction of information. So since the mind does not, and cannot, gain access to the full set of data about the world, it can neither mirror nor construct reality. Instead, for the mind reality is a set of structural *transforms* of primary data taken from the world. This transformation process is hierarchical in that "stronger" structures are formed from "weaker"

23. Polanyi 1963.

24. Editor's note: Stent's paper makes a major excursion here into the uniqueness of discovery. Stent then invokes structuralism to explain uniqueness as well as prematurity.

25. Piaget 1970.

26. Merton 1961.

structures through selective destruction of information. And any set of primary data becomes meaningful only after a series of such operations has so transformed it that it has become isomorphic with a stronger structure preexisting in the mind.

Neurophysiological studies which Stephen Kuffler, David Hubel and Torsten Wiesel have carried out on the process of visual perception in higher mammals have not only shown directly that the brain actually operates according to the tenets of structuralism but also offer an easily understood illustration of those tenets.[27] According to these studies, the primary photoreceptors in the retina report the absolute light intensity that reaches the eye from individual points in the visual field. These primary data are not sent on from the retina to the brain, however. They are first transformed in the retina into information about the light-dark contrast existing at individual points in the visual field, the absolute intensity data having been largely destroyed in the abstraction process. Upon first reaching the brain, the light contrast data for individual points are then transformed into the light contrast data for individual straight edges, or point sets, in the visual field, the information about contrast at individual points being destroyed in that second abstraction process. And at the next level of processing in the brain, the contrast data for individual straight edges are transformed into the corresponding data for sets of parallel edges, or sets of point sets in the visual field, entailing further destruction of information about individual edges. It is not yet clear what transformations take place at the next higher level of processing in the visual pathway, but it is certain that the mind experiences reality without knowing the "real" point-to-point light intensity in its surrounding.

Finally, we may consider the relevance of structuralist philosophy for the problem in the history of science under discussion here. For prematurity, structuralism provides us with an understanding of why a discovery cannot be appreciated until it can be connected logically to contemporary canonical knowledge.

In the parlance of structuralism, canonical knowledge is simply the set of preexisting "strong" structures with which primary scientific data are made isomorphic in the mental abstraction process. Hence, data which cannot be transformed into a structure isomorphic with canonical knowledge are a dead end; in the last analysis, they remain meaningless. They remain meaningless, that is, until a way has been shown how to transform them into a structure that is isomorphic with the canon.

ACKNOWLEDGMENTS

I made an informal presentation of the ideas covered in this essay at a conference on the history of biochemistry and molecular biology, held in May 1970 at the American Academy of Arts and Sciences. I am indebted to the dozen or so con-

---

27. Kuffler 1953; Huble and Wiesel 1968.

ference participants whose vigorous discussion accompanied my presentation and who helped me focus my ideas more sharply. I am particularly grateful to Harriet Zuckerman for calling my attention to Polanyi's paper on prematurity.

BIBLIOGRAPHY

Ames, B. N., and B. Gany. 1959. "Coordinate Repression of the Synthesis of Four Histidine Biosynthetic Enzymes by Histidine." *Proc. Nati. Acad. Sci.* 45:1453–61.

Avery, O. T., C. M. MacLeod, and M. McCarty. 1944. "Studies on the Chemical Nature of the Substance Inducing Transformation in the Pneumococcus." *J. Exp. Med.* 79:137–58.

Baltimore, D. 1970. "Viral RNA-Dependent DNA Polymerase." *Nature* 226:209–11.

Bertani, G. 1953. "Lysogenic versus Lytic Cycle of Phage Multiplication." *Cold Spring Harbor Symp. Quant. Biol.* 18:65–70.

Chargaff, E. 1950. "Chemical Specificity of Nucleic Acids and Mechanism of Their Enzymatic Degradation." *Experientia* 6:201–9.

———. 1968. "A Quick Climb Up Mount Olympus." *Science* 159:1448–49.

———. 1971. "Preface to a Grammar of Biology." *Science* 172:637–42.

Churchman, C. W. 1966. "Perception and Deception." *Science* 153:1088–90.

Dienes, L. 1946. "Complex Reproductive Processes in Bacteria." *Cold Spring Harbor Symp. Quant. Biol.* 11:51–59.

Dunn, L. C. 1951. *Genetics in the 20th Century.* New York: Macmillan.

Feulgen, R., and H. Rossenbeck. 1924. "Mikroskopish-chemischer Nachweis einer Nucleinsäure vom Typus der Thymonocleinsäure und die darauf beruhende elktive Färbung von Zellkernen in mikroskopishen Präparaten." *Hoppe-Seyler's Z. Physiol. Chem.* 135:203–48.

Hershey, A. D., and M. Chase. 1952. "Independent Function of Viral Protein and Nucleic Acid in Growth of Bacteriophage." *J. Gen. Physiol.* 36:39–56.

Hubel, D. T., and T. N. Wiesel. 1968. "Receptive Fields and Functional Architecture of Monkey Striate Cortex." *J. Physiol.* 195:215–43.

Jacob, F., and J. Monod. 1961. "Genetic Regulatory Mechanisms in the Synthesis of Proteins." *J. Mol. Biol.* 3:318–56.

Kuffler, S. W. 1953. "Discharge Patterns and Functional Organization of the Mammalian Retina." *J. Neurophysiol.* 16:37–68.

Lamanna, C. 1968. Letter to the editor. *Science* 160:1397.

Medawar, P. B. 1968. "Lucky Jim." *New York Review of Books* (March 28).

Mendel, G. 1866. "Versuche über Pflanzen-Hybriden." *Verh. naturf. Ver.* Abhandlungen, Brünn 4, pp. 3–47.

Merton, R. K. 1961. "Singletons and Multiples in Scientific Discovery." *Proc. Am. Phil. Soc.* 105:470–86.

Piaget, J. 1970. *Le Structuralisme.* Paris: Presses Universitaires de France.

Polanyi, M. 1963. "Potential Theory of Adsorption." *Science* 141:1010–13.

Quarton, G. C. 1967. "The Enhancement of Learning by Drugs and the Transfer of Learning by Macromolecules." In *The Neurosciences,* ed. G. C. Quarton, T. Melnechuk, and F. O. Schmitt, pp. 744–55. New York: Rockefeller University Press.

Rhine, J. B. 1948. *The Reach of the Mind.* London: Faber and Faber.

Spiegelman, S. 1946. "Nuclear and Cytoplasmic Factors Controlling Enzymatic Constitution." *Cold Spring Harbor Symp. Quant. Biol.* 11:256–77.

Stent, G. S. 1968a. "That Was the Molecular Biology That Was." *Science* 160:390–95.

————. 1968b. Letter to the editor. *Science* 160:1397.

————. 1968c. "What They Are Saying about Honest Jim." *Quarterly Rev. Biol.* 43:179–84.

————. 1972a. "Prematurity and Uniqueness in Scientific Discovery." *Advances in the Biosciences* 8:433–49.

————. 1972b. "Prematurity and Uniqueness in Scientific Discovery." *Scientific American* 227 (December): 84–93.

Temin, H. M., and S. Mizutani. 1970. "RNA-Dependent DNA Polymerase in Virions of Rous Sarcoma Virus." *Nature* 226:1211–13.

Watson, J. D. 1968. *The Double Helix.* New York: Atheneum.

Watson, J. D., and F. H. C. Crick. 1953. "A Structure for Deoxyribonucleic Acid." *Nature* 171:737.

PART TWO

Observer and Participant Accounts

CHAPTER THREE

Prematurity, Nuclear Fission, and the Transuranium Actinide Elements

Glenn T. Seaborg

In 1934, some five years before the discovery of nuclear fission, as a first-year graduate student at Berkeley I began to read the papers coming out of Italy and Germany describing the synthesis and identification of several elements thought to be transuranium elements. In their original work that year, E. Fermi, E. Amaldi, O. D'Agostino, F. Rasetti, and E. Segrè bombarded uranium and other elements with neutrons and obtained a series of beta-particle-emitting radioactivities.[1] On the basis of the periodic table of that day, they believed that the first transuranium element, with atomic number 93, should be chemically like rhenium (and so designated eka-rhenium, or Eka-Re), the next, with atomic number 94, should be like osmium (Eka-Os), and so forth. Therefore they assigned a 13-minute activity associated with a substance that had chemical properties similar to rhenium to element 93, eka-rhenium. A classic paper by Fermi titled "Possible Production of Elements of Atomic Number Higher than 92," which I remember reading at the time, stated, "This negative evidence about the identity of the 13 min-activity from a large number of heavy elements suggests the possibility that the atomic number of the element may be greater than 92. If it were an element 93, it would be chemically homologous with manganese and rhenium. This hypothesis is supported to some extent also by the observed fact that the 13-min activity is carried down by a precipitate of rhenium sulphide insoluble in hydrochloric acid. However, as several

---

Editor's note: This chapter is an edited version of a presentation made in December 1997 at the symposium "Prematurity and Scientific Discovery," on which this volume is based; the chapter was revised by Seaborg in July 1998. Some bibliographic references have kindly been provided by Al Ghiorso, Mary Ann Singleton, and staff at the Lawrence Berkeley Laboratories. Seaborg suffered a stroke in August 1998 and died on 25 February 1999.

1. Fermi et al. 1934a,b.

elements are easily precipitated in this form, this evidence cannot be considered as very strong."[2]

Soon thereafter I read a paper by Ida Noddack titled (in translation) "On Element 93," which took issue with this interpretation and suggested that the radioactivities observed by Fermi and his colleagues might be due to elements with medium atomic numbers: "One could think that in the bombardment of heavy nuclei with neutrons these nuclei disintegrate into several larger fragments which, although they are isotopes of known elements, are not neighbors of the irradiated elements."[3] That is, Noddack was suggesting that, in bombarding the heavy element uranium, Fermi and his colleagues had not produced heavier elements, such as eka-rhenium and eka-osmium, but in fact had split the uranium atom into isotopes of lighter components. One implication of this interpretation was that the apparently new substance with the 13-minute activity, which had been produced by bombardment of uranium and which showed rhenium-like activity (precipitating with rhenium sulfide), was not the hypothesized element eka-rhenium but simply an isotope of the lighter rhenium itself! Thus she intimated that Fermi and his colleagues had split the atom into at least one lighter component—that, in essence, they had achieved nuclear fission! But this paper was not taken seriously. Experiments in Germany during the following years by O. Hahn, L. Meitner, and F. Strassmann appeared to confirm the Italian interpretation, and for several years the "transuranium elements" were the subject of much experimental work and discussion. In a typical paper by Hahn, Meitner, and Strassmann, part of a series they published during 1935 to 1938, they reported a 16-minute 93Eka-Re237, 2.2-minute 93 Eka239, 12-hour 94Eka-Os237, 59-minute 94Eka-Os239, and 3-day 95Eka-Ir239.[4]

In 1938, I. Curie and P. Savitch, working in Paris, found a product with a half-life of 3.5 hours that seemed to have the chemical properties of a rare earth, but they could not give an interpretation of this astonishing discovery. Their paper, titled (as I translate it) "On the Nature of a Radioactive Element with 3.5 Hour Half-Life Produced in the Neutron Irradiation of Uranium," included the following: "We have shown that[,] in the neutron irradiation of uranium[,] a radioactive element with a half-life of 3.5 hours is produced, with chemical properties similar to those of rare earths. In the following we will refer to it as R3.5h. . . . R3.5h separates cleanly from Ac by going to the 'head' [beginning of the fractionation] while Ac goes to the 'tail' [end]. It seems, therefore, that this species cannot be but a transuranic element having properties very different from those of the other known transuranic elements, a hypothesis that raises interpretational difficulties."[5]

2. Fermi 1934.

3. Noddack 1934. Editor's note: Chapter 10 in this volume elaborates on the reception of Noddack's suggestion.

4. Hahn et al. 1936.

5. Curie and Savitch 1938.

Then early in 1939, on the basis of work performed in December 1938, Hahn and Strassmann described experiments in which they had observed barium isotopes as the result of bombardment of uranium with neutrons. This historic paper, titled (as I translate it) "On the Identification and Behavior of Rare Earth Metals Produced in the Neutron Irradiation of Uranium," concluded, "We, as chemists, based on the briefly described experiments, should rename the abovementioned scheme and replace Ra, Ac, Th with the symbols Ba, La, Ce. As nuclear chemists, being in some respects close to physics, we have not yet been able to take this leap, which contradicts all previous experiences in nuclear physics. It could be that a series of strange coincidences could have mimicked our results."[6]

Subsequent interpretive work published shortly afterward by Meitner, who had been forced to leave Germany, and further work by her nephew Frisch, explained the apparent observations that Hahn and Strassmann found so anomalous that they were willing to postulate a series of strange coincidences rather than make the "leap" to believing they had produced much lighter elements and thus nuclear fission.[7] The Meitner-Frisch interpretation was confirmed shortly afterward by work in many laboratories showing that the radioactivities previously ascribed to transuranium elements were actually due to uranium fission products. Hundreds of radioactive fission products of uranium have since been identified.

Thus, in early 1939 there were again, as five years earlier, no known transuranium elements. During these five years I developed an increasing interest in the transuranium situation. When, in 1936, as a graduate student I gave my required annual talk at the weekly Research Conference at the College of Chemistry in Berkeley, I chose the transuranium elements as my topic, describing the work of Hahn, Meitner, and Strassmann.

During the two years following that talk, and before the discovery of fission, my interest in the neutron-induced radioactivities in uranium continued unabated and, in fact, increased. I read and reread every article published on the subject. I was puzzled by the situation—both intrigued by the concept of the transuranium interpretation of the experimental results and disturbed by the inconsistencies in this interpretation. I remember discussing the problem with Joe Kennedy, a colleague in research, by the hour, often in the early hours of the morning at the old Varsity Coffee Shop on the corner of Telegraph Avenue and Bancroft Way near the Berkeley campus, where we often went for a cup of coffee and a bite to eat after an evening spent in the laboratory.

I first learned of the correct interpretation of these experiments—that neutrons split uranium into two large pieces in the fission reaction—at the weekly Monday night seminar in nuclear physics conducted by Ernest O. Lawrence in Le Conte

6. Hahn and Strassmann 1939. Editor's note: See also Hook, chapter 10 in this volume for further comments.

7. Meitner and Frisch 1939; Frisch 1939.

Hall. On this exciting night in January 1939, we heard the news of Hahn and Strass-mann's beautiful chemical experiments. I recall that at first the fission interpreta-tion was greeted with some skepticism by a number of those present. But, as a chemist with a particular appreciation for Hahn and Strassmann's experiments, I felt that this fission interpretation just had to be accepted. I remember walking the streets of Berkeley for hours after this seminar, both exhilarated by the beauty of the work and disgusted at my inability to have arrived at this interpretation myself despite years of contemplation on the subject.

Products of uranium bombardment by neutrons were actually radioactive isotopes of lighter elements and thus were fission-product elements, such as barium, lan-thanum, iodine, tellurium, or molybdenum. Subsequently, during an investigation of the fission process, Edwin M. McMillan discovered a radioisotope with a half-life of 2.3 days. Working at the University of California, Berkeley, in the spring of 1939, he was trying to measure the energies of the two main recoiling fragments from the neu-tron-induced fission of uranium. He used the 60-inch cyclotron as a source of neu-trons from the reaction of 16 MeV deuterons with beryllium. He placed a thin layer of uranium oxide on one piece of paper, and next to this he stacked very thin paper sheets to stop and collect the fission fragments from uranium. The paper he used was ordinary cigarette paper, the kind used by people who rolled their own cigarettes.[8]

In the course of these studies, he found that the 2.3-day activity did not recoil sufficiently to escape. This activity was further investigated by Emilio Segrè, whose lack of chemical sophistication led him to identify the product as a lanthanide el-ement.[9] But in fact it was not. Further work by McMillan and Philip H. Abelson later in 1940 identified it as the first true transuranium element. Its atomic num-ber was 93; they named it neptunium.[10]

Shortly after the discovery of neptunium, McMillan, Joseph W. Kennedy, Arthur C. Wahl, and I discovered plutonium (atomic number 94) in late 1940 and early 1941, also at the University of California, Berkeley.[11] Even more surprises were to come. One expected these new elements, neptunium and plutonium, to have chem-ical properties similar to those immediately below them in the same columns of the periodic table, rhenium and osmium respectively. But the tracer-chemical experi-ments with neptunium and plutonium showed that their chemical properties were much like those of uranium and not at all like those of rhenium and osmium! That is, they had nothing like the properties predicted from their extrapolated position as "eka" analogs of the lighter elements apparently in the same place in the peri-odic table. The pre–World War 2 periodic table had misled Fermi and Hahn and their coworkers into thinking they had found "eka," or transuranium, elements when they had actually discovered fission. See fig. 3.1.

8. McMillan 1939.
9. Segrè 1940.
10. McMillan and Abelson 1940.
11. Seaborg et al. 1946.

1 H																	2 He
3 Li	4 Be											5 B	6 C	7 N	8 O	9 F	10 Ne
11 Na	12 Mg											13 Al	14 Si	15 P	16 S	17 Cl	18 Ar
19 K	20 Ca	21 Sc	22 Ti	23 V	24 Cr	25 Mn	26 Fe	27 Co	28 Ni	29 Cu	30 Zn	31 Ga	32 Ge	33 As	34 Se	35 Br	36 Kr
37 Rb	38 Sr	39 Y	40 Zr	41 Nb	42 Mo	(43)	44 Ru	45 Rh	46 Pd	47 Ag	48 Cd	49 In	50 Sn	51 Sb	52 Te	53 I	54 Xe
55 Cs	56 Ba	57–71 La– Lu	72 Hf	73 Ta	74 W	75 Re	76 Os	77 Ir	78 Pt	79 Au	80 Hg	81 Tl	82 Pb	83 Bi	84 Po	(85)	86 Rn
(87)	88 Ra	89 Ac	90 Th	91 Pa	92 U	(93)	(94)	(95)	(96)	(97)	(98)	(99)	(100)				

57 La	58 Ce	59 Pr	60 Nd	(61)	62 Sm	63 Eu	64 Gd	65 Tb	66 Dy	67 Ho	68 Er	69 Tm	70 Yb	71 Lu

Figure 3.1. Periodic table before World War 2. Note: An element above another in the figure is lower in atomic number, and so denoted in the text as "below" the other.

For a few years following this, uranium, neptunium, and plutonium were considered to be sort of cousins in the periodic table. It was thought that elements 95 and 96 should be much like them in their chemical properties. Thus it was thought that these and the following elements formed a 14-member "uranide" (chemically similar to uranium) group in which the *f* electron shell filled as the atomic number increased. These assumptions proved to be wrong, and the results of the experiments directed toward the discovery of elements 95 and 96 apparently refused to fit the pattern indicated by the periodic table of 1944.

In 1944, I conceived the idea that perhaps all the known elements heavier than actinium (atomic number 89) were misplaced on the periodic table. I advanced the theory that these elements heavier than actinium might constitute a second series analogous to the series of rare-earth or lanthanide elements.[12] The lanthanides are chemically very similar to each other and usually are listed in a separate row below the main part of the periodic table. This would mean that all these heavier elements really belonged with actinium—directly after radium in the period table—just as the known lanthanides fit in with lanthanum between barium and hafnium. See fig. 3.2.

The new concept meant that elements 95 and 96 should have some properties in common with actinium and some in common with their rare-earth "sisters," europium and gadolinium (i.e., those immediately below them in the rare-earth side columns of what would be a revised periodic table), especially with respect to the difficulty of oxidation above the III state. When experiments were designed according to this new concept, elements 95 and 96 were soon discovered (in 1944 and 1945) at the wartime Metallurgical Laboratory at the University of Chicago—that is, they were synthesized and chemically identified.[13]

The discovery of elements 95 and 96 was initially classified as secret. I revealed it publicly for the first time informally on a nationally broadcast radio program, the *Quiz Kids*, on which I appeared as a guest on 11 November 1945. The discovery information had already been declassified for presentation at an American Chemical Society meeting to be held at Northwestern University the following Friday. Participating on the program with me were "quiz kids" Sheila and Patrick Conlan, Robert Burke, Harvey Fishman, and Richard Williams. When Richard asked me if any new elements had been discovered in the course of research on transuranium elements during the war, I revealed the discovery of elements 95 and 96. Apparently, many kids in America told their teachers about it the next day, and, judging from some of the letters I received from such youngsters, they were not entirely successful in convincing their teachers.

The revised periodic table listed the heaviest elements as a second rare-earth series. These heaviest elements, named actinide elements, were paired off with those

12. Seaborg 1945.

13. Cunningham 1945, pp. 5–6; Seaborg, James, and Morgan 1949; Seaborg, James, and Ghiorso 1949; Ghiorso et al. 1950.

1 H 1.008																1 H 1.008	2 He 4.003
3 Li 6.940	4 Be 9.02											5 B 10.82	6 C 12.010	7 N 14.008	8 O 16.000	9 F 19.00	10 Ne 20.183
11 Na 22. 997	12 Mg 24.32	13 Al 26.97										13 Al 26.97	14 Si 28.06	15 P 30.98	16 S 32.06	17 Cl 35.457	18 A 39.944
19 K 39.096	20 Ca 40.08	21 Sc 45.10	22 Ti 47 90	23 V 50 95	24 Cr 52.01	25 Mn 54 93	26 Fe 55.85	27 Co 58 94	28 Ni 5869	29 Cu 63.57	30 Zn 65 38	31 Ga 69.72	32 Ge 72.60	33 As 74.91	34 Se 78.96	35 Br 79.916	36 Kr 83.7
37 Rb 85.48	38 Sr 87.63	39 Y 88.92	40 Zr 91.22	41 Cb 92.91	42 Mo 95.95	43	44 Ru 101.7	45 Rh 102.91	46 Pd 106.7	47 Ag 107.880	48 Cd 112.41	49 In 114.76	50 Sn 118. 70	51 Sb 121 76	52 Te 127 61	53 I 126.92	54 Xe 131.3
55 Cs 132.91	56 Ba 137.36	57 LA 138.92 / 58-71 SEE La SERIES	72 Hf 178.6	73 Ta 180.88	74 W 183 92	75 Re 186.31	76 Os 190 2	77 Ir 193.1	78 Pt 195.23	79 Au 197.2	80 Hg 200.61	81 Tl 204.39	82 Pb 207.21	83 Bi 209.00	84 Po	85	86 Rn 222
87	88 Ra	89 AC / SEE Ac SERIES															

LANTHANIDE SERIES	57 La 138.92	58 Ce 140.13	59 Pr 140.92	60 Nd 144 27	61	62 Sm 150.43	63 Eu 152 0	64 Gd 156.9	65 Tb 159.2	66 Dy 162.46	67 Ho 163.5	68 Er 167.2	69 Tm 169.4	70 Yb 173.04	71 Lu 174.99
ACTINIDE SERIES	89 Ac	90 Th 232.12	91 Pa 231	92 U 238.07	93 Np 237	94 Pu	95	96							

Figure 3.2. Periodic table showing heavy elements as members of an actinide series (arrangement by Glenn T. Seaborg, 1945).

in the already known lanthanide rare-earth series in a table published in *Chemistry and Engineering News,* 10 December 1945, under the title "Chemical and Radioactive Properties of the Heavy Elements."[14] This actinide concept was initially received with much skepticism. Moving all these elements from the main body of the periodic table to a place below—that is, outside the main body—was too much to accept. Two leading inorganic chemists, Wendell Latimer at the University of California at Berkeley and Don Yost at the California Institute of Technology, both said it was wrong and cautioned me that its publication would ruin my scientific reputation. I replied that I had no scientific reputation to ruin and went ahead and published it.

This new understanding of the actinide series led not only to the elements americium and curium (95 and 96) but, subsequently, to the synthesis and identification of berkelium and californium (97 and 98) in 1949 and 1950, einsteinium and fermium (99 and 100) in 1952 and 1953, mendelevium (101) in 1955, and nobelium (102) in 1958. Discovery in 1961 of lawrencium (103), the last element predicted from extrapolation from the rare earths, signaled the end of the actinide series. Elements discovered subsequently (104 and above) are in the main body of the table.

BIBLIOGRAPHY

Cunningham, B. B. 1945. *Metallurgical Laboratory Report.* CS-3312. Chicago: University of Chicago.

Curie, I., and P. Savitch. 1938. "Sur la nature du radioélément de période 3–5 heures formé dans l'uranium irradié par les neutrons." *Comptes Rendus* 206:1643–44. Translated as "Concerning the Nature of the Radioactive Element with 3–5 Hour Half-Life, Formed from Uranium Irradiated by Neutrons," in *The Discovery of Nuclear Fission: A Documentary History,* ed. H. G. Graetzer and D. L. Anderson, pp. 37–38 (New York: Arno Press).

Fermi, E. 1934. "Possible Production of Elements of Atomic Number Higher Than 92." *Nature* 133:898–99.

Fermi, E., O. Amaldi, R. F. D'Agostino, F. Rasetti, and E. Segrè. 1934a. "Artificial Radioactivity Produced by Neutron Bombardment." *Proc. Roy. Soc.* (London), ser. A, 146:483–500.

———. 1934b. "Radioattiva provocata da bombardemento di neutrini III." *Ric Scientifica* 5:452–53.

Frisch, O. R. 1939. "Physical Evidence for the Division of Heavy Nuclei under Neutron Bombardment." *Nature* 143:276.

Ghiorso, A., R. A. James, L. O. Morgan, and G. T. Seaborg. 1950. "Preparation of Transplutonium Isotopes by Neutron Irradiation." *Phys. Rev.* 78:472.

Hahn, O., L. Meitner, and F. Strassmann. 1936. "Neue Umwandlungsprozesse bei Bestrahlung des Urans; Elemente jenseits Uran." *Ber. Dt. Chem Ges.* 69:905–19.

Hahn, O., and F. Strassmann. 1939. "Über den Nachweis und das Verhalten der bei der Bestrahlung des Uransmittels Neutronen entstehenden Erdalkalimetallen." *Naturwissenschaften* 27:11–15. Translated as "Concerning the Existence of Alkaline Earth Metals

14. Seaborg 1945.

Resulting from the Neutron Irradiation of Uranium," trans. H. G. Graetzer, *Am. J. Phys.* 32 (1964): 10–14.

McMillan, E. M. 1939. "Radioactive Recoils from Uranium Activated by Neutrons." *Phys. Rev.* 55:510.

McMillan, E. M., and P. H. Abelson. 1940. "Radioactive Element 93." *Phys. Rev.* 57:1185–86.

Meitner, L., and O. R. Frisch. 1939. "Disintegration of Uranium by Neutrons: A New Type of Nuclear Reaction." *Nature* 143:239–40.

Noddack, I. 1934. "Über das element 93." *Angewandte Chemie* 47:653–55.

Seaborg, G. T. 1945. "Chemical and Radioactive Properties of the Heavy Elements." *Chem. Eng. News.* 23:2190.

Seaborg, G. T., R. A. James, and A. Ghiorso. 1949. "The New Element Curium (Atomic Number 96)." (National Nuclear Energy Series, Plutonium Project Record, vol. 14B, Paper No. 22.2.) In *The Tranuranium Elements: Research Papers,* pp. 1554–71. New York: McGraw-Hill.

Seaborg, G. T., R. A. James, and L. O. Morgan. 1949. "The New Element Americium (Atomic Number 95)." National Nuclear Energy Series, Plutonium Project Record, vol. 14B, Paper No. 22.1. In *The Tranuranium Elements: Research Papers,* pp. 1525–53. New York: McGraw-Hill.

Seaborg, G. T., E. M. McMillan, J. W. Kennedy and A. C. Wahl. 1946. "Radioactive Element 94 from Deuterons on Uranium." *Phys. Rev.* 69:366–67.

Seaborg, G. T., A. C. Wahl, and J. W. Kennedy. 1946. "Radioactive Element 94 from Deuterons on Uranium." *Phys. Rev.* 69:367.

Segrè, E. 1939. "An Unsuccessful Search for Transuranium Elements." *Phys. Rev.* 55:1104–5.

CHAPTER FOUR

Resistance to Change and New Ideas in Physics

A Personal Perspective

Charles H. Townes

I will try to illustrate the problems which Gunther Stent and others have raised about the hang-ups in science; what we do wrong and what we miss. Striking examples have occurred in radio astronomy and in development of the maser and laser. As I've had some contact with these fields, I'll develop these ideas from a very personal point of view. That may help to provide some specific insights.

RADIO ASTRONOMY

Radio waves were detected in outer space by Karl Jansky, an engineer at Bell Labs who was assigned the idea of finding out from where radio noise was coming.[1] He was a good engineer and found some strange noise which, he determined with the help of astronomers, arose from the center of our galaxy, the Milky Way. That was a tremendous discovery. He was very interested, and continued to work on it. Astronomers essentially did very little follow-up work, yet it was the first detection of anything at the center of our galaxy. Astronomers, who normally worked with visible light, had never seen anything there because it is surrounded by dust clouds. Obviously the first radiation detected coming from the center of the galaxy would seem highly important. I read about this when I was an undergraduate student and felt it was tremendous. And nobody could explain the source of such radiation. Another engineer, Grote Reber, built a radio telescope in his backyard to detect this radiation.[2] He commented, "Well, it sounds like waves beating on the shore." While he obviously didn't get much of the astronomy, he was making measurements and the astronomers weren't.

1. Jansky 1932.
2. Reber 1940.

As I see it, there are a number of reasons why things don't develop, or are premature. One is that people may not see the significance of a new idea or how deeply it can penetrate. Its significance may only become clear later, after subsequent things have come along. One may have a great idea that can lead somewhere, but if nobody can see where it leads, then it can't go anywhere until there is a subsidiary idea that is applied well. That's one type of reason. The "not invented here" syndrome provides another explanation for why things get neglected. In the case of radio waves, a radio engineer discovered something bearing on astronomy. The astronomers were working with light telescopes, and they didn't know anything about radio waves. This, they said to themselves, was a funny thing that some engineer at Bell Labs found and not us. We don't see how it connects. This type of "not invented here" reaction accounts for a good deal of lack of interest in important developments. Another problem is that while one may have a promising technological idea, the technology may just not be ready yet. People don't know quite how to do it. To really make it pay off one must have further technological developments. Only when the additional technology comes along can the field become tremendously important and hot. Still another phenomena that gets us in trouble, I think, comes from focusing too strongly on one particular channel or one particular idea. We make strong assumptions and get frozen into them. It's not only the fixed assumptions that we make; it's that we just have a habit of thought going down a particular path. And if that is successful, and popular, and everybody's doing it, then we ignore another channel leading off somewhere perhaps even more interesting.

Clearly astronomers didn't know much about radio waves in the 1930s, and they didn't see how they could use them. Moreover, the field was technically not quite ready in a sense, although one could do some things, and really hard work then would have done quite a lot. However, World War II brought the development of radar with magnetrons and klystrons—new ways of making radio waves, and also new ways of detecting them. With this background, engineers began to take advantage of Jansky's discovery, both measuring and trying to explain the radio waves coming in from space. Eventually, the astronomers began to wake up to possibilities after engineers and physicists made clearly important discoveries.

I was at Bell Telephone Labs during World War II, and after the war, I thought it would be a great place to do radio astronomy, so I went off to see a former professor of mine, Professor I. S. Bowen at Cal Tech. He had been one of my favorite professors at Cal Tech; he'd always been very kind to me, and was by then head of both the Mount Wilson Observatory and the Palomar Observatory—a big shot in astronomy. I came out West and said to him, "Look, I've gotten interested in going into radio astronomy. It seems to me it's a good thing to do at Bell Labs. What do you think are the most important things to do?" Bowen looked at me and said, "Well, you know, I'm sorry to disappoint you, but I don't really think radio waves will do anything for astronomy." I was at Bell Labs, all equipped and ready to work in the field, but he concluded, "No, I'm really sorry . . . " This was 1945, right after the War. I felt he was probably wrong, but I didn't know what to do. I was not much of an astronomer. I thought I knew what to do in another direction, so I went off in the other direction and did work which also turned out to be rewarding.

While American astronomers didn't push that field at all after the war, even though we had all the technology, the British and Australians pushed it very hard. One reason, I think, was that in contrast to the U.S. then they didn't have much money, but *did* have radar equipment and could use it. It was a function of what could be done with available money. Also Lawrence Bragg, the head then at the Cavendish Laboratory of Cambridge University in England, had brought a great impetus to crystallography and interferometry when he succeeded Ernest Rutherford there. I talked to Bragg some time before his death, and he commented about the relationship between crystallography and radio astronomy. He said, "You know, my science has always been optics. And really x-ray crystallography and radio astronomy are both interferometry on different wavelength scales."

The Dutch also did significant work in this field, I think largely because of one or two particular astronomers in the Netherlands who recognized its importance, so the field developed there as well as in Australia and Britain after the war. Professor Bart Bok, a Dutch astronomer, was at that time professor in the Harvard University department of astronomy, and a good friend of mine. He would come down to visit occasionally at Columbia University, where I was. He said there were very few people with whom he could talk at Harvard, and the Harvard Astronomy Department was not enthusiastic about his work because he had turned from optical astronomy to radio astronomy. They didn't think there was much in it. He eventually resigned and went to Australia because at the time he found it a hard life doing radio astronomy at Harvard, even though he already had a full professorship. The situation gradually changed. One big impetus came about in the following way. Professor Jan Oort was a very famous theoretical astronomer in the Netherlands. The country has been outstanding in astronomy for a century or more for special historical reasons, although they frequently have a cloudy sky. He was asked to give a very special honorary lecture in London. He talked about radio astronomy. I remember my astronomer friends in the U.S. saying, "Gee, Oort talked about radio astronomy! That was a very important event, and he talked about radio astronomy. Maybe there's something in it." After all, Jan Oort was one of the great astronomers of the day.[3] They began to get interested.

The technology came along too. The British pushed hard on interferometry, which allows high angular resolution. That illustrates one of the difficulties which put off astronomers. With long waves, as radio waves are, it was hard to get high angular resolution. Yet while high angular resolution is important, it was by no means critically important to advances.

Arno Penzias, one of my former students, had gone to Bell Labs. He and Robert Wilson looked carefully for background "noise" and found radio frequency radia-

---

3. E. Robert Pail (1997) has written that Oort during his lifetime "was considered by many as the greatest living astronomer of the twentieth century."

tion coming from all directions. They realized they had observed something which was hard to explain. The two were able to recognize, with the help of others, that this was residual noise due to the origin of the universe—one "Big Bang." What could be more fundamental and more important to astronomy than to discover the origin of the universe? And actually, the Big Bang radiation had been detected before Penzias and Wilson.[4] But it was regarded simply as some strange noise coming in from all directions, and not important enough to be followed up carefully. There had also been theoretical papers published saying that if there had been a Big Bang, then there ought to be residual radiation noise.[5] But theoretical scientists didn't know much about microwaves, so perhaps they didn't think it could be detected. Almost any radio engineer would have known that if it were as much as predicted, they could detect it. But the theoretical paper wasn't read by engineers or by people who were in radio astronomy.

Another interesting aspect of the field was the discovery of molecules in space with radio astronomy. I was asked by a Dutch astronomer, when I was on sabbatical in Europe in 1955, to give a talk to an international meeting in England. He knew I was interested in molecular spectra and somewhat interested in astronomy. And I had been talking informally with him about the possibility of detecting molecules, that is, finding radio radiation emanating from molecules in space. In my talk I outlined various molecules I felt might be there, their frequencies and how to detect them with radio astronomy. I thought somebody ought to look. He congratulated me on a very interesting talk. But nobody paid any attention. No astronomers ever tried to do it. However, Alan Barrett, my student, got interested.

After getting his degree with me, Alan worked at the problem. He looked for the molecule OH, which I thought was a good choice as it was similar in some properties to the three molecules in space that had been found and identified by methods of optical astronomy in the late 1930s and early 1940s.[6] These were CH, CH+, and CN. They're free radicals, very interactive unstable molecules, so their existence is temporary, in interstellar regions highly excited by radiation. So OH, another simple free radical, seemed a natural choice for which to search. Alan looked for it but failed at first. He then got a job at the University of Michigan in the astronomy department. His boss called me at one time, saying, "Look, you recommended Alan to me, but he's locked in on OH, and he isn't getting anywhere, I worry about him. I don't think he's doing much." I said, "Well, I thought OH was a sensible thing to do." Fortunately Alan then got a job in the engineering department at MIT. He kept on. He got an engineer to help him and he did eventually find OH. This was the first spectrum of a molecule found by using radio waves. That was a major discovery.

Astronomers were very happy about the discovery of OH and began work on it, but then the work slowed down. I had made my first suggestion in 1955. Alan found

---

4. McKellar et al. 1941; DeGrasse et al. 1959; Ohm 1961; Herzberg 1950.
5. Gamow 1948; Alpher and Herman 1949; Dorosch Kevich and Novikov 1964.
6. Swings and Rosenfeld 1937; Douglas and Herzberg 1941.

OH after several tries in 1963. Then nothing else happened. I moved to Berkeley in 1967. I said to myself, "Well, this is a time I'm going to really go into astronomy, and I'm going to try to look for more molecules." While four unstable molecules had been found, astronomers were sure stable and more complex molecules weren't there. They had theoretical reasons. The gas in interstellar space is so rarefied they believed that atoms there wouldn't be sticking together to form stable molecules. Now, there was actually some contrary evidence, but it was not obvious. It was in the astronomical literature, but astronomers tended to ignore it. There were certain interstellar clouds in which atomic hydrogen was not found. This provided a puzzle which only a few astronomers thought was worth a comment.[7] But it suggested to me that it might be a sign that hydrogen in these clouds was in molecular, not atomic, form, and thus it could be worthwhile to look for stable molecules with hydrogen. So when I arrived in Berkeley in 1967, I said to myself, "I think we ought to look and give it a try." Jack Welch, a radio astronomer at Berkeley, agreed to work with me, and he was open-minded about it. But several astronomers in the department disagreed. In fact, George Field, the former chairman, has written a book stating that he had argued that the molecule I wanted to look for couldn't be there, and told me so.[8] His conclusion was theorists should never advise experimentalists because the experimentalists won't believe them anyhow!

We started by looking for ammonia (NH_3), and there it was. We looked for water (H_2O), and there it was too. Now it happened that just about nine months before that, two young postdocs at the Center for Radio Astronomy in West Virginia had asked for time on the radio telescope to look for water. A peer committee reviewed it and said, "That's crazy. That would be a waste of time with the telescope. We mustn't let them try to do that." Six months later after we found water, they immediately gave the two young postdocs time on the telescope, of course, and they found lots more molecules. Soon, everybody else was finding molecules, and it became a very rich field.[9] So in radio astronomy a lot of resistance originated from the fact that radio waves involved a field with which the people in mainstream astronomy were not familiar. And they reacted perhaps with a "not invented here" syndrome. After all, they had a lot of important things to do and on which their attention was fixed, and work with radio waves didn't make sense to them.

THE MASER

The maser (an acronym for "microwave amplification by stimulated emission of radiation") is based on physical principles well-known since the mid 1920s to physicists familiar with quantum mechanics. The only new things about them lay in putting together things in the right way and recognizing that it was important to do

7. Garzoli and Varsavsky 1966.
8. Field and Chaisson 1985.
9. Rank et al. 1971.

that. Why were they missed? I think in part the physicists weren't terribly interested in making oscillators. Another aspect was that by the time quantum mechanics came in strongly, say by the 1930s, research interest in optics was dying out.

Optics had been a very hot field during the first part of this century but then became regarded as an old classical field. People felt most of the important things had been done, that they understood waves and optics rather fully. With some new technology they recognized one could do things a little bit better. But they ignored or were getting rid of their spectrometers, which had been installed in basements of the physics buildings in the earlier part of the century when optics and spectroscopy were hot fields. Physicists were moving into nuclear physics. Optics was old stuff.

I wasn't thinking in that direction either. From immediately after World War II until the late 1950s, I was working on microwave spectroscopy of molecules with waves generated by klystrons and magnetrons. Those were devices built basically by electrical engineers. I was using them, and trying to get shorter and shorter waves. As one gets shorter and shorter waves, the field of molecular spectroscopy gets richer and richer. More and more spectral lines appear and absorption is stronger. We could then make waves of about a half a centimeter, which was as short a wavelength as could be produced by these electronic oscillators. I wanted to make a real change, to get on down to below a millimeter wavelength.

I tried all kinds of methods for producing radiation. They weren't successful enough to encourage me to develop them further. The Navy even asked me to form a committee to examine how to get down to shorter waves. They didn't know what they could use them for, but they wanted to develop the technology. The committee held meetings for a year or two, exploring all kinds of things, and visiting laboratories. We never came up with any great ideas. The last meeting we had was in Washington. I was chairman, and I was concerned about it. I got up early in the morning, wondering why we hadn't been able to make any progress. I went out in a nearby park, sat down, and mulled over the situation. I knew we had tried this and that, and these things couldn't work for such and such a reason. We can't make an oscillating electronic tube because it has to be very small to couple with the wavelength of size like a millimeter, it has to dissipate a lot of energy, and so it gets overheated. We just can't put that much power in it. Well, we really ought to use molecules because they've got resonances in them at just the right frequencies. We ought to use molecules. But, of course, the second law of thermodynamics says that you can't get out of a molecular gas more energy than just the heat radiation. It says there's a limit to the intensity you can get out, and that intensity was very low. I had been through that reasoning before, and said to myself, "Well, no, you can't use molecules." But suddenly, I thought, "Now, wait a minute! Wait a minute. The second law of thermodynamics doesn't have to apply! That applies when you have things in thermal equilibrium. But we have excited molecules in the laboratory. We can pick out molecules in excited states. They don't obey thermodynamics in the normal sense. And so they don't have to obey that law."

I had all the right background. I was at Columbia where Professors I. I. Rabi and Polykarp Kusch had been working on molecular beams for many years. I'd just heard a talk by a German physicist about how to increase the intensities of a molecular beam. I'd been working with metal cavities resonant at microwave wavelengths, in the laboratory. So I put the ideas all together, I could pick out molecules all in an excited state and in an intense beam. I would then pass them through a cavity and have them radiate. In the cavity, the radiation would bounce back and forth, stimulate the molecules to give off more energy, and energy would build up. I pulled out an envelope and worked it all out and said, "Yes, I think one can get enough molecules to do that. It's kind of a long shot, but it looks like it ought to be possible." I went back to Columbia, and a few months later I had a student who was willing to take a chance on this and try it out as a thesis. People would come through the laboratory and say, "Oh that's a cute idea," and that was all. Nobody else was interested. It wasn't a hot field for anybody. Nobody else tried to do it. I was quite open about the idea, so many people learned about it.

After about two years of work, one day my friends I. I. Rabi and Polykarp Kusch came into my office. They sat down and said, "Look! You know that's not going to work. We know it's not going to work. You really ought to stop. You're just wasting money." Rabi had been chairman of the department, and Kusch was then chairman of the department. I was a young professor. It's very rare that professors come into another's office and say, "You really ought to stop that work because it's silly and you're wasting money. You'd better stop." Fortunately, I had tenure. I was an associate professor by then. Tenure is an important thing. I said, "I think it has a reasonable chance of working." They left after a little huffing and puffing. In another three months we had it working. Kusch came around later and said, "Well, I guess I ought to know you know more about what you're doing than I do." Of course now we had it working. It was contrary to the general mode of thought of the time, but not contrary to physics that many good physicists knew. Almost all well-trained physicists knew something about the laws of physics which were used. The electrical engineering involved was also well known to engineers—how resonant cavities work, how to make resonances, how to have low loss in the cavities, etcetera. All those things were known. It's just putting it all together that created a new field.

But there was at least one aspect that puzzled people. In fact, L. H. Thomas, a famous theorist at Columbia at that time, argued with me a great deal about it. "It can't do that, you know. You're saying it's going to oscillate on a fairly pure frequency. No, it can't do that. It violates the uncertainty principle. The molecules go through the cavity in a short time. The uncertainty principle means you can't define their frequency that accurately." I argued with him about it. I carefully looked up in my notes from my quantum mechanics course back in the 1930s and found yes, the theory was all there. It said that radiation would have to be coherent, that is, at exactly the same frequency. Each molecule would emit exactly the same frequency with which it was stimulated. Then we got it to work, and, sure enough it did. We built two of them. They beat together. We could hear them beating together at a very pure frequency. After it worked, Professor Thomas never spoke to me again about it.

I had a young postdoc who also argued it couldn't work. He bet me a bottle of Scotch it wouldn't. And he paid up. But, more surprising even than that, after I had the device working, I was walking on the street with Niels Bohr. Of course, he asked me what I was doing. I told him, and he said, "Now, wait a minute. That's not possible. No. That isn't. Molecules can't do that. They can't give you pure frequencies." And I explained, "Well, they can and we have shown that they do." I don't think Niels Bohr ever really quite understood it. Perhaps he was just being nice to a young man when he said, "Well, maybe you're right." Then there was the reaction of John von Neumann, the very famous mathematician, and a wonderful theorist who did all kinds of things. I ran into him at a cocktail party at Princeton. And again, as is common in such situations, he said, "Well, what are you doing?" I told him. He exclaimed, "Oh, that can't be right. No, no, you can't get a pure frequency out of that." And I said, "Well, yes, and we have it." He said, "You're fooling yourself somehow," and went off and got another cocktail. And about fifteen minutes later he came back, and said, "You're right. You're right." And then he asked me if I could do the same thing with a semiconductor. I said, "Well, my primary aim was to get down into the infrared wavelengths down below a millimeter, and semiconductors might be good for that. I don't really see at the moment how I can do it with a semiconductor but, yes, in principle I could." He wanted to talk all about that.

I later found that von Neumann had written a letter to Edward Teller a couple of years earlier saying he thought that maybe one could take a semiconductor, excite the electrons into upper states with neutrons, and from that get a very powerful beam of light. That's exactly what I was talking about. Von Neumann didn't say, "Well, I thought of this myself." Obviously, he had. But he didn't realize that the light would be coherent. He didn't realize that it would come out in a pure frequency and in a very pure wave. He clearly didn't because he was arguing against it when we first discussed it. Teller apparently never answered his letter. So his idea died.

If you look back on the record, you'll find about a half a dozen people who had suggested this previously. One German, Houtermans, told me, "I had that idea. You know, somebody was running a hydrogen discharge and they came to me and said that they had very intense light of a peculiar kind, and they didn't understand it." I thought about it and said, "Oh it must be stimulated emission." It was perhaps getting atoms into excited states, releasing their energy due to the wave passing by them, and thus providing more and more energy. But then they came back a couple days later and said they had an explanation which was much more prosaic, and, yes, I felt that was the right explanation, so I never did anything about it and never wrote it up." Well, he did write it up after the laser came, the whole history, how he'd had this idea.[10]

10. Houtermans 1960.

Many of us have ideas and toss them off, and don't take them all that seriously, don't realize their importance until later after someone else has done something with one of them. But in retrospect, yes, many people later can claim the idea. Tolman, for instance, a theoretical chemist back in the mid-1920s, wrote that if you can get more molecules in the excited than in the lower state, there can be negative absorption (more energy generated than lost). He was explaining absorption of light by molecules. And if you have a molecule in the lower state, it will absorb as it gets knocked up to a higher state. A molecule in a higher state will emit energy by falling down to a lower state. It is really the balance between those two that produces absorption, and with ordinary gases, there's always more in the lower state, so there's always absorption. But he pointed out that if there are more in the upper state, this would give you negative absorption, so you'd have more energy.[11]

A Russian named V. A. Fabrikant actually tried to find such an effect. I don't know how many others there were. Every once in a while somebody new says to me, "I remember thinking about this idea." But they missed the technology and they missed the fine points of why it was important. I had thought about it myself a couple of years before I recognized its importance and started work. I felt that stimulated emission was well-established theory, but that, as no one had observed it correctly, it would be nice to try an experiment and observe it. But I also thought it would be hard work and didn't really see any strong reason then to do the experiment, because I felt sure the theory was correct. Even in ordinary gases, we have some molecules going up, and some of them coming down. They have to be coming down and giving up energy. So there's no doubt the effect is there, hence why do a hard experiment just to show that it is true?

Now, on the other hand, to use that effect to produce high frequencies to do spectroscopy and other science, for that I had motivation. And I had enough engineering background to see how to do it. I also knew the physics, which most engineers didn't. The engineers, at that time, knew relatively little quantum mechanics. And the physicists weren't very concerned with oscillators, resonators, or feedback, which were familiar to the engineers. I happened to have the right mixture in my background. So once having the motivation, the background, and being aware of molecular beams, I used a molecular beam mechanism. There were a lot of other mechanisms that would do it. But the idea grew out of the field of microwave spectroscopy. And it's now clear that masers and lasers had to come out of microwave spectroscopy for the following reasons. There were three legitimate, independent ideas at about that time. One from myself. One came from the Russians N. G. Basov and A. M. Prokhorov, who later shared the Nobel Prize with me. They were microwave spectroscopists. They proposed using a different gas. It wasn't as usable, but they had the idea. Their theory was all right. They understood coherence. They understood feedback. Still another initiation of the general idea came from Joe Weber at the University of Maryland. He had been an engineer who con-

11. Tolman 1924.

verted to a microwave spectroscopist. He gave a talk at the Electrical Engineering Society in 1952 because, he told me later, he wanted engineers to recognize that there were new and other ways of getting amplification. It was a theoretical talk, and he didn't take it seriously enough to do anything in the laboratory about it. In fact, his numbers were off by about a factor of 10 million. But, nevertheless, he had the concept. He had it right. So two other initial ideas came from people with exactly the same background as mine. These ideas all occurred with in a couple of years of each other, but the people involved weren't in close enough touch that we knew just what the others were doing.

The maser soon became a hot topic. People became excited about using it. It gave amplification, provided excellent frequency standards, and it was the best amplifier in the world, a hundred times more sensitive than previous amplifiers. The field became so hot that the *Physical Review* editor even announced that he wasn't going to take any more papers on the maser because he was getting swamped with them. I believe that's the only time the *Physical Review* has made such a rule.

THE LASER

After the maser was developed, nobody was thinking of doing the same thing at wavelengths as short as that of light. There were various reasons why. At very short wavelengths, the molecules don't stay in excited states very long. They usually drop down fast, so you can't expect to have a large number in an excited state. Any number of people said, "You can't really get to very short wavelengths with the maser idea." I thought we could get into the infrared at least. In the summer of 1957, I was in an Air Force study group advising about the use of electronics in Air Force technology for the next 25 years. We decided masers would probably be important to the future of the Air Force. I persuaded the group to put into the report that one should also try to get down into the infrared. I thought one can get at least as far as 10 microns, and that ought to be developed.

The report was blacked out that year and redone the next. I decided not to be part of the new team. They then threw out the idea of pushing down towards shorter wavelengths. This, they thought, is just a crazy Townes idea. Masers were okay, but they didn't think one could get any shorter wavelengths than those in the microwave region. I did think it would be hard. I was waiting for a good idea, but since I didn't have a good idea, I eventually just sat down at my desk to figure the best way I could see of doing it. As I wrote down the equations and theoretical relations, I suddenly saw from them that it's not all that hard to get right on down to optical wavelengths. It appeared almost as easy as the long infrared wavelengths. And so I immediately started mapping out a system how to do it.

I was a consultant at Bell Labs at that time. They said my job was just to cruise around the laboratory and talk to people and make contacts. My brother-in-law was there, Arthur Schawlow, who had worked with me as a postdoc and married my younger sister. Of course, I went and talked with him and told him what I was doing. He was interested. And he had a very good idea to complement what I was suggesting. So we put it together, and wrote a paper on the laser. I told Art, "I guess this is a Bell Labs thing, so it's proprietary. We'd better wait until Bell Labs gets it patented before we publish or say anything about it." So for about ten months we wrote the paper, working it all out and reviewing material for the patent people so they could patent it. And we told nobody else about it. Previously I had always been quite open, telling everybody what my ideas were and what ought to be done. Nobody had said anything about getting to shorter wavelengths. Earlier, when I first worked on masers, nobody paid any attention to my efforts. But after the masers worked and became hot, then people paid more attention to whatever I was doing in the maser business. However, they didn't know I was working on short waves. No one wrote or said anything about getting to shorter wavelengths at the time when Art and I were preparing a paper discussing how it could be done. After we published the paper, because I already had a reputation and masers were hot, many people dashed into the field and tried to get to the shorter wavelengths which would make a laser. The first working laser was actually built at Hughes Research Laboratories by Ted Maiman. Practically all the lasers were built in industry. Once industry saw it was important, they hired physicists who came from universities and were in this kind of spectroscopy. They put a lot of manpower into the effort. The effort succeeded and the laser bloomed.

I thought I could see a number of possible applications, but in new fields you cannot see all of them. It's just impossible. I certainly did not foresee all the technical and industrial applications of the laser. Some others initially saw even less potential than I did. A favorite comment of some of my friends at the time was, "The laser is a solution looking for a problem. It's a fantastic device, but what can it do?" I would reply, "The laser marries electronics and light. You know both of those have a large number of applications. And if you get a marriage between them, which empowers both, there's bound to be a lot of applications." And I could think of some of them. In fact, Bell Labs didn't want to patent the laser until we had an application to communication. They seemed to think otherwise it was no more than just interesting physics. Bell Labs' patent lawyers didn't really understand it very well. So we devised a method of communication, and the patent was titled "Optical Masers and Communication." It's very easy to use a light beam to communicate, but the lawyers were doubtful because Alexander Graham Bell himself had tried light as a communication means and decided it didn't work very well.

I had a lot of further ideas as to what to do with lasers. I recognized that with very good intensity you burn tiny, tiny holes, and there's no material that can withstand it. And I could see how it might be used to weld and to cut things. One of the first applications, a medical one that pleased me highly, was for reattaching a detached retina. I had never before heard of a detached retina, but found it very moving to have the laser used on a friend who was prevented from going blind.

Various people, I remember, said, "Well, the laser's never going to get very much power because it's very inefficient. It has an efficiency of maybe one part in 10,000."

I said, "But wait a minute. There's no fundamental reason why it can't have just as much efficiency as any other power transfer." In fact some lasers are now 50 percent efficient. But I didn't know for sure they could reach that at the time. I had just wanted to get a few milliwatts of power from the laser to do spectroscopy and further scientific work. And rather low-powered lasers have indeed become wonderful scientific tools. But by now we are about to get a million billion watts of power in laser light pulses, more power than everything else used on the surface of this planet. Of course, one can only afford to have this power on for a short time, perhaps a thousandth of a billionth of a second.

And who can predict what there is still to be achieved?

CONCLUSION

The well-known physicist Richard Feynman once said to me, "You know how to tell a really good idea? If you hear about it and say, 'Gee, I could have thought of that,' then it's good." Similarly, Albert Szent-Györgi said, "Discovery is seeing what everybody else has seen and thinking what nobody else has thought."

Many things are out there, but we're just not looking at them in the right way. We're not following the right track. We're following what may be important tracks, but we miss many possibilities. Ideas come up from time to time. But what makes them really significant is seeing what's important and demonstrating them in a way that convinces the scientific or technical community.

BIBLIOGRAPHY

Alpher, R. A., and R. C. Herman. 1949. "Remarks on the Evolution of the Expanding Universe." *Phys. Rev.* 25:1089–95.

DeGrasse, R. W., D. C. Hogg, E. A. Ohm, and H. E. D. Scovil. 1959. "Ultra-Low-Noise Antenna and Receiver Combination for Satellite or Space Communication." *Proc. of the Nat. Electronics Conf.* 15:370–79.

Dorosch Kevich, A. G., and I. D. Novikov. 1964. "The Mean Density of Radiation in the Relativistic Cosmology." *Doklady Akademii Nauk SSSR.* 154:809–11.

Douglas, A. E., and G. Herzberg. 1941. "CH+ in Interstellar Space in the Laboratory." *Astrophys. J.* 94:381.

Field, G. B., and E. J. Chaisson. 1985. *The Invisible Universe: Probing the Frontiers of Astrophysics.* Boston: Birkhauser.

Gamow, G. 1948. "The Evolution of the Universe." *Nature* 162:680–82.

Garzoli, S. L., and C. M. Varsavsky. 1966. "The Distribution of Hydrogen in a Region in Taurus." *Astrophys. J.* 145:79–83.

Herzberg, G. 1950. *Spectra of Diatomic Molecules.* New York: Van Nostrand.

Houtermans, F. C. 1960. "Über Maser—Wirkung im Optischen Spektralgebiet und die Möglichkeit Absolut Negativer Absorption für einige Fälle von Molekulspectren." *Helvetica Physica Acta* 33:933–40.

Jansky, K. 1932. "Directional Studies of Atmospherics at High Frequencies." *Institute of Radio Engineers* 20:1920–32.

McKellar, A. 1941. *Publ. Dominion Astrophys. Obs. Victoria, B.C.* 7:251.

Ohm, E. A. 1961. "Receiving Systems." *Bell System Tech. J.* 40:1065–94.

Pail, E. 1997. "Oort, Jan Hendrik (1900–1992)." In *History of Astronomy: An Encyclopedia,* ed. J. Lankford, pp. 374–75. New York: Garland.

Rank, D. M., C. H. Townes, and W. J. Welch. 1971. "Interstellar Molecules and Dense Clouds." *Science* 174:1083–1101.

Reber, G. 1942. "Cosmic Static." *Institute of Radio Engineers* 30:367–78.

Swings, P., and L. Rosenfeld. 1937. "Consideration Regarding Interstellar Molecules." *Astrophys. J.* 86:483–96.

Tolman, R. C. 1924. "Duration of Molecules in Upper Quantum States." *Phys. Rev.* 23:693–709.

CHAPTER FIVE

The Timeliness of the Discoveries of the Three Modes of Gene Transfer in Bacteria

Norton D. Zinder

There are three modes of gene transfer in bacteria: conjugation, transduction, and transformation.[1] I was involved in one way or another with all three and thus feel qualified to comment on the zeitgeist surrounding these discoveries.

Stent's definition of a premature discovery as one that cannot be connected to the canonical knowledge of the time by a sequence of logical steps means one and only one thing to me: All true discoveries are premature; all other "discoveries" are at best just clever, logical extrapolations, although occasionally they also entail brilliant technical innovation.[2] They could be called postmature, a term that Harriet Zuckerman and Joshua Lederberg used to describe one of the gene transfer systems,[3] but that statement implies to me that the discoveries are there "for the grabbing," and that just ain't so. The important part of a scientific discovery in almost any aspect of science is the reception it receives, and this is in large measure a social phenomenon not always based on scientific criteria. In his opening remarks at the 1961 Cold Spring Harbor Symposium, but not in the written text, Francois Jacob explicitly stated this viewpoint, which I paraphrase here: "In science one gets an idea, does some experiments that support it and, with further experiments, convinces oneself that it is true; then one goes out and sells it to one's colleagues. I am doing so today." Of course, he was talking about the messenger RNA hypothesis.

CONJUGATION

Let us consider some of these issues with regard to the discoveries of genetic transfer in bacteria. As he has since written in several accounts, in 1946 Joshua

---

1. Neidhardt 1996.
2. For an explicit statement of Stent's formulation, see chapter 2 in this volume.
3. Zuckerman and Lederberg 1986.

Lederberg, a medical student at Columbia University, at Francis Ryan's suggestion went to Edward Tatum's lab to search for genetic recombination in *E. coli* bacteria.[4] Zuckerman and Lederberg refer to his subsequent finding of genetic conjugation as a postmature discovery.[5] But sex, that is, genetic exchange, was already known to exist in most life-forms other than bacteria, so the finding of conjugation and genetic exchange—that is, recombination in bacteria—was not a true discovery as I have defined it. There was, nevertheless, genius in the technique Lederberg used, which itself was of enormous significance. As Luria said shortly afterward: "The discovery *[sic]* of biochemical mutations in bacteria with production of specific growth factor deficiencies permitted Lederberg and Tatum to demonstrate by a brilliant technique the recombination of characters in mixed cultures of different mutants. These studies, still in the preliminary stage, appear to be among the most fundamental advances in the whole history of bacteriological science."[6]

It was more than ten years before the intimate details of *E. coli* mating were worked out,[7] and Lederberg was among the last to accept some of them. But all the while, *E. coli* conjugation was used by many bacteriologists to show that bacterial genes were organized in a larger structure—a chromosome—and to provide, among other things, a ready explanation for the promiscuous transfer of resistance to antibiotics from bacterium to bacterium. It provided, too, the tools for the eventual analysis of gene expression and the understanding of how the viruses of bacteria, that is, bacteriophages (and by implication some animal viruses), can live in cooperation with their hosts as temperate phages and not kill them, a process termed "lysogeny." Such temperate phages can also integrate into the genome of the host and act like any other gene segment. The existence of such phages and their study allowed for the discovery of transduction, the transfer by phages of genes from one bacterium to another, a true discovery, one whose origins I consider next.

TRANSDUCTION

In 1951, working in Lederberg's lab with salmonella, bacteria closely related to *E. coli,* I discovered genetic transduction in bacteria. I was attempting to extend conjugation as found in *E. coli* but found a different gene transfer process, one that, when ultimately resolved, revealed that bacteriophages could carry genes from one bacterial host to another.[8] The major difference between this and conjugation was

4. Lederberg 1986 and 1987.
5. Zuckerman and Lederberg 1986.
6. Luria 1947.
7. Hayes 1953; Wollman and Jacob 1958.
8. Zinder and Lederberg 1952.

that bacterial contact was unnecessary and only one or two traits were transferred in each event. This was a true discovery because at the time there was no real understanding of bacterial gene organization, or of the nature of viruses per se, or of lysogeny and the temperate viruses. The pope of bacteriophages, Max Delbrück, had said that lysogeny did not exist, and his followers followed. It was only in 1951 that A. Lwoff and A. Gutman provided clear evidence for lysogeny, and in 1952 that A. Hershey and M. Chase showed that viruses were genomes in a package.[9]

Study of transduction led to the analysis of the fine structure of the gene.[10] In transduction one can also see the roots of viral oncogenesis and recombinant DNA technology.[11] Transduction was quickly absorbed into "normal" science.[12] One of the reasons we succeeded so well in convincing people of our discovery, especially the usually skeptical medical community (see discussion of transformation below), was that salmonella came in a variety of serotypes with somatic or cell antigens and flagellar antigens by which the organisms were typed. Strains, or species as they were designated, were characterized by particular antigens. The organism that caused typhoid fever, *Salmonella typhi*, was known to have somatic antigens 9 and 12 and flagellar antigen *d*, while *S. typhimurium*, which infects mice, had somatic antigens 1, 4, 5, 12 and flagellar antigen *i*. By selecting with the antibody against motility of *S. typhi* with the anti-*d* antibody and treating with phages from *S. typhimurium* (which has flagellar antigen *i*), we obtained 9, 12, *i* strains, which did not exist in nature.[13] Whatever our interpretation for what we had done, the scientific community conceded that we had done something important. I later proved that the transducing activity was the result of phage particles by correlating the properties of the two activities, such as their relative physical size and their sensitivity to antibodies and toxic substances. With the exception of ultraviolet sensitivity, the transducing and plaque-forming activities of any solution were always in the same ratio.[14] Moreover, in each bacteriophage growth cycle the spectrum of bacterial genes that could be transduced reflected the genome of the donor. Q.E.D. So much for transduction.

TRANSFORMATION

I have deferred until last a consideration of transformation because it is the most ambiguous and most complex of the findings. Pneumonia, "the old man's friend" and the scourge of children, was known since the early days of the twentieth century to be caused by the pneumococcus bacterium. Also long known as an absolute correlate of virulence was the presence of a polysaccharide capsule on the

9. Lwoff and Gutman 1950; Hershey and Chase 1952.

10. Zinder 1953.

11. On viral oncogenesis, see Zinder 1959; on recombinant DNA technology, see Cohen et al. 1973.

12. Kuhn 1970.

13. Zinder and Lederberg 1952.

14. See Zinder 1953.

bacterium. Noncapsulated strains were avirulent, while the virulent strains had a capsule. The sugary capsules came in a number of "flavors," which were distinguished by investigators serologically and known as types (Types I, II, . . . up to some number X, with X continuously increasing).

Phase one of this story began in 1928, when the British microbiologist F. Griffith noted that injecting very large numbers of noncapsulated ("rough," or R) avirulent pneumococci into mice occasionally led to their death by pneumonia, and that from the dead mice one could isolate the capsulated strain of the original ("smooth," or S) type, from which the rough bacteria had originated.[15] Mutations and other sources of within-species variation being then unknown in bacteriology, he assumed that each noncapsulated bacterium had a tiny residue of S (capsule) substances, and that, when enough of the susceptible bacteria lysed, the residue could be pooled onto a remaining organism in an amount sufficient to convert it to a virulent organism of the original type. It was known that even a single complete pneumococcus could kill a mouse. So, since the residue of each killed rough organism could contribute a small amount of virulent capsular material to a previously avirulent surviving organism, eventually one might gain sufficient capsular material to achieve virulence by means of this pabulum effect. In an era when bacteria were not believed to have genes, this idea made some sense.

He inferred from this, and corroborated it by experiment, that he could replicate the process more efficiently with far fewer avirulent noncapsulated bacteria by injecting heat-killed fully capsulated bacteria. When the dead bacteria lysed after injection they released a substance that accumulated on a few noncapsulated bacteria and reinstated a full capsule and restored virulence. (Template-dependent growth of cell walls is known even today with such varied organisms as *E. coli* and paramecium.)[16] Thus, when the amount of the virulent pabulum was increased, the efficiency of the reaction went up. He obtained much higher and reproducible levels of mouse killing. Moreover, the capsule type recovered was that of the donor; clearly the pabulum explanation was correct! With the results quickly repeated by Neufeld and Levinthal, the story was undoubted.[17]

At the Rockefeller Institute, O. T. Avery's lab had been studying pneumococci for years and preparing rabbit sera against the various serotypes for use in treatment of pneumonia. Young colleagues in Avery's lab repeated Griffith's experiments in the early 1930s. M. H. Dawson and R. H. P. Sia reported producing the effect in vitro by mixing dead virulent (S) bacteria with live avirulent (R) bacteria in a test tube, adding anti-R serum and getting in-vitro growth of the capsulated and virulent bacteria.[18] (The anti-R sera allowed the few transformed smooth bacteria to grow out and be detected.)

15. Griffith 1928.
16. Arndt and Zinder 1956.
17. Neufeld and Levinthal 1928.
18. Dawson and Sia 1931.

A year later another junior colleague, J. L. Alloway, published an account of his success with extracts of "donor" bacteria.[19] Avery's laboratory was on its way. But the test tube experiments used ascitic fluids of varying accessibility and required all kinds of tender loving care. They were not very reproducible.

By 1937, C. M. MacLeod, who had moved to the Rockefeller Institute in 1935, had taken over the experiments. But because the sulfonamides appeared to be useful for the treatment of pneumococcal infection, MacLeod turned his attention to them.[20] Indeed, Avery's unit had previously invested much time and effort preparing various strain-specific rabbit antiserum to pneumococci for use in treatment of human disease, but this work was now given lower priority because of the advent of sulfa drugs. Ironically, sulfa-resistant mutant strains were known by this time but never used in attempts to transform pneumococci. Virulence, the primary reason for the team's interest in the pneumococci, seemed to guide their choice of assay of effects.

M. McCarty arrived about 1940 and took up what appeared to be the less urgent quest for further purification of the transforming principle. He used new and more useful avirulent R strains and larger quantities of donor bacteria to purify material. He also found ways to make extracts more stable. By 1943 purified active extracts seemed to contain solely DNA because, among other reasons, crude DNase—but not protein-destroying enzymes—destroyed the activity. The paper by Avery, MacLeod, and McCarty appeared in 1944.[21]

This story raises significant questions about the nature of discovery. Did Griffith discover anything? By his lights, his experiments, though based on a wrong hypothesis, came out just right. Certainly his version was connected to the canonical knowledge of the time, and most bacteriologists, if provoked, probably would have agreed. There is no evidence of dissent. It was exactly this that spurred Avery on, for, as he said, he felt that by understanding control of capsular substance one would understand virulence.[22] Then began the slow but steady analysis of the remaining details. But what exactly did Avery, MacLeod, and McCarty do? Firstly, they found that Griffith's interpretation, plausible as it was, was wrong. Whatever the active substance, it was not acting as a pabulum. Bacteria whose capsules were dissolved by enzymes yielded a substance that donated just as well.

Then they transferred the phenomenon to the test tube, isolated the transforming agent, and analyzed it in considerable detail. But though they felt the phenomenon was genetic, they made no effort to generalize it. Moreover, by calling it a transforming "principle," they led eminent geneticists—such as T. Dobzhansky, to his dying day—to refer to the phenomenon as a "directed mutation."[23] Another

19. Alloway 1932.
20. McCarty 1985.
21. See McCarty 1985; Avery et al. 1944; and Dubos 1976.
22. McCarty 1985.
23. Dobzhansky 1947.

scientist who did so was A. Boivin, who supposedly repeated the experiment with *E. coli* in 1947.[24] It was "heritable" and therefore had a genetic component, but what did that imply at that time and to whom? It seemed to be caused by DNA. But many did not regard that claim as completely nailed down. DNA was, after all, not supposed to have any biological specificity. At the Rockefeller Institute, some years before, P. A. Levene had proposed that DNA was only a repeating tetranucleotide. No significant objection to this doctrine appeared, and it became widely accepted, but at the same time it made it difficult to imagine DNA having a structure sufficiently complex that it would differ among organisms and pertain to genetic transformation.[25] Despite his professed interest in the project, MacLeod left the Rockefeller to lead microbiology at New York University.

McCarty gave a few talks on the subject to medical people, but because he emphasized neither the genetics nor the DNA chemistry, these talks did not make much impact. In 1946, he presented a paper at a genetics symposium (see the discussion of the presentation of conjugation below) held at the Cold Spring Harbor Laboratory titled "Biochemical Studies of Environmental Factors Essential in Transformation."[26] Those present, including the cognoscenti of the new genetics of the time, probably did not appreciate the possible significance of the paper. Everyone else at the meeting talked about mutation and recombination.[27] Shortly afterward, McCarty left the field and became head of the rheumatic fever division at Rockefeller.[28]

Alfred Mirsky, who was no fool, realized the work was highly important but objected to the interpretation that DNA had been proven the vector of transformation. His vocal objections and alternative explanations, while eventually found to be incorrect, nevertheless brought the subject into public discussion. Mirsky was connected to the geneticists of the time and was in contact with the powerful biology department at Columbia (in which, incidentally, I was then an undergraduate). Mirsky's point was that, although there was little protein or any material other than DNA detected in the solution, there were so few transformations per unit mass and Avogadro's number (the number of molecules in one gram-molecular weight) was so large, viz. 6.022×10^{23}, that one could not be sure all protein molecules had been removed by the purification.[29] There might still be trace amounts of an active protein in the purified material, sheltered from protein-destroying enzymes by a tight surrounding shell of DNA, but which was denatured when the DNase disrupted its cover.

I turned up a set of letters that I exchanged with Rollin Hotchkiss in the spring of 1952, when we were discussing my coming to his laboratory. Several points are

24. Boivin 1947.
25. Levene 1931.
26. McCarty et al. 1946.
27. Ibid.
28. McCarty 1985.
29. Mirsky 1947.

worth mentioning. I had concentrated the purified transducing agent and could not find any material substance, although I had about 10^5 transductions per trait per milliliter. Moreover, the procedure for the purification I used was essentially that for purifying pneumococcal DNA. Obviously, I just had to work harder, and Hotchkiss's lab would be the place to do it. Now, it is clear that 10^5 transductions is equivalent to 10^{10} P22 phages, and that that equals just 1.0 microgram of material—which is hard to detect. Thus, if I had added this material to pneumococcus-transforming preparations, and had not added some salmonella mutants, the latter would not be found there. Partly on this basis, I suggested a new interpretation: the Transforming Principle is inactivated by the products of DNA hydrolysis (DNA breakdown), that is, not by DNase directly. Hotchkiss responded that it was the first good idea he had heard in a bit and did the experiment. How different the history of DNA would have been had this experiment worked.

HIDDEN PROPERTIES OF EXPERIMENTAL SYSTEMS

The experiment just suggested raises questions about the experimental system, and it may also reveal hidden properties in it. In every experiment there are always unknown aspects and attributes. It is often these that lead to discovery. I am sure others who have reflected upon discovery have come to similar conclusions. Let me cite two examples from my own work that stimulated these thoughts.

The presence of a temperate phages in some of the bacteria that I used was totally unknown to me yet was indispensable to the discovery of transduction. All other things being equal, *ceteris paribus,* no such phage no discovery.

Often in microbial genetics one establishes a screen mechanism to select and concentrate organisms with a particular mutant phenotype. One almost always gets what one selected for, but precisely what one selected for—that is, its real, underlying nature can be determined only afterward. Once, when screening for a mutation in a particular gene, I chose as the signal the failure to complement (rescue) a known mutant in that gene. The only apparent phenotype of the mutant I obtained was its failure to complement the known mutant. By itself it had an apparently wild-type phenotype. It took great effort to determine the complex protein-protein interaction that underlay the failure of the complementation—which is now known as an example of a negative-dominant mutant. Though it turned out to be irrelevant for future work, in neither instance did I know what I was really doing.

DIFFERENCES IN THE RECEPTION OF GENETIC EXCHANGE
Conjugation and Transduction

I now turn to an analysis of the differences in the reception of the various instances of genetic exchange. Conjugation and transduction were discovered at least five years apart, and there were minor differences in their reception, but to expedite matters I'll treat them together. At the time of Lederberg's findings in 1946, there was a small coterie of "new" geneticists, including, among others, Delbrück, Luria, Beadle, Tatum, Sonneborn, Spiegelman, and of course, Hermann Muller, who since 1920 was the geneticist's geneticist. It was a strange group in the sense that Beadle and Tatum were clearly chemistry oriented, while the majority did not care about chemistry. Ironically, members of the Delbrück phage school (which, incidentally, Stent later joined and James Watson affiliated with as a graduate student of Luria's) were antichemistry. Chemistry in their view just did not lead to the discovery of new laws of physics that explained life.

Still, this group of "new" geneticists, many of whom were located in the Midwest—Lederberg was in Wisconsin, Novick and Szilard in Chicago, Luria and Spiegelman in Illinois, Sonneborn in Indiana, Levinthal in Michigan—would meet regularly and discuss the new science. Whatever they found they quickly assimilated, and this resulted in wide propagation of the new ideas. True, most of the old-line zoology and botany departments resisted, but the intelligence and dominating presence of the "new" geneticists (and their East and West Coast friends) quickly absorbed conjugation and transduction into the body of science, just as they had done earlier with Beadle and Tatum's one gene, one enzyme, hypothesis. Since the experiments were so easily repeated, within a few years they became classic—cited simply by name.

The public presentations of conjugation and transduction also were remarkably similar. Conjugation was discovered in May 1946. Tatum managed to get a paper inserted into the following month's Cold Spring Harbor Symposium. Lederberg did further linkage studies and published a full paper in the journal *Genetics* in 1947. R. A. Brink, a corn geneticist at the University of Wisconsin, in a brilliant move hired the twenty-two-year-old Lederberg as an assistant professor in 1947. Papers, reviews, and talks streamed from Lederberg, and scientists received the relevant strains and repeated and extended the experiments.

I arrived at the University of Wisconsin in the summer of 1948, all of nineteen years old, sent from Columbia by Lederberg's other mentor, Francis Ryan. I immediately discovered the penicillin enrichment procedure for the isolation of biochemical auxotrophic mutants. This provided an infinity of genetic markers and opened the whole of intermediary metabolism in bacteria to genetic analysis. The summer of 1949 I spent at Cold Spring Harbor taking the phage course, teaching Demerec how to mate *E. coli*, and mapping the streptomycin resistance loci. With transduction already a year old by 1952, Tatum got me inserted into a symposium held by the Society of American Bacteriologists. In the next few years I gave more than a half dozen invited talks about "genes in a bottle," including a plenary lecture at the International Genetics Congress. Few understood completely what we were saying, but few objected. Only Delbrück had reservations about conjugation, but even he conceded transduction.

Transformation

Transformation had a different fate. At the Rockefeller Institute, Avery, MacLeod, and McCarty were surrounded by a group of protein chemists, physical chemists (also involved with proteins), and old-line microbiologists and virologists. Avery, MacLeod, and McCarty had little support in their quest—only T. Shedlovsky helped, by showing in an ultracentrifuge that the transforming principle was about 500,000 molecular weight units—and they were not very good as propagandists. They had no connection with an elite group of scientists, and the experiment itself was so difficult that no one, unless trained at Rockefeller, could repeat it—and perhaps no one tried. Still the experiment had significant impact (see below), especially after the war was over and science was in recovery. But they published in the wrong journals and were shy of fighting for their phenomenon. They assumed that, no matter what the objections or the neglect, sooner or later what was observed in a test tube would be accepted and understood. A battle for acceptance of the interpretation was not necessary.

In 1994 we celebrated at Rockefeller University the fiftieth anniversary of the Avery, MacLeod, and McCarty paper. I organized the event and was able to secure presentations from some of the most important of the early players and observers: McCarty, Chargaff, Hotchkiss, Hershey, Seymour Cohen, and Lederberg. Each related the consequences of the paper upon his own work. Lederberg described how, on hearing from Mirsky about transformation, he was additionally stimulated to look for bacterial recombination; Chargaff described how he dropped his blood work and turned to working out the procedures for DNA nucleotide analysis that eventually resulted in his famous rules; Cohen found RNA in tobacco mosaic virus, joined Chargaff, and then, despite the jeers from Delbrück's phage group, looked at the biochemical-DNA, RNA, and protein-events that followed phage infection; Hershey, also after looking at DNA metabolism following phage infection, was led eventually to undertake the famous Hershey-Chase experiment. Hotchkiss improved the sensitivity of protein detection and then tried transformation with markers other than the capsule, and succeeded. Indeed, Hotchkiss, who is still at Rockefeller University, was the focus of a subsequent endeavor on transformation. However, not unlike Barbara McClintock, he never published real papers on some of his most important work. In fact, for years many of the developments at the Rockefeller appeared only in the annual reports of the Rockefeller Institute for Medical Research (1933–44, 1946–51), which nobody read.

The resistance to accepting DNA as genetic material was clearly a result of Avery, MacLeod, and McCarty's diffidence on both the genetic and chemical sides, their lack of public relations efforts, and the lack of definitive and corroborative experiments, perhaps because they were too hard to do. In addition, there was no elite group like the Midwest phage scientists to propagate the story for them.

At least in the short term, the details of the social reactions to the presentation of a new scientific fact, finding, or discovery are, for many scientists, what decides who wins and who loses, independently of whether the finding is premature, postmature, or just right.

BIBLIOGRAPHY

Alloway, J. L. 1932. "The Transformation *in Vitro* of R Pneumococci into S forms of Different Specific Types by the Use of Filtered Pneumococcus Extracts." *J. Exp. Med.* 55:91–99.

Arndt, W., and N. D. Zinder. 1956. "Production of Protoplasts of *Escherichia coli* by Lysozyme Treatment." *Proc. Soc. Nat'l Acad. Sci.* 42:586–90.

Avery, O. T., C. M. MacLeod, and M. McCarty. 1944. "Studies on the Chemical Transformation of Pneumococcal Types." *J. Exp. Med.* 79:137–58.

Boivin, A. 1947. "Directed Mutation in Colon Bacilli, by an Inducing Principle of Deoxyribonucleic Nature: Its Meaning for the General Biochemistry of Hereditary." *Cold Spring Harbor Symp. Quant. Biol.* 12:7–17.

Cohen, S. N., A. C. Chang, H. W. Boyer, and R. R. Helling. 1973. "Construction of Biological Functional Bacterial Plasmids in Vitro." *Proc. Soc. Nat'l. Acad. Sci.* 70:3240–44.

Dawson, M. H., and R. H. P. Sia. 1931. "In Vitro Transformation of Pneumococcal Types." Pts. 1–2. *J. Exp. Med.* 54:681–99, 701–10.

Dobzhansky, T. 1947. *Genetics and the Origin of Species.* New York: Columbia University Press.

Dubos, R. L. 1976. *The Professor, the Institute, and DNA.* New York: Rockefeller University Press.

Griffith, F. 1928. "The Significance of Pneumococcal Types." *J. Hyg.* 27:113–59.

Hayes, W. 1953. "The Mechanism of Genetic Recombination in *Escherichia coli.*" *Cold Spring Harbor Symp. Quant. Biol.* 18:75–93.

Hershey, A., and M. Chase. 1952. "Independent Functions of Viral Proteins and Nucleic Acid in Growth of Bacteriophage." *J. Gen. Physiol.* 36:39–56.

Kuhn, T. 1970. *The Structure of Scientific Revolutions.* Chicago: University of Chicago Press.

Lederberg, J. L. 1986. "Forty Years of Genetic Recombination in Bacteria: A Fortieth Anniversary Reminiscence." *Nature* 324:627–28.

———. 1987. "Genetic Recombination in Bacteria: A Discovery Account." *Ann. Rev. Genet.* 21:23–46.

Levene, P. A. 1931. *Nucleic Acids.* New York: Chemical Laboratory Company.

Luria, S. E. 1947. "Recent Advances in Bacterial Genetics." *Bact. Rev.* 2:1.

Lwoff, A., and A. Gutman. 1950. "Recherches sur un B. megatherium lysogene." *Ann. Ist. Pasteur.* 78:711–39.

McCarty, M. 1985. *The Transforming Principle.* New York: W. W. Norton.

McCarty, M., H. E. Taylor, and O. T. Avery. 1946. "Biochemical Studies of Environmental Factors Essential in Transformation of Pneumococcal Types." *Cold Spring Harbor Symp. Quant. Biol.* 11:177–83.

Mirsky, A. E. 1947. "Chemical Properties of Isolated Chromosomes." *Cold Spring Harbor Symp. Quant. Biol.* 12:142–47.

Neidhardt, F. C. 1996. *Escherichia coli and Salmonella.* Washington, D.C.: ASM Press.

Neufeld, F., and W. Levinthal. 1928. "Beiträge zur Variabilität der Pneumokokken." *Zeit. für Immunität* 55:324–40.

Stent, G. S. 1972. "Prematurity and Uniqueness in Scientific Discovery." *Scientific American* 227 (December): 84–93.

Wollman, E., and F. Jacob. 1958. "Sur les processus de conjugasion et de recombinaison chez E. coli." *Ann. Inst. Pasteur* 95:641–46.

Zinder, N. D. 1953. "Infective Heredity in Bacteria." *Cold Spring Harbor Symp. Quant. Biol.* 18:261–70.

———. 1959. "Virology, 1959." *American Scientist* 48:608–16.

Zinder, N. D., and J. Lederberg. 1952. "Genetic Change in Salmonella." *J. Bact.* 64:679–99.

Zuckerman, H., and J. Lederberg,. 1986. "Forty Years of Genetic Recombination in Bacteria: A Postmature Scientific Discovery?" *Nature* 324:629–31.

CHAPTER SIX

Scotoma

Forgetting and Neglect in Science

Oliver Sacks

I

We may look at the history of ideas backward or forward—we can trace the earlier stages, the intimations, and the anticipations of what we think now; or we can concentrate on the evolution, the effects and influences of what we once thought. Either way, we may imagine that history will be revealed as a continuum, an advance, an opening like the tree of life. What one often finds, however, is very far from a majestic unfolding, and very far from being a continuum in any sense. This is a conclusion that I will try to illustrate by some stories (which might be multiplied a hundredfold) of how odd, complex, contradictory, and irrational the processes of scientific discovery can be. And yet, beyond the twists and anachronisms in the history of science, beyond the vicissitudes and fortuities, perhaps there is an overall pattern to be discerned.

I began to realize how elusive scientific history can be when I became involved with my first love, chemistry. I vividly remember, as a boy, reading a history of chemistry by F. P. Armitage, a former master at my school, and learning that oxygen had been all but discovered in the 1670s by John Mayow, along with a theory of combustion and respiration. But Mayow's work was then forgotten and concealed from view by a century of obscurantism (and the phlogiston theory), and oxygen was only rediscovered a hundred years later by Lavoisier. Mayow died at thirty-four: "Had he lived but a little longer," Armitage adds, "it can scarcely be doubted that he would have forestalled the revolutionary work of Lavoisier, and stifled the theory of phlogiston at its birth." Was this a romantic exaltation of John Mayow, or a romantic misreading of the structure of the scientific en-

Editor's note: This is an edited and updated revision of a lengthier essay published in Silvers 1995.

terprise, or could the history of chemistry have been wholly different, as Armitage suggests?[1]

I thought of this history in the mid-sixties, when I was a young neurologist just starting work in a headache clinic. My job was to make a diagnosis—migraine, tension headache, whatever—and prescribe treatment. But I could never confine myself to this, nor could many of the patients I saw. They would often tell me, or I would observe, other phenomena: sometimes distressing, sometimes intriguing, but not strictly part of the medical picture—not needed, at least, to make the diagnosis.

Often in a classical migraine there is an aura, so-called, where the patient may see scintillating zigzags slowly traversing the field of vision. These are well described and understood. But sometimes, more rarely, patients would tell me of more complex geometrical patterns that appeared in place of, or in addition to, the zigzags: lattices, whorls, funnels, and webs, all shifting, gyrating, and modulating constantly. When I searched the current literature, I could find no mention of these. Puzzled, I decided to go back and look at nineteenth-century accounts, which tend to be much fuller, much more vivid, much richer in description, than modern ones.

My first discovery was in the rare book section of our college library (everything written before 1900 counted as "rare")—an extraordinary book on migraine written by a Victorian physician, Edward Liveing, in the 1860s. It had a wonderful, lengthy title, *On Megrim, Sick-Headache, and Some Allied Disorders: A Contribution to the Pathology of Nerve Storms,* and it was a grand, meandering sort of book, clearly written in an age far more leisurely, less rigidly constrained, than ours.[2] It touched briefly on the complex geometrical patterns I had been told of, and it referred me to a paper written a few years before, "On Sensorial Vision," by John Frederick Herschel, son of Frederick Herschel (both father and son, as well as being eminent astronomers, had "visual" migraines and wrote about them). I felt I had struck paydirt at last. The younger Herschel gave meticulous, elaborate descriptions of exactly the phenomena my patients had described; he had experienced them himself, and he ventured some deep speculations about their possible nature and origin. He thought they might represent "a sort of kaleidoscopic power" in the sensorium, a primitive, pre-personal generating power in the mind, the earliest stages, even precursors, of perception.[3]

I could find no adequate description of these "Geometrical Spectra," as Herschel called them, in the entire hundred-year period between his observations and

1. Armitage's book was written in 1905 to stimulate the enthusiasm of Edwardian schoolboys, and it seems to me now, with different eyes, that it had a somewhat romantic and jingoistic ring, an insistence that it was the English, not the French, who discovered oxygen. William Brock, in his *Norton History of Chemistry* (1993), writes: "Early historians of chemistry liked to find a close resemblance between Mayow's explanation and the later oxygen theory of calcination." But such resemblances, Brock stresses, "are superficial, for Mayow's theory was a mechanical, not a chemical, theory of combustion. [Moreover,] . . . it marked a return to a dualistic world of principles and occult powers."

2. Liveing 1873.

3. Herschel 1858.

my own, and yet it was clear to me that at least one person in twenty affected with migraine experienced them on occasion. How had these phenomena—startling, highly characteristic, unmistakable hallucinatory patterns—evaded notice for so long? In the first place, someone must make an observation and report. A few years after Herschel reported his spectres, G. B. A. Duchenne, in France, described a case of muscular dystrophy.[4] But here the stories diverge. As soon as Duchenne's observations were published, physicians started "seeing" the dystrophy everywhere, and within a few years, scores of further cases were reported and described. The disorder had always existed, ubiquitous and unmistakable. Why did we need Duchenne to open our eyes? His observations entered the mainstream of clinical perception at once, as a syndrome, a disorder of great importance.

Herschel's paper, by contrast, sank without a trace. He was not a physician making medical observations but an independent observer of great curiosity. He considered himself an astronomer even in regard to his own hallucinations and, indeed, called himself "an astronomer of the inward." Herschel suspected that his observations had scientific importance, that such phenomena could lead to deep insights about the brain, but whether they had medical importance too was not in his mind. Since migraine was usually defined as a "medical" condition, Herschel's observations had no professional status; they were seen as irrelevant and, after a brief mention in Liveing's book, were forgotten, ignored by the profession. If they were to point to new scientific ideas about the mind and brain, there was no way of making the connection in the 1850s; the necessary concepts only emerged 120 years later.

These necessary concepts emerged in conjunction with the recent development of chaos theory, which shows that, while it is impossible to predict in detail the individual disposition of each element in a system, when there are a large number of elements in interaction (as, for example, with the million-odd nerve cells in the primary visual cortex), patterns can be discerned at a higher level by using recently developed methods of mathematical and computer analyses. There are "universal behaviors" that emerge in such interactions, behaviors that represent the ways such dynamic, nonlinear systems organize themselves. They tend to take the form of complex, reiterative patterns in space and time—indeed the very sort of networks, whorls, spirals, and webs that one sees in the geometrical hallucinations of migraine.

Such chaotic behaviors have now been recognized in a vast range of natural systems, from the eccentric motions of Pluto to the striking patterns that appear in the course of certain chemical reactions, to the multiplication of slime fungi, and the vagaries of the weather. With this, a hitherto insignificant or unregarded phenomenon like the geometrical patterns of migraine aura suddenly assumes a new importance. It shows us, in the form of a hallucinatory display, not only an ele-

4. Duchenne 1868.

mental activity of the cerebral cortex but an entire self-organizing system, a universal behavior, at work.[5]

II

With migraine, I had to go back to an earlier, forgotten medical literature—a literature that most of my colleagues saw as superseded or obsolete. I also found myself in a similar position with Tourette's Syndrome, the *maladie des tics* described in the 1880s by Georges Gilles de la Tourette. My interest in it had been kindled in 1969 when I was able to "awaken" a number of *encephalitis lethargica* patients with L-DOPA and saw how many of them rapidly swung from motionless, trance-like states through a tantalizing brief "normality" and then to the opposite extreme—violently hyperkinetic tic-ridden states very similar to the half-mythical "Tourette's syndrome." I say "half-mythical" because no one in the 1960s spoke much about Tourette's; it was considered extremely rare and possibly factitious. I had only vaguely heard of it. Things were soon to change: in the 1970s, Tourette's was rediscovered and found to be a thousand times commoner than suspected; there was a surge of interest in the syndrome and research into its course.

But this surge of interest, this rediscovery, followed a silence and neglect of sixty years or more, during which the syndrome was rarely discussed or even diagnosed. Indeed, when I started to think about it, in 1969, as my own patients were becoming palpably Tourettic, I had difficulty finding any current references whatever, and once again had to go back to the literature of the previous century: to Gilles de la Tourette's original papers in 1885 and 1886 and to the dozen or so reports that followed them.[6] It was an era of superb, mostly French, descriptions of the varieties of tic behavior, which culminated (and terminated) in the book on tics, first published in 1902 by Henry Meige and E. Feindel.[7] Yet between 1903 and 1970, the syndrome itself seemed almost to have disappeared.

Why? One must wonder whether this neglect was not caused by the growing pressures at the beginning of the new century to try to explain scientific phenomena, following a time when it was enough to *describe* them. And Tourette's was peculiarly difficulty to explain. In its most complex forms it could express itself not only as convulsive movements and noises but also as tics, compulsions, obsessions, and tendencies to make jokes and puns, play with boundaries, and engage in social provocations and elaborate fantasies. Though there were attempts to explain the syndrome in psychoanalytic terms, these, while casting light on some of the

---

5. I described the phenomena of migraine aura in the original edition of Sacks 1970, but could only say that they were "inexplicable" by then existing concepts. I have discussed them in the new light of chaos theory, in collaboration with my colleague Dr. Ralph Siegel, in an additional chapter in the revised edition, Sacks 1992.

6. Tourette 1885.

7. See Meige and Feindel 1907 for the English translation.

phenomena, were impotent to explain others; there were clearly organic components as well. In 1960, the finding that a drug, haloperidol, which counters the effects of dopamine, could extinguish many of the phenomena of Tourette's, generated a much more tractable hypothesis—that Tourette's was, essentially, a chemical disease caused by an excess of (or an excessive sensitivity to) the neurotransmitter dopamine in the brain.

With this comfortable, reductive explanation to hand, the syndrome suddenly sprang into prominence again, and indeed seemed to multiply its incidence a thousandfold. There is now a very intensive investigation of Tourette's Syndrome, but it is an investigation almost confined to its molecular and genetic aspects. And while these may explain some of the overall excitability of Tourette's, they may do little to illuminate the particular forms of the Tourettic disposition to engage in comedy, fantasy, mimicry, mockery, dream, exhibition, provocation, and play. Thus while we have moved from an era of pure description to one of active investigation and explanation, Tourette's itself has been fragmented in the process and is no longer seen as a whole.[8]

This sort of fragmentation is perhaps typical of a certain stage in science—the stage that follows pure description. But the fragments must somehow, sometime, be gathered together and presented once more as a coherent whole. This will require an understanding of determinants at *every* level, from the neurophysiological to the psychological to the sociological—and of their continuous and intricate interaction.[9]

III

I had spent fifteen years as a physician making neurological observations, but in 1974 I had a neurological experience of my own—experienced, so to speak, the "inside" of a neuropsychological syndrome. I had severely injured the nerves and muscles of my left leg while climbing in a remote part of Norway. I needed surgery to connect the muscle tendons and time to allow the healing of nerves. During the two-week period in which the leg was denervated and immobilized in a cast,

8. I have written about Tourette's Syndrome, and the history of our ideas about it, in Sacks 1987a.

9. A somewhat similar sequence has occurred in "medical" psychiatry. If one looks at the charts of patients institutionalized in asylums and state hospitals in the 1920s and 1930s, one finds extremely detailed clinical and phenomenological observations, often embedded in narratives of an almost novelistic richness and density (as in the "classical" descriptions of Kraepelin and others at the turn of the century). With the institution of rigid diagnostic criteria and manuals (the "diagnostic and statistical manuals" called DSM-III and DSM-IV), this richness and detail and phenomenological openness have disappeared, and one finds instead meager notes that give no real picture of the patient or his world but reduce him and his disease to a list of "major" and "minor" diagnostic criteria. Present-day psychiatric charts in hospitals are almost completely devoid of the depth and density of information one finds in the older charts, and will be of little use in helping us to bring about the synthesis of neuroscience with psychiatric knowledge that we also need. The "old" case histories and charts, however, will be invaluable.

it was not only bereft of movement and sensation, it ceased to feel like a part of me. It seemed to have become a lifeless, almost inorganic object, not real, not mine, inconceivably alien and strange. But when I tried to communicate the experience to my surgeon, he said, "Sacks, you're unique. I've never heard of anything like this from a patient before."

I found this absurd. How could I be "unique"? There must be other cases, I thought, even if my surgeon had not heard of them. As soon as I was mobile enough, I started to talk to my fellow patients, and many of them, I found, had similar experience of "alien" limbs. Some had found the experience so uncanny and fearful that they had tried to put it out of their minds; others had worried about it secretly, but not tried to communicate it.

When I left the hospital, I went to the library determined to seek out the literature on the subject. For three years I found nothing. Then I came across an account by Silas Weir Mitchell, the great nineteenth-century American neurologist, fully and carefully describing phantom limbs ("sensory ghosts," as he called them). Mitchell also wrote of "negative phantoms," experiences of the subjective annihilation and alienation of limbs following severe injury and surgery. He encountered vast numbers of such cases during the Civil War and was so struck by them that he at once published a special circular on the matter *(Reflex Paralysis),* which was distributed by the Surgeon General's office in 1864. His observations aroused brief interest, then disappeared.[10] More than fifty years had to pass before the syndrome was rediscovered. This again occurred during wartime, when thousands of new cases of neurological trauma were seen at the front. In 1917, the eminent neurologist J. Babinski published (with J. Froment) a monograph entitled *Syndrome Physiopathique,* in which, apparently ignorant of Mitchell's report, he described the syndrome I had. Once again the observations sank without a trace. (When, in 1975, I finally came upon the book in our library, I found I was the first person to have borrowed it since 1918.) During World War II, the syndrome was fully and richly described for the third time by two Soviet neurologists, A. N. Leont'ev and A. V. Zaporozhets, again in ignorance of their predecessors. Though their book, *Rehabilitation of Hand Function,* was translated into English in 1960, their observations completely failed to enter the consciousness of either neurologists or rehabilitation specialists.[11]

As I pieced together this extraordinary, even bizarre, story, I felt more sympathy with my surgeon and his saying he had never heard of anything like my symptoms before. And yet the syndrome is not that uncommon: it occurs whenever there is a significant dissolution of body image. But why is it so difficult to put this on record and to give the syndrome its due place in our neurological knowledge and consciousness?

10. Mitchell 1872.
11. Leont'ev and Zaporozhets 1960.

The term *scotoma* (darkness, shadow)—as used by neurologists—denotes a disconnection or hiatus in perception, essentially a gap in consciousness produced by a neurological lesion. Such lesions may be at any level, from the peripheral nerves, as in my own case, to the sensory cortex of the brain. It is therefore extremely difficult for a patient with such a scotoma to be able to communicate what is happening. He himself, so to speak, scotomizes the experience. It is equally difficult for his physician and his listeners to take in what he is saying, for they, in turn, tend to scotomize what they are hearing. Such scotoma is literally unimaginable unless one is actually experiencing it (this is why I suggest, only half-jocularly, that people read *A Leg to Stand On* while under spinal anesthesia, so that they will know in their own persons what I am talking about).[12]

If, somehow, by an almost superhuman effort, these barriers to communication are transcended—as they were by Mitchell, Babinski, and Leont'ev and Zaporozhets—no one seems to read or remember what they have written. There is a historical or cultural scotoma, a "memory hole," as Orwell would say.

IV

Let us return from this uncanny realm to a more positive (but still strangely neglected or scotomized) phenomenon, in particular that of total colorblindness following a cerebral injury or lesion—a so-called acquired cerebral achromatopsia. (This is, of course, a completely different condition from common colorblindness, which is caused by a deficiency of one or more color receptors in the retina.) I select this example because I have explored it in some detail, but I learned of it quite by accident, when a patient with the condition wrote to me, asking if I had ever encountered it before. My friend and colleague Dr. Robert Wasserman and I spent a great deal of time with this extraordinary patient, and the original report appeared in the *New York Review of Books* in 1987.[13]

But when we looked into the history of this condition, we soon encountered a remarkable gap or anachronism. Acquired cerebral achromatopsia—and even more dramatically, hemiachromatopsia, the loss of color perception in only one-half of the visual field, coming on suddenly as a consequence of a stroke—had been described in exemplary fashion by a Swiss neurologist, Louis Verrey, in 1888. When his patient subsequently died, Verrey was able to delineate the exact area of the visual cortex that had been damaged by her stroke. Here, he said, "the center for chromatic sense will be found."[14] Within a few years of Verrey's report, there were other careful reports of similar problems with color perception and the lesions that caused them, and achromatopsia and its neural basis seemed firmly established. But then, strangely, there were no more reports—not a single full case

---

12. Sacks 1984. I have discussed this more fully in the afterword in Sacks 1998.
13. Sacks 1987b. An expanded and revised version of the article also appears in Sacks 1995.
14. Verrey 1888.

report for seventy-five years, between the last nineteenth-century report in 1899 and the "rediscovery" of achromatopsia in 1974.

This story has been discussed, with great scholarship and acumen, by two colleagues of mine: Semir Zeki of the University of London and Antonio Damasio of the University of Iowa.[15] Zeki, remarking that resistance to Louis Verrey's findings started the instant they were published, sees their virtual denial and dismissal as springing from a deep and perhaps unconscious philosophical attitude, the then prevailing belief in the seamlessness of vision. The notion that we are given the visual world as a datum, an image, complete with color, form, movement, and depth, is a natural and intuitive one that had seemingly been given scientific and philosophical legitimation by Newtonian optics and Lockean sensationalism. The invention of the camera lucida, and later of photography, seemed to exemplify such a mechanical model of perception. Why should the brain behave any differently? Color, it was obvious, was an integral part of the visual image and not to be dissociated from it. The ideas of an isolated loss of heretofore normal color perception and of a center for chromatic sensation in the brain were thought self-evident nonsense. Verrey had to be wrong; such absurd notions had to be dismissed out of hand. So they were, and achromatopsia "disappeared."

Darwin often remarked that no man could be a good observer unless he were an active theorizer as well, and Darwin himself, his son Francis wrote, seemed "charged with theorizing power" that would animate and illuminate all his observations, even the most trivial ones. But, Francis added, this power was always balanced by skepticism and caution and, above all, by experiments that as often as not demolished the new theory. Theory, though, can be a great enemy of honest observation and thought as well, especially when it forgets that it *is* theory or model and hardens into unstated, perhaps unconscious, dogma or assumption. Mistaken assumptions killed Verrey's observation, killed the entire subject for three-quarters of a century.[16]

The notion of perception as "given" in some seamless, overall way was finally shaken to its foundations by the findings of David Hubel and Torsten Wiesel in the late fifties and early sixties that there were cells and columns of cells in the visual cortex that acted as "feature detectors" specifically sensitive to horizontals, verticals, edges, alignments, and other features of the visual field.[17] The idea began

15. Zeki 1993; Damasio 1985.

16. And yet there seem to have been other factors at work as well. Damasio describes how, when the renowned neurologist Gordon Holmes published his findings on 200 cases of war injuries to the visual cortex in 1919, he stated summarily that none of these showed isolated deficiencies in color perception, and that his research gave no support to any notion of a color center in the brain. See Holmes 1918 and 1919. Holmes was a man of formidable authority and power in the neurological world, and his empirically based antagonism to the notion of an isolated cerebral color defect or color center, reiterated with increasing force for over 30 years, was a major factor, Damasio feels, in actually preventing the clinical recognition of the syndrome.

17. See, for example, Hubel and Wiesel 1962.

to develop that vision had components, that visual representations were in no sense "given," like optical images or photographs, but were *constructed* by an enormously complex and intricate correlation of different processes. Perception was now seen as composite, as modular: the interaction of a huge number of components. The seamlessness of perception was not "given" but had to be *achieved* in the brain.

It thus became clear in the 1960s that vision was an analytic process depending on the differing sensitivities of a large number of cerebral (and retinal) systems, each "tuned" to respond to different components of perception. It was in this atmosphere of hospitality to subsystems and their integration that Zeki discovered specific cells sensitive to wavelength and color in the visual cortex of the monkey; and he found them in much the same area as Verrey had suggested as a color center eighty-five years before. Zeki's discovery seemed to release clinical neurologists from their almost century-long inhibition. Within a few years, scores of new cases of achromatopsia were found, and it was at last legitimized as a valid neurological condition.[18]

V

Can we draw any lessons from the examples I have been discussing? I believe we can. One might first invoke the concept of prematurity here—and see the nineteenth-century observations of Herschel, Mitchell, Gilles de la Tourette, and Verrey as having come before their times, so that they could not be integrated into contemporary conceptions. Gunther Stent, writing about prematurity in scientific discovery, says, "A discovery is premature if its implications cannot be connected by a series of simple logical steps to contemporary canonical [or generally accepted] knowledge."[19] Prematurity, I think, though relatively rare in science, may be much commoner in medicine, partly because medicine does not have to do elaborate experiments but, in the first place, can simply describe.

But scotoma involves more than prematurity, it involves the *deletion* of what was originally perceived, a loss of knowledge, a loss of insight, a forgetting of insights

---

18. That a conceptual bias was responsible for the dismissal and "disappearance" of achromatopsia is confirmed by the completely opposite history of central motion blindness (akinetopsia), which was described in a single case, in Zihl et al. 1983. This patient could see people, or cars, at rest, but they disappeared as soon as they moved, and then reappeared, motionless, in different places. Zihl's case, Zeki notes, was "immediately accepted by the neurological . . . and neurobiological world, without a murmur of dissent . . . in contrast to the more turbulent history of achromatopsia." This dramatic difference relates to timing, to the profound change in intellectual climate that had come about in the years immediately before. In the early 1970s it had been shown that there was a specialized area of motion-sensitive cells in the prestriate cortex of monkeys, and the *idea* of functional specialization was fully accepted within a decade. Thus by 1983, in Zeki's words, "all conceptual difficulties had been removed." There was no longer any conceptual reason for rejecting Zihl's findings—indeed, quite the contrary: they were embraced with delight, as a superb piece of clinical evidence in consonance with the new climate.

19. See chapter 2 in this volume.

that once seemed clearly established, a regression to less perceptive explanations. All these not only beset neurology but are surprisingly common in all fields of science. They raise the deepest questions about why such lapses occur. What makes an observation or a new idea acceptable, discussible, memorable? What may prevent it from being so, despite its clear importance and value?

Freud would answer this question by emphasizing resistance: the new idea is deeply threatening or repugnant, and hence is denied full access to the mind. This doubtless is often true, but it reduces everything to psychodynamics and motivation; and even in psychiatry this is not enough.

It is not enough to apprehend something, to "get" something, in a flash. The mind must be able to accommodate it, to retain it. This process of accommodation, of being able to create a mental space, a category with potential connections—and the readiness to do this—seems to me crucial in determining whether an idea or discovery will take hold and bear fruit, or whether it will be forgotten, fade and die without issue. The first difficulty, the first barrier, lies in one's own mind, in allowing oneself to encounter new ideas and then to bring them into full and stable consciousness, and to give them conceptual form, holding them in mind even if they do not fit, or they contradict, one's existing concepts, beliefs, or categories. Darwin remarks on the importance of "negative instances" or "exceptions" and how crucial it is to make immediate note of them, for otherwise they are "sure to be forgotten."

That it is crucially important to take note of exceptions, and not forget them or dismiss them as trivial, was brought out in "On Unnoticed Sensations and Errors of Judgment," the first paper written by Wolfgang Köhler, before his pioneer work in Gestalt psychology. Köhler spoke here of premature simplifications and systematizations in science, psychology in particular, and how they could blind one, ossify science, and prevent its vital growth.

"Each science," he wrote, "has a sort of attic into which things are almost automatically pushed that cannot be used at the moment, that do not quite fit.... We are constantly putting aside, unused, a wealth of valuable material [which leads to] the blocking of scientific progress" (1913).

Thus, at the time Köhler wrote this, visual illusions were seen as "errors of judgement"—trivial, of no relevance to the workings of the mind-brain. But Köhler would soon show that the opposite was the case, that such illusions constituted the clearest evidence that perception does not just passively process sensory stimuli but actively creates large configurations or "gestalts" which organize the entire perceptual field. These insights now lie at the heart of our present understanding of the brain as dynamic and constructive. But it was first necessary to seize on an "anomaly," a phenomenon contrary to the accepted frame of reference, and by according it attention, to enlarge it, to revolutionize it entirely.

But if exceptions, anomalies, promise a transition to a larger mental space, they may do so through a very painful, even terrifying, process of undermining one's existing beliefs and theories—painful because our mental lives are sustained, consciously or unconsciously, by theories, sometimes invested with the force of ideology or delusion. And to these inner conflicts are added the external ones—the acceptance or contesting or dismissal of new ideas by one's contemporaries.

The history of science and medicine, in a sort of Darwinian way, has taken much of its shape from intellectual and personal competitions that force people to confront both anomalies and deeply held ideologies; and competition, in the form of open debate and trial, is therefore essential to its progress. When these debates are open and straightforward, a rapid resolution and advance can sometimes be achieved.

VI

But perhaps it is equally important to look at discoveries and ideas not only by taking account of their acceptance or dismissal by contemporaries but also by seeing them with respect to the history of ideas, as some of the greatest scientists have done. Einstein entitled his own idiosyncratic book *The Evolution of Physics,*[20] and the story he tells is not just one of emergence but one of radical discontinuity too. Thus part I is entitled "Rise of the Mechanical View." The mechanical worldview, as he sees it, had to collapse and leave a rather frightening intellectual vacuum before a radically new concept could be born. The conception of a "field" of forces—which was a prerequisite of the theory of relativity—in no way emerges or evolves *from* a mechanical one. Thus it is less evolution than revolution that Einstein speaks of—a revolution that he himself, of course, took to heretofore unimaginable heights.

But, most importantly, Einstein is at pains to say that the new theory does not destroy the old, does not invalidate it or supersede it, but rather "allows us to regain our old concepts from a higher level." Einstein expands this notion in a famous simile: "To use a comparison, we could say that creating a new theory is not like destroying an old barn and erecting a skyscraper in its place. It is rather like climbing a mountain, gaining new and wider views, discovering unexpected connections between our starting point and its rich environment. But the point from which we started out still exists and can be seen, although it appears smaller and forms a tiny part of our broad view gained by the mastery of the obstacles on our adventurous way up."

Hermann von Helmholtz, in his partly autobiographical memoir, "On Thought in Medicine," also uses the image of a mountain climb (he himself was an ardent Alpinist) but describes the climb as anything but linear. One cannot see in advance, he says, a way up the mountain; it can only be climbed by trial and error. The intellectual mountaineer makes false starts, gets stuck, gets into blind alleys and culde-sacs, finds himself in untenable positions, has to backtrack, has to descend and to start again. Thus, slowly and painfully, with innumerable errors and corrections,

20. This book was written in collaboration with Einstein's friend and colleague Leopold Infeld, but its thoughts and tone are pure Einstein.

he makes his zigzag way up the mountain. It is only when he reaches the summit or the height he desires that he will see that there was, in fact, a direct route, a "royal road," to it. In his publications, Helmholtz says, he takes his readers along this royal road, but this bears no resemblance to the crooked and tortuous processes by which he constructed a path for himself.[21]

In such accounts we find a common theme—that there is some vision, intuitive and inchoate, of what must be done, and that it is this, once glimpsed, that drives the intellect forward. Thus Einstein at fifteen had fantasies about riding a light-beam and ten years later developed the theory of special relativity, going from a boy's dream to the grandest of theories. Was the achievement of the theory of special relativity, and then of general relativity, "inevitable," part of an ongoing historical process? Or the result of a singularity, the advent of a unique genius? Would relativity have been conceived in Einstein's absence? (And how quickly would relativity have been accepted had it not been for the solar eclipse of 1919, which, by a rare chance, allowed the theory to be confirmed by accurate observation of the effect of the sun's gravity on light?) Neither "historical process" nor "genius" is an adequate explanation—each glosses over the complexity, the chancy nature, of reality. What emerges from a close study of a life such as Einstein's is the immense role that fortuity played in his life and the fact that this or that technical achievement was available to be used—the Michelson-Morley experiment, for example. If Georg Riemann and other mathematicians had not developed non-Euclidean geometries, Einstein would not have had the intellectual techniques available to move from a vague vision to a fully developed theory, which requires the concepts of non-Euclidean geometries. He was, of course, intensely alert, primed to perceive and seize whatever he could use. It was a particularly happy coincidence that non-Euclidean geometries had been developed at this time. They had been worked out as pure abstract constructions, with no notion that they might be appropriate to any physical model of the world.

A huge number of isolated, autonomous individual factors must be present before the seemingly magical act of a creative advance, and the absence (or insufficient development) of any one may suffice to prevent it. The huge role of contingency, of sheer luck (good or bad), it seems to me, is never emphasized enough. And this is much more so in medicine even than in science, for medicine often depends crucially on rare and unusual, perhaps unique, cases being encountered by the "right" person, at the right time.

Could the history of science—like life—be rerun quite differently? Does the evolution of ideas resemble the evolution of life? Assuredly we see sudden explosions of activity, when enormous advances are made in a very short time—this was so for molecular biology in the 1950s and 1960s; for quantum physics in the 1920s; and a similar burst of fundamental, Kuhnian, work seems to be occurring in neuro-

21. Helmholtz 1877.

science now. Sudden bursts of discovery change the face of science, and these are often followed by long periods of consolidation and, in a sense, stasis. One cannot but be reminded of the picture of "punctuated equilibrium" given us by Niles Eldredge and Stephen Jay Gould and wonder if there is at least an analogy here to a natural evolutionary process.[22]

And yet, even if this holds as a general pattern, the specifics, one feels, could be very different, for ideas seem to arise, flourish, go in all directions, or abort and become extinct, in completely unpredictable ways. Gould is fond of saying that if the tape of life on earth could be replayed, it would be wholly different the second time around. Suppose that John Mayow had indeed discovered oxygen in the 1670s; that Babbage's Difference Engine—a computer—had been built in the last century: might the course of science have been quite different?[23] This is the stuff of fantasy, of course, but fantasy that brings home the sense that science is not an ineluctable process but, in its details, contingent in the extreme.

BIBLIOGRAPHY

Armitage, F. P. 1906. *A History of Chemistry.* New York: Longmans, Green.

Brock, W. H. 1993. *The Norton History of Chemistry.* New York: W. W. Norton.

Damasio, A. R. 1985. "Disorders in Visual Processing." In *Principles of Behavioral Neurology,* ed. M. M. Mesulam, pp. 259–88. Philadelphia: F. A. Davis.

Duchenne, G. B. A. 1868. "Recherches sur la paralysie musculaire pseudohypertrophique ou paralysie myosclerotique." *Archives Générales de Médicine* 11, no. 5: 178, 305, 421, 552.

Einstein, A., and L. Infeld. 1961. *The Evolution of Physics.* New York: Simon and Schuster.

Eldredge, N., and S. J. Gould. 1972. "Punctuated Equilibrium: An Alternative to Phyletic Gradualism." In *Models in Paleobiology,* ed. T. J. M. Schopf, pp. 82–115. San Francisco: Freeman Cooper.

Gibson, W. 1991. *The Difference Engine.* New York: Bantam Books.

Helmholtz, H. von. 1877. "On Thought in Medicine." In *Science and Culture: Popular and Philosophical Essays,* ed. D. Cahan, pp. 309–27. Chicago: University of Chicago Press, 1995.

Herschel, J. F. 1858. "On Sensorial Vision." Reprinted in *Familiar Lectures on Scientific Subjects.* London: Alexander Strahan, 1866.

Holmes, G. 1918. "Disturbances of Visual Orientation." *Br. J. Ophthalmol.* 2:449–86, 506–16.

———. 1919. "Disturbances of Spatial Orientation and Visual Attention with Loss of Stereoscopic Vision." *Arch. Neurol. Psychiatry* 1:385.

Hubel, D., and T. Wiesel. 1962. "Receptive Fields, Binocular Interactions, and Functional Architecture in the Cat's Visual Cortex." *J. Physiol. Lon.* 160:106–54.

22. Eldredge and Gould 1972.

23. Some of these as-if fantasies have been powerfully explored by science fiction writers. Thus William Gibson and Bruce Sterling in the 1991 novel *The Difference Engine* imagine science and the world set on a different course with the actual construction of Charles Babbage's Difference Engine (and the start of a computer era) in the 1850s. (Intriguingly, Babbage's Analytical and Difference Engines have now been made, exactly as he specified them, and are on display in the Science Museum in London. They work, and could in fact have been made a century and a half ago, although the cost would have been prodigious.)

Leont'ev, A. N., and A. V. Zaporozhets. 1960. *Rehabilitation of Hand Function.* Trans. B. Haigh. Ed. W. R. Russell. New York: Pergamon Press.

Liveing, E. 1873. *On Megrim, Sick-Headache, and Some Allied Disorders: A Contribution to the Pathology of Nerve Storms.* London: Churchill.

Meige, H., and E. Feindel. 1907. *Tics and Their Treatment.* Trans. S. A. K. Wilson. London: Sidney Appleton, 1907.

Mitchell, S. W. [1872] 1965. *Injuries of Nerves.* New York: Dover Press.

Sacks, O. [1970] 1992. *Migraine.* Rev. ed. Berkeley and Los Angeles: University of California Press.

————. 1984. *A Leg to Stand On.* New York: Summit Books.

————. [1984] 1998. *A Leg to Stand On.* Rev. ed. New York: Simon and Schuster.

————. 1987a. "Tics." *New York Review of Books* (29 January 1987): 37–41.

————. 1987b. "The Case of the Colorblind Painter." *New York Review of Books* (19 November 1987): 25–34.

————. 1995. *An Anthropologist on Mars.* New York: Knopf.

Silvers, R. 1995. *Hidden Histories of Science.* New York: New York Review of Books.

Tourette, G. G. de la. 1885. "Étude sur une affection nerveuse caractérisée par de l'incoordination motrice accompagnée d'echalalie et de copralalie." *Arch. Neur.* (Paris) 9. A partial translation is found in C. G. Goetz and H. L. Klawans, "Gilles de la Tourette on Tourette Syndrome," in *Advances in Neurology,* ed. A. J. Friedhoff and T. N. Chas, vol. 35: *Gilles de la Tourette Syndrome* (New York: Raven Press, 1982).

Verrey, L. 1888b. "Hemiachromatopsie droite absolue." *Arch. Opthalmol.* (Paris) 8:289–300.

Zeki, S. 1993. *A Vision of the Brain.* Oxford: Blackwell's Scientific Publications.

Zihl, J., D. von Cramon, and N. Mai. 1983. "Selective Disturbance of Movement Vision after Bilateral Brain Damage." *Brain* 106, no. 2:313–40.

PART THREE

Historical Perspectives

Relatively Unproblematic Examples

CHAPTER SEVEN

Prematurity and Delay in the Prevention of Scurvy

Kenneth J. Carpenter

Gunther Stent has used the concept of prematurity to explain missed opportunities, where knowledge was available but not used. In my own reading the most striking example of such a phenomenon is the failure of the British navy from 1755 to 1795 to use lemon juice to prevent disabling outbreaks of scurvy in their ships kept at sea for long periods.[1] It appears that this came about because there was no appealing theory to support its use, while there was an attractive theory, with prestigious backing, that supported an alternative measure even though the alternative had never been put directly to the test.

Scurvy (or scorbutus) was probably a much older problem for poor people in northern countries, but it came to the attention of governments and medical writers in the 1500s, after sailors from western Europe had begun to make ocean voyages lasting ten weeks or more. From expedition after expedition it was reported that after about two months at sea the men began to weaken and develop blood blisters. Their thighs became stiff and painful to the touch, their teeth became loose, and finally the men would "die suddenly—in the middle of a sentence"[2]—from, we now realize, the bursting of a main blood vessel.

By the late 1500s it had been discovered that sour fruits, including lemons and oranges, were effective cures and preventives. However, these could not be stored for long periods, under the damp conditions on ships, without going moldy. Even stoppered bottles of lemon juice would go moldy in a few weeks of warm weather.

In London, the College of Physicians was consulted. They "reasoned" that healthy bodies were in balance between acid and alkaline principles. Since scurvy was cured by sour—that is, acidic—fruits, it must be the consequence of the body becoming imbalanced to the alkaline side. What was needed therefore was an acid

1. Carpenter 1986, pp. 91–97.
2. Wagner 1929, pp. 244–46.

that was stable and could be included in the ship's surgeon's medicine chest in concentrated form. They recommended "elixir of vitriol," that is to say, sulfuric acid with a little flavoring.[3]

The British navy followed this advice, and by 1747 it had been a standard navy issue for about a hundred years. In that year James Lind, a thirty-year-old ship's surgeon, carried out what is now praised as the first "controlled clinical trial." In his sick bay he had 12 men with scurvy "as similar as I could have them," kept them all on the standard "sick" diet, and tested six different treatments—with two men in each treatment group—for a period of two weeks. I will list just three.

1. Twenty-five drops of elixir of vitriol (diluted) three times per day.
2. Two spoonfuls of vinegar, three times per day.
3. Two oranges and one lemon per day—for six days only, when the supply ran out.

The result was that those receiving the fruit were cured, and those receiving the other two acidic materials were *not* cured, or even improved.

Six years later (in 1753), after Lind had obtained his M.D. degree at Edinburgh and was licensed as a physician, he published his finding demonstrating that it was not "acidity" as such that prevented or cured scurvy, but some special property of citrus and other fruits.[4] However, his book was ignored, and citrus fruits were not supplied to sailors as standard practice for the next 40 years, even when a fleet had ready access to them, as in the West Indies. Then the practice *was* adopted and was of great importance in enabling the British fleets to keep at sea for long periods during the Napoleonic wars.

Herbert Spencer, the Victorian economist-philosopher who disapproved of state interventions, was to write with hindsight that "it was an amazing perversity of officialdom that the Admiralty had not adopted a procedure after its chief medical officer had given conclusive evidence of its worth."[5] (Actually, Lind *was* put in charge of the navy's largest hospital in the 1750s but was never its chief medical officer.)

I believe that Lind was ignored in large part because the theory he proposed to explain his observation did not prove sufficiently attractive to compete with some other new ideas. His idea was that, because of the moist atmosphere at sea, the pores in human skin became blocked. This hindered perspiration, which was believed to be the route by which toxins were excreted from the body. When these toxins accumulated in the body, scurvy developed. Lemon juice had a detergent action that cleared the obstruction. This theory failed to explain some well-established examples of scurvy appearing on land, particularly in besieged cities.

The alternative "new" idea was based on the work of Sir John Pringle, a medical scientist who had considerable influence in his position as president of the Royal Society. He had studied factors that affected the rate of putrefaction of pieces of meat

3. Lloyd and Coulter 1961, p. 294.
4. Lind 1753, pp. 145–48.
5. Spencer 1879, pp. 162–63.

placed in small tubes with added water. He judged its extent by the intensity of the smell when he removed the cork and sniffed over the open tube, and concluded that adding bread to the meat "after a period checked the putrefaction with release of much air." In his opinion this was of medical relevance since "excessive putrefaction led to putrid diseases of which scurvy was one."[6] (With hindsight, we would probably attribute the observation to the acidic fermentation of the bread resulting in the volatile "putrid-smelling" amines, forming water-soluble, nonvolatile salts.)

David MacBride, a medical man in Dublin with scientific interests, followed up on Pringle's idea. He pointed out, correctly, that nonacidic fresh vegetables had also been found to cure scurvy, and he concluded that their important common property was that they readily fermented. He recommended, for preventing scurvy at sea, the use of malted barley. This was barley that had been moistened, so that it had begun to sprout and the starch had been partially hydrolyzed to sugars, and then oven-dried so as to stop the process. This material, when suspended in hot water to give a "sweet wort," served as the easily fermentable material for brewing beer. MacBride reasoned that such malted barley could be stored indefinitely on board ship, and if a sailor developing scurvy were then given freshly prepared wort, it would ferment rapidly in his tissues and inhibit putrefaction, just as it did in Pringle's test tubes.[7]

Captain James Cook's famous first voyage round the world began in 1768. The naval authorities supplied him and instructed him to evaluate a whole range of "antiscorbutics," that is, products claimed to prevent scurvy. These included sauerkraut, "rob [concentrated juice] of oranges and lemons," sugar, and vinegar. He was also supplied with a large quantity of malt, together with copies of MacBride's *Experimental Essays,* and advised that "there was great reason to believe that malt made into wort may be of great benefit to seamen in scorbutic and other putrid diseases." There is no evidence that the authorities supplied him with Lind's *Treatise of the Scurvy.* As we know from Cook's diary, he also took pains to gather green leafy material or yams and coconuts whenever he came to land.

His completion of a three-year voyage without having lost a single man from scurvy was unprecedented. But with so many antiscorbutics in use, it was difficult to place a particular value on any one of them. William Perry, his surgeon's mate, wrote, "They were of such infinite service . . . that the use of malt, with respect to necessity, was almost entirely precluded. Nevertheless from its mode of operation, from Mr. MacBride's reasoning, I shall not hesitate a moment before declaring my opinion that the malt is the best medicine I know."[8]

Now, a surgeon's mate was not at that time a university-educated man, but we also have the diary of the scientist (or "naturalist," as he was called at the time) on the expedition. This was the twenty-five-year-old Joseph Banks, later to become "Sir Joseph" and to preside over the Royal Society. Seven months into the voyage he

6. Pringle 1750, pp. 553–55.
7. MacBride 1767, pp. 32–33; Scott 1970, pp. 46–50.
8. Beaglehole 1955, pp. 632–33.

wrote, "I drank a pint or more of wort every evening [as a pleasant substitute for sauerkraut] but this did not check the distemper [scurvy] entirely. . . . About a fortnight ago my gums swelled and some small pimples rose on the inside of my mouth and threatened to become ulcers. I then flew to the lemon juice 'in my private store, preserved with brandy.' . . . In less than a week my gums became as firm as ever."[9]

A year later (still on the ship) he wrote, "Our malt having turned out so indifferent[,] the surgeon made little use of it." They therefore boiled their breakfast wheat in wort, and Banks wrote again, "I thought I received great benefit from use of this mix in banishing costiveness [constipation]. Whether that is a more beneficial method of administering wort must be left especially to that excellent surgeon Mr. MacBride, whose ingenious treatise on the sea scurvy can never be sufficiently commended."[10] Despite these commendations of malt, there had been no actual demonstration of its value during Cook's voyage or any other.

That the British navy did finally adopt lemon (or lime) juice as a regular issue was largely the work of Sir Gilbert Blane. He was a physician of high social standing and a friend of Admiral George Rodney, who took him to the West Indies in 1780 as "physician to the fleet." There he collected statistics and made the significant point that losses of effective men from disease were many times greater than from war wounds. Turning to scurvy particularly, he was to write in 1788 that "every fifty oranges or lemons might be considered as a hand to the fleet inasmuch as the health, and perhaps the life, of a man would thereby be saved." Admirals who took his advice kept their fleets free from scurvy, and finally, in 1795, as a member of the Sick and Hurt Board in London, Blane persuaded the Admiralty "to sanction a daily issue of $^3/_4$ ounce of lemon, 2 ounces of sugar mixed in a man's grog [rum and water]."[11] Admiral Horatio Nelson obtained a good supply of lemons from Sicily in the Napoleonic wars, and his ability to keep the fleet at sea for long periods contributed to his successes.[12]

To jump forward 200 years, it is now accepted that scurvy results from the continued consumption of a diet containing little vitamin C, and analyses have shown that citrus fruits are rich in the vitamin. The vitamin content of stored lemon juice does slowly decline, but the loss is not accelerated by the presence of added brandy, which is useful in delaying the growth of mold.[13] When barley is sprouted, vitamin C is synthesized, but commercial drying then destroys almost all of it, so that dried malt and wort contain little or none, and we can understand now that its value was never demonstrated.[14] But then, as now, we are happier to accept an idea that

9. Banks 1896, pp. 71–72.

10. Ibid., pp. 258–59.

11. Lloyd 1961, pp. 129–31.

12. Acerra 1981, p. 74. The subtitle of her article, translated from French, is "Was It the Power of Their Lemon Juice That Won the Battles of Aboukir and Trafalgar for the English Navy?"

13. Chick and Hume 1917.

14. Hughes 1975.

fits with our background of ideas, than one that does not, even if we have seen little or no evidence to support it. In my example the use of wort as an antiscorbutic, for which there was no experience, did fit, and lemon juice, for which there had been a long experience, did not.

To a modern reader the comments of Perry and of Banks may seem merely ridiculous. But that reaction comes, in part anyway, from Pringle's theory of putrefactive disease seeming so bizarre to us. At the time it was clearly extremely attractive and appeared to be based on experimental evidence. MacBride had also made a good point, verified by experience, that nonacidic fresh vegetable materials, such as green leaves, could also be antiscorbutic. To us, his chain of reasoning has obvious gaps, but these may not have been evident to someone hearing "on good authority" of this new treatment based on the most modern scientific advances. I do not believe that human nature has changed. We should therefore take this example to heart as the kind of thing that could happen at any time.

BIBLIOGRAPHY

Acerra, M-M. 1981. "Le Scorbut: La Peste du Marin." *L'Histoire* 36 (July–August): 74–75.

Banks, J. 1896. *Journal during Captain Cook's First Voyage, 1768–1771*. London: Macmillan.

Beaglehole, J. C. 1955. *The Journals of Captain James Cook: The Voyage of the Endeavour.* Cambridge: Hakluyt Society.

Carpenter, K. J. 1986. *The History of Scurvy and Vitamin C.* New York: Cambridge University Press.

Chick, H., and E. M. Hume. 1917. "The Distribution among Foodstuffs of the Substances Required for the Prevention of Beriberi and Scurvy." *Trans. Soc. Trop. Med. Hyg.* 10:141–78.

Hughes, R. E. 1975. "James Lind and the Case of Scurvy: An Experimental Approach." *Med. Hist.* 19:342–51.

Lind, J. 1753. *A Treatise of the Scurvy.* Edinburgh: Millar.

Lloyd, C. 1961. "The Introduction of Lemon Juice as a Cure for Scurvy." *Bull. Hist. Med.* 35:123–32.

Lloyd, C., and J. L. S. Coulter. 1961. *Medicine and the Navy, 1200–1900: 1714–1815.* Vol. 3. Edinburgh: Livingstone.

MacBride, D. 1767. *An Historical Account of a New Method of Treating Scurvy at Sea.* London: Thomas Ewing.

Pringle, J. 1750. "Some Experiments on Substances Resisting Putrefaction." *Philos. Trans. R. Soc. London.* 46:480–88, 525–34, 550–58.

Scott, E. L. 1970. "The Macbridean Doctrine of Air: An Eighteenth Century Explanation of Some Biochemical Processes Including Photosynthesis." *Ambix* 17:43–57.

Spencer, H. 1879. *The Study of Sociology.* New York: Appleton.

Wagner, H. R. 1929. *Spanish Voyages to the North West Coast of America in the Sixteenth Century.* San Francisco: California Historical Society.

CHAPTER EIGHT

A Triptych to Serendip

Prematurity and Resistance to Discovery in the Earth Sciences

William Glen

The global-warming debate, now "perhaps the most pressing and urgent environmental issue on the world's agenda," invites comparison with earlier and contemporary theoretical debates that have triggered upheavals in science during the past half century.[1] Many of the templates I fashioned in the past from historical studies of the debate over continental drift/plate tectonics and the meteorite impact/volcanism/mass-extinction conflict seem also to fit the global-warming controversy, which has been well delineated by Spencer Weart.[2] I compare here the three controversies from various vantage points and bolster illations of both commonalities and differences among them by referring to episodes of discovery in still other areas of science. The history of the global-warming issue is less widely known than the other cases and thus will be treated more fully.

The potential problem of global warming was first recognized in 1896 by Svante Arrhenius. He explained in detail how carbon dioxide being put into the atmosphere through human industry might result in global warming. But surprisingly, almost nothing of significance was learned about global warming until the 1950s—it took a full half century for science to finally come to regard it as a potential danger. That realization triggered an upheaval.

The reasons for such a seemingly delayed response to Arrhenius's discovery are curious, but even more interesting is a comparison of global warming to other theoretical conflicts. Such comparisons, made here, include the notion of prematurity of novel ideas; the tendency for science to resist new ideas—especially ideas that threaten a reigning paradigm; and the unplanned, fortuitous character of the respective upheaval-triggering discoveries.

---

1. The quote is from National Environmental Research Council 1998 *fide* Stanhill 1999. On the theoretical debates, see Glen 1990, 1994, 1996, 1998.
2. Glen 1981, 1982, 1985; Weart 1997, 1998.

Just a few years after Arrhenius opened the global-warming question, he was attacked both for the methods he used to judge the effect of CO_2 on solar radiation and because it was commonly held that water vapor essentially made the atmosphere opaque, so that all the long-wave radiation that might be absorbed by additional CO_2 was already absorbed by water vapor. Such a virtual canon made it appear that adding CO_2 to the atmosphere was of no consequence. That fallacious, encumbering idea was prevalent as late as the early 1950s.[3]

Warming of the climate was apparent to both scientists and the public during the first half of the twentieth century. In the Northern Hemisphere, the familiar white Christmas seemed to be disappearing, real blizzards were becoming infrequent, hockey players waited in vain for the rivers to freeze over, glaciers were retreating, and other signs of warming were widely discussed.[4] However, no one seemed concerned, and some even thought that warming would be beneficial—a longer growing season, not drought, is what first sprang to the farmer's mind.

The warming was attributed to benign weather cycles: cyclic natural phenomena that were not threatening. Indeed, cycles have been the firmest launching platform for prediction since the rise of science. No one dreamed that humankind could influence the climate globally. The effects of human industry on climate were assumed to be locally confined and brief.[5] Only Stewart Callender, a lone voice in Britain, warned in 1938 that carbon dioxide in the atmosphere had increased by 10 percent since the 1890s, and that that could be the reason for climatic warming over that interval.

Callender's data were thought unreliable, and he was ignored by most meteorologists. His idea was also rejected because it was thought at the time that the oceans act to equalize, or balance out, carbon dioxide fluctuations. Therefore, virtually no one thought there was cause for alarm. Why worry about seemingly minor atmospheric CO_2 perturbations when the oceans contain 50 times as much CO_2 as the atmosphere? Very few reliable data were known, and Callender even had to cite 30-year-old studies in his paper of 1938! Data that could guide reliable interpretation did not exist.

The subject of climate change went largely uninvestigated until the mid-1950s, partly because climatology remained a minor branch of meteorology, which was mainly a data-gathering service. Climatology was then a field that was almost purely descriptive and unlikely to attract gifted students of theoretical research.

It was not until the late 1950s that some scientists began to regard the warming trend as something that was not an innocuous part of a natural cycle. The 1960s marked a shift in understanding and interpretation of warming. Weart believes that shift may eventually be regarded as "one of our century's pivotal scientific developments." He also believes that this scientific turn was mainly a lucky break, be-

3. Weber and Randall 1932; Elder and Strong 1953.
4. Carson 1951.
5. Blair 1942, pp. 90, 101.

cause it resulted from research that was aimed at entirely different questions—in short, the rediscovery of the global-warming problem was serendipitous. Shortly, we will see other cases of serendipity in different disciplines that resulted in similar unpredictable advances. Among them is the evidence, found in 1965, that finally convinced the geologic community that continents were not fixed but were laterally mobile; and the discovery of unearthly evidence that resurrected the centuries-old, languishing, meteorite-impact theory of mass extinction.[6]

The first compelling evidence to eventually figure in the global-warming question came from research during World War 2 and shortly afterward. It showed that CO_2 could affect the absorption of radiation quite apart from the effect of water vapor. The idea that CO_2 did not count—which was used to reject Arrhenius's claim in 1896 and similar claims of others afterward—was finally recognized as wrong. Here we see an old canon discarded, but only after many years during which the canon—really an unfounded assumption—essentially precluded progress across a broad potential front of research.

By 1955, Gilbert Plass, using digital computers, demonstrated that adding CO_2 to the atmosphere would increase the interception of infrared radiation. In the same year, Hans Suess found fossil carbon in the atmosphere that had been produced by burning fossil fuels. But whether the oceans could harmlessly soak up all that carbon was still not known. That was because the oceans were an extremely complex chemical system that was still little understood.

The fate of a molecule of carbon dioxide in the atmosphere or in the sea was still enigmatic. Roger Revelle had long been interested in the problem and was motivated further by the need to learn the fate of radioactive fallout from atomic tests in the mid-1950s. He attacked the problem by researching carbon-14 with Hans Suess. At the same time, Harmon Craig and also the team of James Arnold and Ernest Anderson approached the problem similarly through carbon-14 work. They all agreed to share priority by publishing simultaneously.

Their findings were quite similar: a CO_2 molecule spends about ten years in the atmosphere before entering the sea, and the oceans turn over completely in several hundred years. Such rates were thought fast enough to carry CO_2 into the ocean depths.

After studying the carbon isotope data, Revelle and Suess initially stated that the oceans have probably absorbed all the CO_2 released from artificial fuels since the start of the industrial revolution—a quieting conclusion for all. However, Revelle's other work on the chemistry of seawater had left him uneasy; it had engendered more sophisticated questions for him and others about what happened to a molecule of CO_2 that went into the ocean. Would it stay there or escape back into the atmosphere? And how much additional CO_2 could seawater hold? Revelle was an extremely charismatic and powerfully influential figure: it is thus important to ex-

6. On the lateral mobility of continents, see Glen 1982; on the meteorite-impact theory of mass extinction, see Glen 1994.

amine the sequence of events that shaped his opinions on the global-warming question—opinions that weighed heavily in the formulation of research policy globally. Weart writes:

> January 1956: Revelle apparently tells Arnold that he suspects 80% of the CO_2 in the atmosphere will stay there. This can only be based on guesswork. March 1956: Revelle announces publicly, in testimony to Congress, that serious global warming effects could occur within 50 years. August 1956: Revelle and Suess publish a paper which says that most CO_2 will stay in the atmosphere and so a global warming world "experiment" may be underway—by now Revelle has a solid calculation, but it gives a centuries time scale, and the result is an afterthought, not prominently featured in his paper. The science remains uncertain and poorly understood. 1959: Bolin and Eriksson explain the science fully (only at this point was the lack of ocean uptake of CO_2 made entirely plausible). Finally it should be noted that Revelle talked about the risk of global warming in the later 1950s in further testimony to Congress and in conversation with science journalists, some of whom wrote about it (Revelle went beyond what he could prove in his public warning about the dangerous potential of global warming—nobody, except perhaps Plass, was more vocal on that point than Revelle).[7]

Gilbert Plass's *Scientific American* article of 1959 reflected the growing concern about the threat of global warming. Although data were still wanting and approximations uncertain, Plass predicted that the global temperature would climb by 3°F by the year 2000. His article catapulted industrial pollution and the degradation of the environment to the forefront of both science and the media.

The demand to know the exact level of CO_2 in the atmosphere had led Revelle to hire Charles Keeling, whose two-year-long, painstaking measurements showed in 1960 that CO_2 was rising ominously. Keeling's easily comprehensible CO_2 curve was arresting; it thus became the icon and logo of the greenhouse effect. It triggered an upheaval in climate studies. Efforts to understand and predict climate warming were energetically launched and have continued to broaden and mount, seemingly exponentially, up to the present.

This is clearly not a story of step-by-step accumulation of facts gathered in a conscious attempt to answer a specific larger question. We certainly do not find a mapped path toward a sought solution. Weart's history shows that the vital things learned did not appear in a "linear and logical sequence." After Callender's work in 1938, each new discovery that helped to eventually raise the global-warming question to prominence was made in a different subject area that was completely distant from climate studies. None of the travelers had purposefully set out to derive data aimed at the global warming question. None sensed that they were on the road to Serendip.

Data on infrared absorption came from studies of war weapons; on spectral line widths from high-altitude research; on carbon-14 from attempts to fathom

---

7. Weart, personal communication, July 1998 (cited with permission of Spencer Weart).

Egyptian chronology; and on the movement and residence of CO_2 from concerns about nuclear-waste disposal. Data on seawater chemistry and other information came from still other fields that were not addressing the global-warming question.[8]

In addition to the lack of evidence of global warming—evidence that would appear only long after Arrhenius's alert in 1896—there were further encumbrances to progress in the form of suppositious generalizations emplaced as canons, which forestalled even other theoretical debates. Among those canons were the principle of uniformitarianism, belief in the repeatability of certain natural cycles that permit prediction, and the notion of a self-balancing homeostatic Earth system.

The principle of uniformitarianism (uniformity) as espoused by James Hutton and Charles Lyell—which has long been the philosophical mainstay of the earth sciences but is now being revised to accommodate a rising neocatastrophism—holds that past is prologue: that we can understand Earth's history and predict its future by peering through the lens of the present.[9] It is the belief that Earth has been shaped slowly and gradually during its history by the uniform operation of the forces, processes, and principles of nature that we can now observe.

Uniformitarianism displaced the earlier philosophy of biblically anchored global catastrophism during a strenuous struggle in the nineteenth century. Uniformitarianism's more obvious development can be seen in the late eighteenth century, when the rational conception of the universe came to be expressed increasingly in mathematical language, and knowledge systems entered the logician's paradise of classical, precise determination. Uniformity's ascendance was complete before the end of the nineteenth century. With the rise of uniformity, large-scale catastrophes and contingencies came to fit less and less with the new paradigm that saw the universe as a stable system in a state of equilibrium.

The extent to which the principle of uniformitarianism pervaded and influenced natural philosophy is seldom realized even by scientists who practice in the domains where the idea has ruled up to the present. The principle has shaped resistance to any idea of large-scale, rapid change in the natural order. Uniformitarianism has traveled with the companion idea that the universe and Earth could always regain their balanced states however great the disturbance.

And uniformitarianism has worked against the idea that human enterprise could affect the global workings and balance of nature. During its century-long emplacement, uniformity has simply disallowed global catastrophes—especially sudden ones—in explaining the past or predicting the future. Uniformity is an intellectual-emotional cofferdam against the specter of the unpredictability and uncertainty of a hostile natural world.

The notion of catastrophism—which plays a role antithetical to uniformity—asks us to accept the unpredictable as part of the regular order, and as a tool by

---

8. Weart 1997, 1998.

9. On uniformitarianism, see Hutton 1788; and Lyell 1831; on neocatastrophism, see Glen 1994b,c.

which to retrodict the past and predict the future. Science is mainly charged with making more and more of nature predictable, whether it is a massive volcanic explosion, a climate change that is lethally stressful to humanity, or a devastating meteorite impact.

How such philosophical underpinnings have played roles in scientific controversies can be seen in the case of the Alvarez-Berkeley group meteorite-impact hypothesis, which sought to explain the mass extinction of life 65 million years ago that killed all the dinosaurs and 75 percent of species in the sea.[10] That great mass extinction marks the K-T boundary between the Cretaceous (K) and Tertiary (T) periods. The impact hypothesis was based on unearthly chemical evidence that was discovered, serendipitously, precisely in that boundary.

The find was made while searching *not* for evidence of an instantaneous catastrophe but rather—as Walter Alvarez has noted—for information that would determine the ratio between the constant rate of infall of cosmic iridium to the Earth and the variable rate of earthly sedimentation, in an attempt to estimate the time represented by the K-T extinction.[11] He was doing a study of the stratigraphic section in which the uncannily anomalous iridium find was made, and was not possessed of a catastrophic mind-set when he started to address questions about the time frame for the K-T extinction. Before the Berkeley group discovered the chemical anomaly at the boundary, neither he nor any other member of the group envisioned the kind of catastrophe that the concentration of iridium came to suggest.[12] The iridium anomaly drove the formulation of their theory of catastrophe. They could not have suspected that what they would find would come to impugn the principle of uniformitarianism and trigger an upheaval across several sciences.

The ruling plate-tectonics paradigm, in which the slow rates of process and change—notably crustal deformation—perfectly accommodated uniformitarianism, had already come to form the basis for well-received, Earth-based, mass-extinction and faunal-change hypotheses.[13] Uniformity-friendly plate tectonics of the 1970s was accepted as a cause of mass extinctions. Mass extinctions were then thought to have taken a few million years, which is the time frame for mountain building (orogeny), a process with which mass extinctions had been associated for almost a century.[14] For that reason especially, uniformitarianism contributed to the poor reception of the impact hypothesis, which, by its demand for a geologically instantaneous catastrophe of extraterrestrial origin, flew in the face of uniformitarianism.

Impact theory's reception was further chilled by other factors: the theory was based on evidence (iridium) unfamiliar to the paleontologic community charged with its appraisal; its mechanism of extinction cause was improbable in terms of

10. Alvarez et al. 1980.
11. Alvarez 1984; Glen 1994b.
12. Alvarez 1984; Asaro 1984.
13. Valentine and Moores 1970; Glen 1996.
14. Holmes 1978.

canonical knowledge; and it was authored by a physicist, nuclear chemists, and a geologist, none of whom belonged to the paleontologic community, which was specifically charged by convention with fathoming the cause of mass extinctions. The Alvarez-Berkeley team had formulated a theory with revolutionary bearing on an alien domain; thus they suffered immediate indifference or strong criticism.[15]

But as various forms of evidence for meteorite impact were uncovered over the decade that followed, most earth scientists came to believe that impact(s) had triggered the mass extinction 65 million years ago. However, most of the paleontologists—with the exception of certain specialists whose taxonomic groups had suffered severely right at the Cretaceous-Tertiary boundary—did not embrace impact(s) as the major cause of the extinction. That differing viewpoint was based on the specific nature of the fossil record that they treated and on the canons of their discipline. Students of the severely reduced planktonic foraminifera were the best example of this.[16]

The evolution of life—including the mammals that culminated in humankind—had been taught for a century as having been accomplished within the framework of uniformitarianism. It is a gradualist or uniformitarian view of organic evolution leading to higher, more complex and successful life-forms. That notion of evolutionary progression was read from the fossil record of Earth's crust. Such interpretation of evolution's gradual and uniformitarian character was sustained over the course of a century, even as evidence of several mass extinctions accumulated.

Mass extinctions, however puzzling their time span, were *assumed* to have taken millions of years. Those few great mass extinctions served as the benchmarks for the major subdivisions of the geological time scale. And it was an accepted idea that significant intervals of geologic time had gone unrecorded at such boundaries. It was also understood—long before the end of the last century—that the sedimentary rock record with its contained fossils was profoundly fragmentary; and its severe incompleteness was fully explicated in Wyatt Durham's presidential address to the Paleontological Society in 1967.[17] That knowledge of incompleteness further underpinned the mind-set that permitted massive taxonomic death—even at knifeblade-sharp horizons within the sequence of rock layers—to be accommodated within a uniformitarian framework. Such boundaries, marked by breaks in the continuity of the layered rocks and fossil groups, were assumed to have resulted from chance, rare combinations of uniformitarian normal forces and processes.

The prevailing paleontologic gestalt permitted discontinuities in the rock layers representing unrecorded time intervals to be accommodated within the framework of uniformitarianism; therefore missing fossil taxa were assumed to have gone ex-

15. Glen 1994a,b.
16. Glen 1990, 1994b, 1994c.
17. Durham 1967.

tinct at a normal pace within those unrecorded time gaps.[18] The explanation of missing time intervals provided a refuge from the need to invoke catastrophic forces or near-instantaneous extinction rates. The paleontologists held that, without evidence to the contrary, one should not invoke a giant instantaneous catastrophe to explain the disappearance of many species at a single horizon in the stratigraphic record; mass disappearances were "read" as significant lost time intervals in the book of the rocks.[19]

The newly ascendant idea that a meteorite impact could cause a geologically sudden mass extinction and thus reset the time scale and schedule for evolution has forced evolutionary biologists to rethink the question of whether evolution is as much a matter of "good genes as good luck."[20]

The success of impact theory is forcing a redefinition of uniformitarianism and a concomitant rise of neocatastrophism. This view is conditioning thinking about a number of larger questions, including global warming, impact-induced continental drift, and impact-induced massive volcanic episodes. Simultaneously, uniformity's companion idea of a strict determination of nature by mathematical science has been opened to question by several discoveries, especially chaos theory's nonlinear dynamics, which focuses on the daunting complexity of fluid systems such as those that make up global climate.

Chaos theory holds that a fluid dynamic system can never be predicted; therefore, nothing, in principle, will ever give us certainty in predicting global climate.[21] Chaos theory instructs that small changes in initial conditions of a dynamical system may produce grossly disproportionate changes as the system evolves. Thus any small change in any of the factors in the system may cause a catastrophic, unpredictable change down the line.

Any one of the several anthropogenic contributions to the atmosphere may have already altered the system's course such that a large-scale, unpredictable, destructive change is imminent. We really do not know. But isotopic data, such as Hans Oeschger's Dye 3 ice core from Greenland, are alarming. He documented a rise of 13°F in average annual temperature within a 10–20 year period about 35,000 years ago—a rate and magnitude of rise theretofore inconceivable—which was then sustained for a millennium, before falling back just as rapidly. Moreover, the atmospheric carbon dioxide of the time seemed to fluctuate directly with the alternating air temperatures.[22] That should have been enough to alert us to the possibility of a precipitous, catastrophic climate change in our unforeseeable future.

18. Newell 1967.
19. Ager 1981.
20. Raup 1991; Sepkoski 1990.
21. Shaw 1994.
22. Oeschger 1984 *fide* Kunzig 1996.

Detailed knowledge of the geologic past is wanting for most intervals, so why should we have supposed that Oeschger had stumbled onto a singular or even rare event? In the last few years, further studies have appeared that support and extend Oeschger's early surmise. Many of those studies were prompted by Oeschger's radical conclusion that the climate had virtually jumped back and forth from cold to warm and then cold again, staying in a given state for about a thousand years. And a group under K. C. Taylor then measured the electrical conductivity of the ice core as a surrogate for the ratio of fallout of acids to that of calcium carbonate–bearing dust and determined that the great temperature transition took place in less than three decades.

Recent correlations between radiocarbon-dated glacial deposits from New Zealand and cores of rapidly deposited sediments from California's Santa Barbara basin indicate that the changes were not only rapid but were spread across both hemispheres. A whole series of temperature-change events that punctuate the period from about 65,000 to 25,000 years ago, known from ice core research, has now been duplicated in ocean sediments from both the northern Atlantic and the Santa Barbara basin.

Many other such events must lie hidden within the murky, still-unexamined parts of the ice and stratigraphic records that tell of Earth's climatic tribulations. Over the years, the late Cesare Emiliani—a pioneer of isotopic interpretation of Earth history—repeatedly remarked to me on our ignorance of the mechanisms that control the possible suddenness of global warming and global cooling. In 2000, shortly before his death, he expressed an even greater sense of urgency about the safety of our climatic future.

Wally Broecker, the distinguished geophysicist, believes that whatever pushed Earth's climate did not lead to smooth changes but rather to jumps from one state of operation to another, and that essentially we do not know what the probability is that, through adding CO_2, we will cause the climate system to jump to one of its alternative modes of operation.[23]

Whether we treat global climate changes, meteorite impacts, earthquakes, landslides, volcanic eruptions, or other phenomena that demonstrate self-similarity at different scales (or what is called fractal scaling), we enjoy an abundance of data for small events with high frequencies of occurrence and suffer scarcity of data for large, infrequent events. Unfamiliarity with great, infrequent events that either are not known or are lost to history helps to sustain the encumbering, archaic, uniformitarian view that has blinkered science for over a century to the role of catastrophe in shaping Earth's history. Derek Ager was among the very few earth scientists aware of fractals who pleaded decades ago for an open-minded reexamination of the stratigraphic record and cautioned against overreliance on uniformitarianism—and he continued to make this plea until his death.[24]

23. Broecker 2000.
24. Ager 1993.

If catastrophic global environmental change can be demonstrated as more than merely a few extremely rare events—such as the ice-core-evidenced massive temperature changes on Earth thousands of years ago or the dinosaur-killing, K-T meteorite impact—but instead can be shown to be generalities of Earth's history, then uniformitarianism will be fully redefined; such appears increasingly likely. In its new form, uniformity will accommodate such gigantic excursions from the norm. Such inordinacies near the ends of fractal scales often escape discrimination because of the gross mismatch of the human and geologic historical time scales.

Recognizing that catastrophes of such sorts are really generalities—rather than singularities that do not figure in prediction making—will reshape our thinking about our place on the planet and in the universe. Catastrophism will be subsumed by the recognition that everything falls within a fractal scale of one sort or another, and that, as one ascends such a scale of phenomenology, not only are effects on the biosphere enlarged, but new effects appear that are qualitatively different from those produced at lower energy levels. That has now been demonstrated for a number of phenomena, including meteorite impacts.[25]

The geologic record instructs us against optimism. Our batting average for predicting infrequent, fractally obscure, catastrophic events is poor. The alarming rate of rise in carbon dioxide and other greenhouse gases must be rethought in terms of what we have learned about catastrophically proportioned chaos-theory responses in dynamic fluid systems: they are mainly and frighteningly unpredictable.

LESSONS FROM OTHER THEORETICAL CONFLICTS

Important ways in which science worked in the global-warming controversy can be identified in other theoretical conflicts and upheavals. Cognizance of the potentially ominous character of CO_2 emissions was delayed not only by the reign of uniformitarianism and its companion ideas but also by the date and intellectual context in which Arrhenius made his global-warming proposal. Arrhenius's idea seems to be a clear case of prematurity in science as formalized by Gunther Stent at Berkeley in 1972. "An idea is premature if its implications cannot be connected by a series of simple logical steps to canonical or generally accepted knowledge." (Stent's choice of the term "canonical" was precise: it has been used to denote a standard of judgment since the opening of the seventeenth century.)[26]

25. Melosh 1989.

26. Those alert to Whiggism and recentism who find Stent's term "premature" objectionable may find satisfaction in a substitute term such as "radical." "Radical" accords etymologically with Stent's taxonomic intent: "radical" pertains to that which proceeds from a new root; it is original, fundamental, and it connotes great change, including that of conceptualization. The radical idea may succeed or fail, but it is met with rejection or ignored at its advent by the discipline for which it was formulated. I propose, in addition, that the radical idea may, however, be favorably received in a discipline other than that for which it was formulated. An example is the meteorite-impact hypothesis of K-T mass

My own detailed historical studies of both the continental-drift debates (which eventuated in the ascendance of plate tectonics theory in the late 1960s) and the current debates triggered by the meteorite-impact theory of mass extinction show that those cases also share commonalties of development with the global-warming conflict—commonalities in addition to their both being premature ideas.[27] Recall that the crucial new comprehension of global warming was mainly a stroke of fortune that did not come from a purposeful effort toward understanding global warming but instead came from a series of unrelated, serendipitous contributions. The several different new discoveries that shed light on global warming came from research programs in completely different areas. Each program that birthed each discovery was not aimed at the global-warming question. Weart made that clear.[28]

Similarly, Alfred Wegener drew his case for continental drift from several elegant lines of empirical evidence.[29] But unfortunately, Wegener's idea of great lateral movements of the continents flew in the face of geology's then-reigning canon—also based on suppositions—that the geographical position of both the continents and ocean basins had always been fixed.[30]

Insurrectionist, paradigm-impugning ideas such as Wegener's continental drift theory have always catapulted their authors onto the hot griddle of orthodoxy. In addition, Wegener was decidedly premature in his idea in the same sense that Arrhenius was.

There was virtually nothing known about the geology of the seafloor in Wegener's time. His most vocal and influential critics, the geophysicists, examined the drift question by inappropriately applying the first-order principles of physics equally to both the continents—whose geology was fairly well understood—and the seafloor, whose geology was a complete mystery.[31] Beyond that, understanding of the subcrustal regions was scant. Geophysicists made surmises about the oceanic crust and the subcrust from the very short-term behavior of earthquake waves. However, such surmises were inappropriate in considering the rheology and deformability of such solids over vast periods of time. Geophysicists concluded—from a highly fragmentary factual basis and immature theoretical surmise—that granitic continents could not move through the unyielding basaltic ocean crust. Only

---

extinction, which was formulated by a physicist, nuclear chemists, and a geologist; it was rejected by most paleontologists—members of the discipline charged by convention with its assessment. However, the theory was embraced by almost all astrophysicists and meteoriticists. "Radical" may be defined such that it both accords with Stent's criteria for the category of ideas called premature and embraces further defining qualities that should preclude allegations of Whiggism or recentism.

27. On the continental-drift debates, see Glen 1982, 1985; on the meteorite-impact theory of mass extinction, see Glen 1990, 1994a,c,d, 1996, 1998.

28. Weart 1997.

29. Wegener 1912a,b, 1915.

30. Van Waterschoot van der Gracht et al. 1928.

31. Jeffreys 1924, 1974.

through painstaking research over the next half century did the nature of the seafloor and mantle below get partially clarified.

Scientists learned that the seafloor's history of formation, structure, volcanicity, magnetization, heat flow, and other characteristics were completely different from anything known from the continents. They also found overwhelming evidence— discussed shortly—that the continents had drifted thousands of kilometers, carried on giant fragments of Earth's outer shell that came to be called tectonic plates.

The various programs that contributed the data used in making the plate tec- tonics revolution were addressing a variety of different problems. Not a single one of those programs was directly examining the continental drift question. In no re- search proposal supporting the various programs *that eventually yielded the confirma- tory forms of evidence for drift* can mention of a test of some aspect of the continental drift question be found. That includes the crucial program that yielded the Eltanin 19 magnetic anomaly profile, which is perhaps the most important single datum in the history of geology.[32]

The invention of the static-mode mass spectrometer by John Reynolds and the development of new argon extraction equipment and methods by Jack Evernden and Garniss Curtis in the 1950s, all at Berkeley, refined radiometric dating by an order of magnitude.[33] That refined dating capability was used to advantage at the U.S. Geological Survey in Menlo Park by Allan Cox, Brent Dalrymple, and Richard Doell—all trained in Evernden and Curtis's department—who became the chief architects of the first time scale of geomagnetic reversals dated by radiometry.

They measured the age and magnetic polarity of hundreds of rocks from around the globe (in a several-years-long experiment suggested by Patrick Blackett in 1956) to find that all rocks of the same age had the same polarity, and thus, with some of the data contributed by other research groups, proved that Earth's magnetic field had repeatedly and randomly reversed during the geologic past. By that demonstration they removed Earth's magnetic polarity from the roster of physical constants.[34]

Simultaneously, they formulated the first time scale of geomagnetic reversals.[35] That scale became the magical key by which the puzzling, zebra-stripe magnetic patterns of seafloor crust recorded by Ronald Mason and Arthur Raff were inter- preted when Walter Pittman discovered the Eltanin 19 magnetic anomaly profile in 1966.[36] It was that magnetic anomaly profile that proved that the Vine-Matthews- Morley hypothesis was correct.[37] This hypothesis had predicted that bilaterally sym- metrical magnetic stripes, whose widths were proportional to the time intervals of the polarity-reversal scale of the Cox group, would be distributed across midocean

32. Glen 1979, 1982.
33. Reynolds 1954; Evernden and Curtis 1961; Evernden and James 1964.
34. Cox et al. 1967; Glen 1982.
35. Cox et al. 1963; Glen 1982.
36. Mason and Raff 1961; Pittman and Heirtzler 1966.
37. Vine and Matthews 1963; Morley and Larochelle 1964; Glen 1982, 1985.

ridges and rises (seafloor spreading centers). The Eltanin 19 magnetic anomaly profile proved, virtually overnight, the theory of seafloor spreading—a theory which subsumed that of continental drift—and triggered the ascent of the more complete plate-tectonics theory (the germ of which had been contributed two years earlier, but ignored, in a landmark paper of 1965 by J. Tuzo Wilson on transform faults).

The development of the static-mode mass spectrometer and refined dating methods, the formulation of the geomagnetic time scale, the discovery of magnetic stripes on the seafloor, and the recording and digitization of the Eltanin 19 magnetic anomaly profile *all came from programs that did not address the continental drift question.*

The *premature* concepts of the transform fault by Wilson (which defined the tectonic plate by providing its third type of boundary structure); nemataths, or thread ridges, defined also by Wilson in 1963 as age-graded, linear volcanic chains formed by the passage of a tectonic plate over a hot spot or fixed lava source; and other ideas directed at the drift question were mainly ignored. Those premature concepts languished until the powerful, conversionary evidence of the magnetization pattern displayed in the Eltanin 19 profile was deciphered—only then were those neglected ideas thrust under immediate, global scrutiny.

The Eltanin profile virtually forced the matching of the seemingly magical numerical ratios between the widths of magnetized blocks of seafloor crust, the recently assembled geomagnetic-polarity-reversal time scale, and the magnetic-polarity intervals found in marine sediment cores. The simple, immediately comprehensible congruence of those different data sets lent itself to presentation in diagrams so visually simple and directly representative that reality seemed barely abstracted or surrogated. Most in earth science were awestruck on first sight. What they saw in those diagrams constituted an interlocked framework of such conceptual cohesion and symmetry that even the authors of the hypothesis that had predicted such results could not have anticipated the near-perfect, universally legible form in which the effects of multiple, independent processes had been written and had revealed themselves.[38] That momentous episode constitutes an exemplar for the instantaneous erection of a conceptual bridge from canons to the Vine-Matthews-Morley hypothesis. Simultaneously, proof of that hypothesis demonstrated the virtues of other neglected, related hypotheses such as those concerning transform faults and nemataths.[39]

These various cases of theoretical conflict have much to teach us. Each seems to be a demonstration of too much faith put in canonical and orthodox knowledge that hid suppositions and uncertainties. It is as if the community of scholars, like the individual mind—as classically explicated by the Overstreets in *The Mature Mind*—is virtually incapable of holding a suspended judgment.[40] Premature ideas are the acid test of the mature mind.

---

38. Vine 1978.
39. Glen 1982.
40. Overstreet and Overstreet 1949.

ACKNOWLEDGMENTS

The U.S. Geological Survey, the National Science Foundation, and the Center for History of Physics provided support during much of the time in which the data for this paper were gathered. Spencer Weart generously shared his knowledge of the global-warming question with me.

BIBLIOGRAPHY

Ager, D. V. 1981. *The Nature of the Stratigraphical Record.* 2d ed. New York: Halstead Press.

————. 1993. *The New Catastrophism: The Importance of the Rare Event in Geological History.* New York: Cambridge University Press.

Alvarez, L. W., W. Alvarez, F. Asaro, and H. V. Michel. 1980. "Extraterrestrial Cause for Cretaceous-Tertiary Extinction." *Science* 208:1095–1108.

Alvarez, W. 1984. Interviews by William Glen. Tape recordings. Archives of the Center for History of Physics, College Park, Md.

Arnold, J. R., and E. C. Anderson. 1957. "The Distribution of Carbon-14 in Nature." *Tellus* 9:28–32.

Arrhenius, S. 1896. "On the Influence of Carbonic Acid in the Air upon the Temperature of the Ground." *Philosophical Magazine* 41:237–76.

Asaro, F. 1984. Interviews by William Glen. Tape recordings. Archives of the Center for History of Physics, College Park, Md.

Blackett, P. M. S. 1956. *Lectures on Rock Magnetism.* Jerusalem: Weizmann Science Press of Israel.

Blair, T. A. 1942. *Climatology, General and Regional.* New York: Prentice-Hall.

Bolin, B., and E. Eriksson. 1959. "Changes in the Carbon Dioxide Content of the Atmosphere and Sea due to Fossil Fuel Combustion." In *The Atmosphere and the Sea in Motion. Scientific Contributions to the Rosby Memorial Volume,* ed. B. Bolin, pp. 130–42. New York: Rockefeller Institute Press and Oxford University Press.

Broecker, W. S. 2000. Interview by William Glen, 7 February 2000. Manuscript.

Callender, G. S. 1938. "The Artificial Production of Carbon Dioxide and Its Influence on Temperature." *Qtly. J. Royal Meteorological Soc.* 64:223–40.

————. 1941. "Infra-red Absorption by Carbon Dioxide, with Special Reference to Atmospheric Radiation." *Qtly. J. Royal Meteorological Soc.* 67:263–75.

————. 1949. "Can Carbon Dioxide Influence Climate?" *Weather* 4:310–14.

Carson, R. 1951. "Why Our Winters Are Getting Warmer." *Popular Science* 159:114.

Chamberlain, T. C. 1897. "A Group of Hypotheses Bearing on Climatic Changes." *Jour. Geology* 5:653–83.

Cox, A., G. B. Dalrymple, and R. R. Doell. 1967. "Reversals of the Earth's Magnetic Field." *Sci. Amer.* 216:44–54.

Cox, A., R. R. Doell, and G. B. Dalrymple. 1963. "Geomagnetic Polarity Epochs and Pleistocene Geochronometry." *Nature* 198:1049–51.

Craig, H. 1957. "The Natural Distribution of Radiocarbon and the Exchange Times of CO_2 between Atmosphere and Sea." *Tellus* 9:1–17.

Dobson, G. M. B. 1942. "Atmospheric Radiation and the Temperature of the Lower Stratosphere." *Qtly. J. Royal Meteorological Soc.* 68:202–4.

Durham, J. W. 1967. "Presidential Address: The Incompleteness of Our Knowledge of the Fossil Record." *Jour. Paleon.* 41:559–65.

Elder, T., and J. Strong. 1953. "The Infrared Transmission of Atmospheric Windows." *Jour. of the Franklin Institute* 255:189–208.

Evernden, J. F., and G. H. Curtis. 1961. "The Present Status of Potassium-Argon Dating of Tertiary and Quaternary Rocks." *Intern. Quaternary Assoc. Proc. 6th Cong.* (Warsaw), 643–52.

Evernden, J. F., and G. T. James. 1964. "Potassium-Argon Dates and the Tertiary Floras of North America." *Amer. Jour. Sci.* 262:945–74.

Glen, W. 1979. Archive of Project in Geomagnetic History, 1975–79. Bancroft Library, University of California at Berkeley. This archive contains copies of proposals to conduct the Eltanin survey and interviews with the Columbia University graduate students and faculty who did that magnetic survey over the East Pacific Rise.

———. 1981. "The First Potassium-Argon Geomagnetic Polarity Reversal Time Scale: A Premature Start by Martin G. Rutten." *Centaurus* 25:222–38.

———. 1982. *The Road to Jaramillo: Critical Years of the Revolution in Earth Science.* Stanford: Stanford University Press.

———. 1985. *Continental Drift and Plate Tectonics.* Columbus, Ohio: Charles E. Merrill Publishing, 1975; reprint, San Mateo: Geo-Resources Associates.

———. 1990. "What Killed the Dinosaurs: A Decade of Debates." *Amer. Scientist* (July-August): 354–70. Reprinted in *Annual Editions: Biology,* 6th ed. (Guilford: Dushkin Publishing Group, 1992), pp. 173–83.

———. 1994a. "What the Impact/Volcanism/Mass-Extinction Debates Are About." In *The Mass Extinction Debates: How Science Works in a Crisis,* ed. William Glen, pp. 7–38. Stanford: Stanford University Press.

———. 1994b. "How Science Works in the Mass-Extinction Debates." In *The Mass Extinction Debates: How Science Works in a Crisis,* ed. William Glen, pp. 39–91. Stanford: Stanford University Press.

———. 1994c. "A Panel Discussion on the Debates." In *The Mass Extinction Debates: How Science Works in a Crisis,* ed. William Glen, pp. 268–86. Stanford: Stanford University Press.

———. 1994d. "How Different Disciplines Have Responded to the Alvarez-Berkeley Group Hypothesis." *Abstracts with Programs, Annual Meeting, Geological Society of America.* A-282.

———. 1996. "Observations on the Mass-Extinction Debates, 1996." In *The Cretaceous-Tertiary Event and Other Catastrophes in Earth History,* ed. G. Ryder, D. Fastovsky, and S. Gartner, pp. 39–53. *Geol. Soc. Amer. Spec. Paper* no. 307. Boulder, Colo.: Geological Society of America.

———. 1998. "A Manifold Current Upheaval in Science." *Earth Sciences History* 17:190–209.

———. Forthcoming. "Myth, Muse, and Mind: Problems in the Rational Reconstruction of Science History: Presented at a Working Conference." In *Interviews in Writing the History of Recent Science.* Cambridge: Harvard University Press. The conference took place 28–30 April 1994 at Stanford University.

———, ed. 1994. *The Mass Extinction Debates: How Science Works in a Crisis.* Stanford: Stanford University Press, pp. 1–370.

Hall, A. R. 1954. *The Scientific Revolution: 1500–1800, the Formation of the Modern Scientific Attitude.* Boston: Beacon Press, pp. 1–390.

Holmes, A. H. 1978. *Principles of Physical Geology.* 3d ed., rev. by D. L. Holmes. New York: John Wiley and Sons.

Hutton, J. 1788. "Theory of the Earth; or, an Investigation of the Laws Observable in the Composition, Dissolution, and Restoration of Land upon the Globe," *Transactions of the Royal Society of Edinburgh* 1, pt. 2:216.

Jeffreys, H. 1924. *The Earth: Its Origin, History, and Physical Constitution.* Cambridge: Cambridge University Press.

———. 1974. "Theoretical Aspects of Continental Drift." In *Plate Tectonics: Assessments and Reassessments,* ed. C. Kahle, pp. 395–405. Vol. 23. Tulsa: Amer. Assoc. of Petroleum Geologists.

Kunzig, R. 1996. "In Deep Water." *Discover* (December): 86–96.

Lyell, C. 1831. *Principles of Geology; Being an Attempt to Explain the Former Changes of the Earth's Surface, by Reference to Causes Now in Operation.* Vol. 1. London: J. Murray.

Mason, R. G., and A. D. Raff. 1961. "A Magnetic Survey off the West Coast of North America 32°N to 42°N." *Geol. Soc. Amer. Bull.* 72:1259–65.

Melosh, J. J. 1989. *Impact Cratering.* New York: Oxford University Press, 1–245.

Morley, L. W., and A. Larochelle. 1964. "Paleomagnetism as a Means of Dating Geological Events." In *Geochronology in Canada,* ed. F. F. Osborne, pp. 39–51. Roy. Soc. Canada Spec. Pub. no. 8. Toronto: University of Toronto Press.

National Environmental Research Council. 1998. *Climate Change: Scientific Certainties and Uncertainties.* Swindon, England: NERC.

Newell, N. D. 1967. "Revolutions in the History of Life." In *Uniformity and Simplicity: A Symposium on the Principle of the Uniformity of Nature,* ed. C. C. Albritton Jr., pp. 63–91. Geol. Soc. Amer. Spec. Paper no. 89. Boulder, Colo.: Geological Society of America.

Overstreet, H. A., and B. Overstreet. 1949. *The Mature Mind.* New York: W. W. Norton.

Pittman, W. C., and J. R. Heirtzler. 1966. "Magnetic Anomalies over the Pacific-Antarctic Ridge." *Science* 154:1164–71.

Plass, G. N. 1956. "Carbon Dioxide and the Climate." *Amer. Scientist* 44:302–16.

———. 1959. "Carbon Dioxide and Climate." *Sci. Amer.* 201:41–47.

Raup, D. M. 1991. *Extinction: Bad Genes or Bad Luck?* New York: W. W. Norton.

Revelle, R., and H. E. Suess. 1957. "Carbon Dioxide Exchange between Atmosphere and Ocean and the Question of an Increase of Atmospheric CO_2 during the Past Decades." *Tellus* 9:18–27.

Reynolds, J. H. 1954. "A High Sensitivity Mass Spectrometer." *Phys. Rev.* 98:283.

Sepkoski, J. J., Jr. 1990. "The Taxonomic Structure of Periodic Extinction." In *Global Catastrophes in Earth History,* ed. V. L. Sharpton and P. D. Ward, pp. 33–44. Geol. Soc. Amer. Spec. Paper. no. 247. Boulder, Colo.: Geological Society of America.

Shapely, H., ed. 1953. *Climatic Change: Evidence, Causes, and Effects.* Cambridge: Harvard University Press.

Shaw, H. R. 1994. *Craters, Cosmos, and Chronicles: A New Theory of the Earth.* Stanford: Stanford University Press.

Stanhill, G. 1999. "Climate Change Science Is Now Big Science." *EOS, Trans. Am. Geophys. Union* 80:396–97.

Stent, G. S. 1972. "Prematurity and Uniqueness in Scientific Discovery." *Sci. Amer.* 227 (December): 84–93.

Suess, H. E. 1955. "Radiocarbon Concentration in Modern Wood." *Science* 122:415–17.

———. 1957. "Residence Time of CO_2 in the Atmosphere from C14 Measurements." In *Proceedings, Conference on Recent Research in Climatology, Scripps Institution of Oceanography, La*

Jolla, California, March 25–26, ed. Craig Harmon, pp. 50–52. University of California Water Resources Center, contrib. no. 8.

Valentine, J. W., and E. M. Moores. 1970. "Plate Tectonics Regulation of Faunal Diversity and Sea Level: A Model." *Nature* 228:657–59.

van Waterschoot van der Gracht, W. A. J. M., B. Willis, R. T. Chamberlain et al. 1928. *Theory of Continental Drift: A Symposium on the Origin and Movement of Land Masses both Inter-Continental and Intra-Continental, as Proposed by Alfred Wegener.* Tulsa: Amer. Assoc. of Petroleum Geologists.

Vine, F. J. Interview by William Glen, 15 May 1978. Tape recording. Archive of the History of Science and Technology Program, Bancroft Library, University of California at Berkeley.

Vine, F. J., and D. H. Matthews. 1963. "Magnetic Anomalies over Ocean Ridges." *Nature* 199:947–49.

Weart, S. 1997. "Global Warming, Cold War, and the Evolution of Research Plans." *Historical Studies in the Physical and Biological Sciences* 27:319–56.

———. 1998. "Climate Change, Post-1940." In *Sciences of the Earth: An Encyclopedia of Events, People, and Phenomena,* ed. G. A. Good. New York: Garland Publishing.

Weber, L. R., and H. M. Randall. 1932. "The Absorption Spectrum of Water Vapor beyond 10 Microns." *Physical Review* 40:835–47.

Wegener, A. 1912a. "Die Entstehung der Kontinente." *Petermanns Geogr. Mitt.* 58:185–95, 253–56, 305–8.

———. 1912b. "Die Entstehung der Kontinente." *Geol. Rundschau.* 3:276–92.

———. 1915. *Die Entstehung der Kontinente.* Braunschweiz: Vieweg.

Wilson, J. T. 1965. "A New Class of Faults and Their Bearing on Continental Drift." *Nature* 207:343–47.

CHAPTER NINE

Theories of an Expanding Universe

Implications of Their Reception for the Concept of Scientific Prematurity

Norriss S. Hetherington

One may regard Gunther Stent's formulation of prematurity in scientific discovery as a work in progress in the best sense of this term, given the wider range of issues discussed in this volume. I explore here the delayed response to theories of an expanding universe. My approach departs radically in some ways from Stent's concept as he presented it originally, but owes much to the stimulation of his 1972 papers, as well, in its revision, as to comments by others advanced at the conference that preceded this volume.[1]

Several mathematical models of an expanding universe appeared in the scientific literature of the 1920s and now are celebrated, retrospectively, in historical accounts of the development of modern cosmology. One early proponent of an expanding universe was Georges Lemaître, a Belgian astrophysicist, Catholic priest, and, from 1960 until his death in 1966, president of the Pontifical Academy of Sciences. Lemaître offered a second chance—however subtle and implicit in its manifestation—for the Catholic church to embrace and be embraced by a second Galileo.[2] An even earlier proponent of an expanding universe was the Russian meteorologist and mathematician Alexander Friedmann. He came to be hailed explicitly, in the years immediately preceding the breakup of the Soviet Union, as an example of great Soviet science. No matter that the difficult conditions in revolutionary Russia in the early 1920s, and Friedmann's early death from typhoid fever, limited markedly his scientific output.[3]

Some may mine, if they choose, considerable polemical value from both Friedmann and Lemaître. Additionally, scientist-historians have a notorious predilection

---

1. On Stent's exposition, see chapter 2 in this volume; on the conference, see the preface.

2. For an example of emphasis on the ways in which Lemaître may have foreseen future developments in science, see Heller 1996; Hetherington 1997.

3. See Troop et al. 1993; Hetherington 1995.

to discover and display anticipations of great scientific ideas in their chronologies of accumulating positive achievement. Too often they do this without regard to historical influence or impact.[4]

Myriad temptations, polemical and pedagogical, to use and misuse history all but guarantee continued emphasis on Friedmann's and Lemaître's early proposals of an expanding universe, notwithstanding the fact that neither man's work received during the 1920s nearly the attention it now does. As the English astronomer Arthur Eddington noted at the 10 January 1930 meeting of the Royal Astronomical Society, up to that time astronomers were looking only for static models of the universe.[5] And the Dutch astronomer Willem de Sitter, looking back from the vantage point of 1932, explained that the universe had been thought to be static and thus only static models had been sought.[6]

The historical problem confronting us is not to document in ever-greater detail Friedmann's and Lemaître's celebrated contributions to science, but to explain why during the 1920s there was so little appreciation of their theories of an expanding universe.[7] Stent's concept may prove a valuable heuristic guide for exploring this problem.

An obvious starting point in attempting to explain why discoveries and theories now honored in retrospect initially went unappreciated is to investigate how widely they were known in their own time. Stent's illustrative historical cases of prematurity involve claims that were published and thus presumably known to at least some of their contemporaries even though the claims were unaccepted or underappreciated, as judged from a later perspective. We are dealing not with the unknown but with the unloved: with ugly ducklings since become swans.

All great discoveries virtually by definition may suffer initial underappreciation, because such discoveries, as a condition of subsequently recognized greatness, must counter prevailing belief and, ultimately, change the scientific canon. Many more discoveries and theories underappreciated in their own time never achieve greatness. Were they also once premature, or is the concept of prematurity to be restricted to discoveries and theories retrospectively recognized as great? Such a state of affairs would be unsatisfactory to historians intent on understanding past science in its own unique temporal context, who sometimes even go to the extreme of deliberately blinding themselves to subsequent developments.

One may raise the specter of Whiggish history (so named after the Whig party in England and historians who attributed to past Whig politicians an effort to realize the unknowable-for-them future).[8] More generally, and especially in its ap-

4. Kuhn 1968, reprinted in Kuhn 1977, pp. 105–26.

5. Eddington 1930.

6. De Sitter 1932, p. 12.

7. Initial neglect of proposals of an expanding universe has been discussed by North 1965, p. 117; Hetherington 1973, pp. 22–28; and Kerszberg 1989, pp. 13–14.

8. Butterfield 1931.

plication to the history of science, Whiggish history has come to mean praise of successful revolutions, an emphasis on progress, ratification of the present, and an absence of sensitivity to what earlier scientists actually thought they were doing. The charge of Whiggism (or the similar but distinct presentism bias) is a potent but indiscriminate club, one too quickly raised by contextualists and prigs against those who would use their own experience in science to help understand and empathize with the intellectual state of past researchers.[9] Historians would be foolish not to take advantage of modern formulations of issues and use them to probe the past— in fact, failure to do so would be every bit as foolish as indiscriminately reading the present back into the past. Pragmatically, if we are to achieve heuristic benefit from the concept of prematurity in science, we must set aside an understandable aversion to Whiggish history and seek instead any advantages that may flow from it.

From current belief that the universe is expanding comes much of the interest in early manifestations of this particular theory, not only in the proposals by Friedmann (in 1922 and 1924) and Lemaître (in 1927) but also by the mathematician H. P. Robertson (in 1928).[10] Friedmann's first paper, which considered the possible case of a nonstationary world, appeared in a major journal, the *Zeitschrift für Physik*. There, it was seen by Einstein, who voiced his suspicion of Friedmann's work in 1922 but within a year accepted it is as a possible solution of his field equations.[11] (Einstein's set of equations accounted simultaneously for the effects of gravitation and how it worked, and established the foundations of general relativity, a science of the universe departing fundamentally from previous cosmological thinking.)[12] Friedmann's 1924 paper titled "Über die Möglichkeit einer Welt mit Konstanter negativer Krümmung des Raumes," adding the possibility of a negative curvature of space, also appeared in the *Zeitschrift für Physik*.[13]

Friedmann's 1922 paper was listed in the annual survey of scientific papers on astronomical topics, the *Astronomischer Jahresbericht,* under the heading "Relativity Theory" by its title. No abstract appeared there to indicate that the paper presented an alternative cosmological model.[14] The 1924 paper was not listed. This omission is explained by a change in editorial policy in 1923, which resulted in the exclusion of articles on relativity theory judged not to be of astronomical interest.[15] Evidently Friedmann's work fell into this category. The editorial change in 1923 perhaps should be understood not as a change from an explicit existing policy, one of including all articles on relativity theory, but as the creation of policy de novo in response to a growing number of papers on relativity theory and hence a need to be selective in listing them.

9. For the anti-anti-Whig case, see Hull 1979; Harrison 1987; and Brush 1995.
10. Friedmann 1922, 1924; Lemaître 1927; Robertson 1928.
11. Einstein 1922, 1923.
12. Kerszberg 1989, 1993.
13. Friedmann 1924.
14. *Astronomischer Jahresbericht* 24 (1922): 61.
15. *Astronomischer Jahresbericht* 25 (1923): iii.

Elihu Gerson has suggested that we might replace the word "prematurity" with "isolation" or "disconnection," and that we should examine the relevance of a discovery or theory to the audience with which it is associated.[16] It may well be that Friedmann's work was of interest to a very few theoreticians pursuing relativity theory and was not relevant to most astronomers. Abstract theoretical speculations would have to be translated into concrete observational consequences before a cosmological model could gain many adherents among astronomers, or even be listed in the *Astronomischer Jahresbericht* after 1923.

Prematurity might also be understood in the sense that scientists can do nothing with a certain discovery or theory, as David Hull suggested.[17] Astronomers need more than abstract theory to direct their observations. Nor were many astronomers mathematically capable of drawing observational consequences from relativity theory. (Eddington, when asked if it were true that only three persons in the whole world really understood Einstein's relativity theory, is said to have paused for a moment and then asked who was the third.)

George Von der Muhll has described analogously how mathematization rendered a field of political science premature (and nearly killed it), in the sense that researchers in the field were unable to participate in it because they were mathematically illiterate.[18] Similarly, while astronomers during the 1920s possessed some limited mathematical ability, few other than Eddington himself could have put relativity theory to use.

Interaction between theory and observation is implicit in and essential to Stent's concept of prematurity, with discoveries rendered premature when they cannot be fit into a theoretical context—into canonical or generally accepted knowledge. Indeed, Stent applies his concept of prematurity only to discoveries, not to theories. One of the most striking results of viewing the cosmology of the 1920s through the lens of prematurity is the resulting focus on the absence of interaction between theory and observation.

An exception who illustrates the rule is George Ellery Hale, founder and first director of the Mount Wilson Observatory. He brought scientists, including the Dutch astronomer J. C. Kapteyn and the American geologist T. C. Chamberlin, for summer visits to the observatory with the hope that their theoretical interests might help guide observational programs. Astronomers at the observatory, however, found little of interest in speculative treatments, and Hale might well have concluded that astronomers in general were narrowly focused on particular observational problems and disinterested in abstract theory.[19]

In a few instances in which theory did guide observation at Mount Wilson, the results were not happy. Led astray by expectation, one astronomer found what he

16. See Gerson, chapter 19 in this volume.
17. See Hull, chapter 22 in this volume.
18. See von der Muhll, chapter 17 in this volume.
19. See Hetherington 1994, pp. 113–23.

expected to find, something we now know did not exist. His purported measurement of the rapid rotation of spiral nebulae meant that they were nearby objects within our galaxy, not similar galaxies beyond the boundaries of our galaxy. Soon Edwin Hubble, also at the Mount Wilson Observatory, obtained equally definitive evidence placing the spiral nebulae at great distances beyond our galaxy. This embarrassing situation emerged during the 1920s.[20]

There was already much reason for astronomers to distrust theory, given the example of Percival Lowell setting out to find signs of intelligent life on Mars and then reporting canals invisible to astronomers at other observatories. In his 1916 obituary of Lowell, the astronomer Henry Norris Russell of Princeton University warned that

> if the observer knows in advance what to expect[,] . . . his judgment of the facts before his eyes will be warped by this knowledge, no matter how faithfully he may try to clear his mind of all prejudice. The preconceived opinion unconsciously, whether he will or not, influences the very report of his senses, and to secure trustworthy observations, it has been recognized everywhere, and for many years, he must keep himself in ignorance of what he might expect to see.[21]

Science never has fit completely the naive Baconian inductive model, in which all facts are collected and only after that stage is complete are inevitable theories then inevitably induced. Instead, in scientific discovery, "the catalytic role of intuition and hypothesis is essential in making sense of disjoint empirical results and in mapping out the search for new data."[22] In the 1920s, however, cosmology was yet to become a modern science characterized by fruitful interaction between theory and observation.

Astronomers, especially in the United States, were largely content to produce observations while leaving theory to theoreticians. In contrast to the lush growth of observational astronomy, "theoretical cosmology failed to germinate on American soil." This may have been due partly to an intellectual climate of skepticism toward theoretical activities, especially those without evident practical benefit. Also, American astronomers could not cope with the mathematics of relativity theory. In England, in contrast, interest in relativity theory flourished, but it was conceived of as a strictly mathematical entity without observational input. Here, too, there was little inclination to mix theory and observation.[23]

Not only would it be anachronistic to assume on the part of astronomers much interest in cosmological theories in general during the 1920s, but the theory of an expanding universe specifically ran directly counter to the then-widely-held belief that the universe was static. Einstein himself had noted in 1917, while working out the consequences of the general theory of relativity, that the universe would not be

20. Hetherington 1972, 1983, 1988.
21. Russell 1916. On the Martian controversy, see Hetherington 1971b, 1976, 1981.
22. Thomson 1983.
23. Gale 1993; Gale and Urani 1993.

static unless he added an otherwise unnecessary constant term to his field equations describing the universe. Eddington called this constant a "hidden hand" preventing the gravitational collapse of the universe otherwise predicted by the theory of general relativity.[24] Einstein, who professed that "the viewpoint of logical simplicity [was] an indispensable and effective tool of [the scientist's] research," was immediately disturbed by his mathematical makeshift and soon expressed doubts about the cosmological constant.[25] Eventually he would term his cosmological constant "a hypothetical term added to the field equations," saying also that it was "not required by the theory as such nor did it seem natural from a theoretical point of view."[26]

The cosmological constant was necessary to satisfy Einstein's sense of physical reality. Replying to Friedmann's 1922 paper, with its nonstatic solution to Einstein's field equations, Einstein wrote, but then crossed out before sending his note to the journal for publication, that "a physical significance can hardly be ascribed to it [Friedmann's nonstatic universe]."[27]

Einstein's private correspondence reveals that he opposed the idea of an expanding universe on aesthetic grounds as well. He wrote to de Sitter that the circumstance of an expanding universe irritated him because it implied that the universe had a beginning, and that "to admit such possibilities [of an expanding universe] seems senseless."[28]

De Sitter found a second static solution to Einstein's field equations. It had a diminishing frequency of light vibrations with increasing distance from the origin of coordinates, which would give the appearance (but not the reality) of a velocity-distance relation (more distant objects apparently receding at greater speeds). The model could not contain matter if it were to remain stable.[29]

In 1917 there was some observational support for Einstein's model. The radius of curvature of space calculated from two different considerations of mass and density came out roughly the same; moreover, these two considerations were not entirely incompatible with a calculation based on the absorption of light in intergalactic space. Though far from a decisive argument for Einstein's model, the agreement on the value of the curvature of space was remarkable. On the other hand, stars thought to be the most distant gave some indication of systematic velocity displacements, thus arguing against Einstein's static model and for de Sitter's. (The displacements are now attributed to entirely different causes.) A more decisive test would be based on objects at greater distances.[30]

24. Kerszberg 1989, p. 6.
25. Ibid., pp. 9, 161, 164.
26. Einstein 1961, p. 133.
27. Kerszberg 1989, p. 335.
28. Quoted in Jastrow 1978, pp. 27–28. See also Kahn and Kahn 1975.
29. De Sitter 1917a,b.
30. Ibid.

Spiral nebulae were considered to be among the most distant objects known, and a few radial velocities of spirals had been determined by 1917. Vesto M. Slipher at the Lowell Observatory in Arizona published his measurements of some 15 spirals in 1915 and 10 more in 1917.[31] Evidently de Sitter did not see these reports, published in the American journal *Popular Astronomy* and in the *Proceedings of the American Philosophical Society*. Neither was a major astronomical journal; in addition, the war may have hindered communication between the United States and neutral Holland. De Sitter instead relied on two council reports of the Royal Astronomical Society.[32] One, by Eddington, mentioned that Slipher had determined the radial velocities of 15 spiral nebulae, but reported only the 2 velocities confirmed by other observers. The other council report supplied a third confirmed velocity. One of the 3 reported velocities was negative (approaching), which de Sitter easily explained away as a large, random negative velocity superimposed on a smaller, theoretical positive (receding) velocity. De Sitter concluded that more observations would be needed to confirm a possible systematic recession.[33] Even after he obtained all 25 velocities, only 3 of them negative, de Sitter still maintained that there was yet no physical criteria for distinguishing between the two models of the universe, his and Einstein's.[34]

Slipher had interpreted his first few measurements, which found approaching nebulae on the south side of our galaxy and receding spirals on the north side, as the manifestation of our galaxy's drift relative to the spiral nebulae. Observations of more spirals soon turned up an anomaly in terms of the drift hypothesis: spirals with positive velocities of recession on the south side of the galaxy. Rather than discard his hastily formulated drift hypothesis, Slipher professed his belief that as even more spirals were observed, a predominance of negative velocities would be found on the south side, in the area toward which he thought our galaxy was moving.[35] But though his drift hypothesis suggested the study of spirals near the supposed apex of the Milky Way's motion, and Slipher explicitly stated his intent to explore group motions, the nontheoretical consideration of brightness and the corresponding convenience of measurement was the primary criterion in his selection of which nebulae to study.[36] Of the 34 nebulae whose velocities Slipher had measured by 1921, only 2 were significantly fainter than the rest and located near the supposed apex of galactic motion. Despite its persistence in Slipher's thoughts, the drift hypothesis provided little if any guidance for his observing program.[37] Slipher's drift hypothesis was a premature theory with which little could be done because of the difficulties of observing faint spiral nebulae.

31. Slipher 1915, 1917.
32. Eddington 1917; Newall 1917.
33. De Sitter 1917a.
34. De Sitter 1921.
35. Slipher 1915.
36. Slipher 1922.
37. Hetherington 1971a.

Slipher measured radial velocities of a few spiral nebulae, but their distances remained unknown. De Sitter did not explicitly attribute to the lack of distance determinations for spiral nebulae his own hesitancy to choose between his and Einstein's models of the universe, but he must have been aware of the importance of such information for testing the hypothesis of a velocity-distance relation.

In this sense, all cosmological models, static and expanding, were premature in the 1920s. F. L. Holmes has suggested that one may, at least in one sense, characterize prematurity as, or as attributable to, an insufficiency of data required to support a conclusion.[38] (This is not Stent's meaning of prematurity, especially since he admits only discoveries and excludes theory.) Philosophers of science, on the other hand, may seize on the notion of prematurity as unverifiability, eschewing, as they do, psychological aspects of discovery and instead devoting their attention to the logic of verification.

The extremely limited extent to which cosmological models during the 1920s were testable or verifiable in the absence of distance determinations for spiral nebulae may not only have prevented a very few scientists, such as de Sitter, from drawing definite conclusions but may also have discouraged most scientists from even exploring the issue, diverting their attention instead to topics more promising of solution. This episode suggests more generally that prematurity is likely to arise in a field in which, or a theory about which, one can "do" hardly anything, at least with regard to pertinent observations.

De Sitter's static model could remain static only if there were no matter in it. This insistence on an absence of matter was objectionable on theoretical as well as observational grounds, because a material postulate of relativity of inertia denied the logical possibility of the existence of a world without matter.[39] For a time de Sitter was able to gloss over this theoretical point with an argument on the possible statistical equilibrium of the universe.[40] To avoid the contradiction inherent in admitting the observed existence of matter while simultaneously maintaining his static model as a possible model of the universe, de Sitter, at least for a few years, could suppose that the density of matter in the universe was approximately zero. His fellow Dutch astronomer Jan Oort, however, produced in 1927 a new estimate of the mass of our galaxy, an estimate that caused de Sitter to reexamine and reject his earlier assumption that the average density of matter in space was zero.[41] There could be no more pretense that de Sitter's static model might correspond to reality.

Nor was Einstein's static model to survive new observational evidence. From 1924 on, Hubble was slowly determining distances to a handful of spiral nebulae, these distances constituting the other half of a possible velocity-distance relation.[42]

38. See Holmes, chapter 12 in this volume.
39. De Sitter 1917b.
40. De Sitter 1921.
41. Oort 1927; De Sitter 1930.
42. Hetherington 1990, 1996.

Initially, though, Hubble showed little interest in de Sitter's or Einstein's cosmological models. Hubble's tentative calculation in a 1926 paper of the dimensions of the universe based on Einstein's static model was taken directly from a textbook on theoretical physics and does not necessarily indicate any comprehension on Hubble's part of the complex theoretical issues involved in the calculation.[43] In 1928, however, after attending the International Astronomical Union meeting in Holland and presumably discussing issues there with de Sitter and others, Hubble returned to Mount Wilson determined to test de Sitter's model of the universe. Hubble had Milton Humason, a meticulous and gifted observer, systematically study faint and more distant nebulae to determine if they had velocities greater than those of closer nebulae.[44]

The result was the famous 1929 paper, now regarded as the first conclusive demonstration of the expansion of the universe.[45] Using the period-luminosity relation for Cepheid-type stars, established earlier by Harlow Shapley at Mount Wilson before he moved to Harvard University, Hubble determined distances to 5 nebulae in which he had found Cepheids, as well as to a sixth nebula, which was a physical companion of one of the first 5. Hubble then calibrated the absolute magnitude of the brightest stars in the 6 nebulae and, from observations of the apparent magnitudes of the brightest stars in 14 more nebulae, estimated their distances. (Absolute, or intrinsic, brightness is diminished by the object's distance from the observer. Knowing the absolute brightness and measuring the apparent, or observed, brightness, one can calculate the distance.) Next Hubble found an average absolute magnitude for all 20 nebulae and compared that value to the apparent magnitudes of 4 nebulae in the Virgo Cluster of galaxies, thus obtaining distances to them. The distances combined with velocities showed a linear velocity-distance relation. For the remaining 22 nebulae that had known radial velocities but unknown distances, and which were too far away for observation of Cepheids or bright stars in them, Hubble measured the apparent magnitude of each nebula; calculated a mean apparent magnitude for all 22 nebulae; compared that value to the mean absolute magnitude for nebulae whose distances were known, obtaining a mean distance for the 22 nebulae; and then showed that the mean distance and the mean velocity of the 22 nebulae agreed well with the velocity-distance relation determined from the first 24 nebulae. The database was skimpy and the interpretation shaky in detail, but it was a brilliant and bold extrapolation outward into space.

Not until the final paragraph of the 1929 paper did Hubble mention either de Sitter or cosmological theory, and then he simply noted that the velocity-distance relation might represent the de Sitter effect and might be of interest for cosmological discussion. (Though de Sitter's prediction of an apparent but not real velocity-distance relation had helped guide Hubble's research, the spectral shifts Hubble was

43. Hetherington 1970, p. 156.
44. Smith 1982.
45. Hubble 1929.

measuring would generally be interpreted as real Doppler velocity shifts in a non-static, expanding universe.) Hubble did not mention Einstein's static model nor its definitive contradiction by the empirical velocity-distance relation. Such was Hubble's understated introduction of the key to the scientific exploration of the universe.

Hubble was pursuing a conscious strategy to convince his scientific peers of the reality of an empirical velocity-distance relation, as is apparent from his otherwise incomprehensible emphasis on a rather insignificant matter, a correction for the solar motion. The velocity-distance relation earlier had come under suspicion when the Polish-American mathematical physicist Ludvik Silberstein in a 1924 paper had attempted to prove the relation using velocities and distances for globular clusters of stars within our galaxy.[46] But he selectively excluded data not in agreement with the hypothesis. Gustaf Strömberg and Knut Lundmark, both then at the Mount Wilson Observatory, quickly revealed this scientific impropriety, noting that the complete body of evidence did not support a velocity-distance relation. They were careful to distinguish between Silberstein's work and the general hypothesis of a velocity-distance relation, which would have to await a conclusive test using the more distant spiral nebulae.[47] Nonetheless, theoretical preconceptions of a velocity-distance relation such as had poisoned Silberstein's analysis were suspect, and an obstacle for Hubble to overcome. His new data, on spiral nebulae, indicated that a linear correlation existed between velocity and distance regardless of whether the velocities were corrected for the solar motion, yet Hubble emphasized the correction, even reporting that he had Strömberg check it. Hubble did not state that Strömberg accepted the velocity-distance relation, though readers could easily have leapt to this conclusion. Lundmark, Silberstein's other critic, who had returned to Sweden, could not be enlisted as easily. But Hubble did cite Lundmark's solution for the solar motion, no matter that it differed from Hubble's, nor that Lundmark had used a nonlinear velocity-distance relation! What was important was the implication that Silberstein's critics endorsed Hubble's result. Certainly Hubble required neither Strömberg's nor Lundmark's assistance to check the straightforward calculation of the solar motion.[48] There is more to the advance of science than new observations and new theories: ultimately, people must be persuaded.

It was now evident that neither Einstein's nor de Sitter's static model was a physically possible model of the universe. Einstein's model was ruled out by the velocity-distance relation and de Sitter's model was eliminated by the existence of matter in the universe. At the 10 January 1930 meeting of the Royal Astronomical Society, de Sitter summed up the dilemma. Eddington then rose to remark, "One puzzling question is why there should be only two solutions. I suppose the trouble is that people look for static solutions."[49]

46. Silberstein 1924.
47. Strömberg 1925; Lundmark 1924.
48. Hetherington 1986.
49. Eddington 1930.

The remark was printed, and seen by Lemaître, a former student of Eddington's. Lemaître wrote to Eddington to remind him of Lemaître's 1927 paper. Published in the *Annals de la Société Scientifique de Bruxelles,* the paper had been easily overlooked by astronomers.[50] Eddington learned of the paper only when Lemaître wrote to him in 1930, and de Sitter first heard about the paper from Eddington.[51] Yet had astronomers been looking for nonstatic solutions before 1930, they could have seen listed in the 1927 *Astronomischer Jahresbericht* Lemaître's paper on a homogeneous universe of constant mass and increasing radius accounting for the radial velocity of the extragalactic nebulae.[52] The next year Robertson published in the far-better-known *Philosophical Magazine* a paper on a Euclidean space with redshifts of spiral nebulae due to real recessional velocities.[53] Eddington read Robertson's paper and began working on the question of whether Einstein's model was stable. But Eddington did not enthusiastically acclaim Robertson's paper as he would Lemaître's upon learning of it. Robertson's was a preliminary sketch, while Lemaître's offered a more thorough development. This may also explain why Lemaître's paper was listed in the *Jahresbericht* and Robertson's was not.

Premature throughout the 1920s, the idea of an expanding universe was, in 1930, an idea whose time had come. It accounted for observational data and provided a satisfactory resolution of the crisis following from the failure of both static models of the universe.

There remained the problem of what could be done with the theory. Hubble's distance determinations coupled with Slipher's initial and Hubble's subsequent radial velocity measurements helped remedy the previous insufficiency of data required to support a conclusion, and more data would soon be obtained at the Mount Wilson Observatory.

Another obstacle to scientific progress was the division between observers and theoreticians, both in interest and in talent. Hubble wrote to de Sitter in 1931 that he had emphasized the empirical features of the velocity-distance correlation, and "the interpretation, we feel, should be left to you and the very few others who are competent to discuss the matter with authority."[54] The intent of Hubble's letter was not to denigrate his own abilities as a theoretical astronomer but to emphasize the importance of careful empirical studies and to secure to the Mount Wilson Observatory credit for the observations. Very quickly Hubble moved on to the interpretation of his observations.

Hubble consciously set out to bridge the gulf between observation and theory, and his effort to interpret the velocity-distance relation is an early example of the now-common joint scientific effort. Cooperation, Hubble wrote in a report, was an

50. Lemaître 1927.
51. Eddington 1931; de Sitter 1917b.
52. *Astronomischer Jahresbericht* 29 (1927): 229.
53. Robertson 1928.
54. Edwin Hubble to Willem de Sitter, 23 September 1931, quoted in Hetherington 1982.

important and distinctive feature of nebular research at the Mount Wilson Observatory. Working in close association with colleagues at the nearby California Institute of Technology, the scientists combined resources in particular investigations and interpreted the results in the light of constructive criticism from the group as a whole.[55]

Particularly complementing Hubble's observational input was theoretical input from Richard Tolman, a theoretical physicist at the California Institute of Technology (Cal Tech) highly knowledgeable regarding the mathematical foundations of relativistic cosmology. In a joint paper, he and Hubble addressed the problem of discriminating between possible models of the universe, precisely calculating theoretical relations and then linking the results to available observations.[56]

Tolman was one of several astronomers, physicists, and mathematicians who, according to Mrs. Hubble's reminiscences, came to her house about every two weeks. "They brought a blackboard from Cal Tech and put it up on the living room wall," she wrote. "In the dining room were sandwiches, beer, whiskey and soda water; they strolled in and helped themselves. Sitting around the fire, smoking pipes, they talked over various approaches to problems, questioned, compared and contrasted their points of view—someone would write equations on the blackboard and talk for a bit, and a discussion would follow."[57] Observation and theory finally were linked in cosmological studies, if not through the efforts of a single individual, at least in a cooperative collective of observers and theoreticians.

More recent appreciation of the role of theory in astronomy perhaps precludes a repetition of the general disdain for cosmological hypotheses that prevailed during the 1920s, and it is doubtful that a theory now would be judged or treated as premature because it lacked observational predictions, or that an absence of needed data would necessarily render an astronomical theory uninteresting. Rather, these conditions might be taken as a challenge: a spur to further work. They might also furnish grounds for funding requests. It is easy, however, to imagine astronomical theories in conflict with philosophical values, cultural values, religious beliefs, or even canonical scientific belief. And consequently it is easy to imagine theories in want of financial support for their testing because they are, in some sense, premature.

BIBLIOGRAPHY

Brush, S. 1995. "Scientists as Historians." *Osiris* 10:215–31.
Butterfield, H. 1931. *The Whig Interpretation of History*. London: G. Bell and Sons.
Eddington, A. S. 1917. "The Motions of Spiral Nebulae." *Monthly Notices of the Royal Astronomical Society* 77:375–77.
———. 1930. "Meeting of the Royal Astronomical Society." *The Observatory* 53:33–44.

55. Hubble 1938.
56. Hubble and Tolman 1935.
57. G. Hubble, undated handwritten note, quoted in Hetherington 1982.

————. 1931. "The Expansion of the Universe." *Monthly Notices of the Royal Astronomical Society* 91:412–16.

Einstein, A. 1922. "Bermerkung zu der Arbeit von A. Friedmann 'Über die Krümmung des Raumes.'" *Zeitschrift für Physik.* 11:326.

————. 1923. "Notiz zu der Arbeit von A. Friedmann 'Über die Krümmung des Raumes.'" *Zeitschrift für Physik* 16:228.

————. 1961. *Relativity: The Special and the General Theory. A Popular Exposition by Albert Einstein.* Trans. R. W. Lawson. New York: Crown Publishers.

Friedmann, A. 1922. "Über die Krümmung des Raumes." *Zeitschrift für Physik* 10:377–86.

————. 1924. "Über die Möglichkeit einer Welt mit Konstanter negativer Krümmundes Raumes." *Zeitschrift für Physik* 21:326–32.

Gale, G. 1993. "Philosophical Aspects of the Origin of Modern Cosmology." In *Encyclopedia of Cosmology: Historical, Philosophical, and Scientific Foundations of Modern Cosmology,* ed. N. Hetherington, pp. 481–95. New York: Garland Publishing.

Gale, G., and J. R. Urani. 1993. "Philosophical Aspects of Cosmology." In *Cosmology: Historical, Literary, Philosophical, Religious, and Scientific Perspectives,* ed. N. Hetherington, pp. 547–68. New York: Garland Publishing.

Harrison, E. 1987. "Whigs, Prigs, and Historians of Science." *Nature* 329:213–24.

Heller, M. 1996. *Lemaître, Big Bang, and the Quantum Universe.* Tucson: Pachart Publishing.

Hetherington, N. 1970. "The Development and Early Application of the Velocity-Distance Relation." (Ph.D. diss., Indiana University.) Abstract in University Microfilms International (1971).

————. 1971a. "The Measurement of Radial Velocities of Spiral Nebulae." *ISIS* 62:309–13.

————. 1971b. "Lowell's Theory of Life on Mars." *Astronomical Society of the Pacific Leaflet* 501:1–8.

————. 1972. "Adriaan van Maanen and Internal Motions in Spiral Nebulae: A Historical Review." *Quarterly Journal of the Royal Astronomical Society* 13:25–39.

————. 1973. "The Delayed Response to Suggestions of an Expanding Universe." *Journal of the British Astronomical Association* 84:22–28.

————. 1976. "Amateur versus Professional: The British Astronomical Association and the Controversy over Canals on Mars." *Journal of the British Astronomical Association* 86:303–8.

————. 1981. "Percival Lowell: Scientist or Interloper?" *Journal of the History of Ideas* 42:159–61.

————. 1982. "Philosophical Values and Observation in Edwin Hubble's Choice of a Model of the Universe." *Historical Studies in the Physical Sciences* 13:41–67.

————. 1983. "Just How Objective Is Science?" *Nature* 306:727–30.

————. 1986. "Edwin Hubble: Legal Eagle." *Nature* 319:189–90.

————. 1988. *Science and Objectivity: Episodes in the History of Astronomy.* Ames: Iowa State University Press.

————. 1990. "Hubble's Cosmology." *American Scientist* 78:142–51.

————. 1993a. *Encyclopedia of Cosmology: Historical, Philosophical, and Scientific Foundations of Modern Cosmology.* New York: Garland Publishing.

————. 1993b. *Cosmology: Historical, Literary, Philosophical, Religious, and Scientific Perspectives.* New York: Garland Publishing.

————. 1994. "Converting an Hypothesis into a Research Program: T. C. Chamberlin, His Planetesimal Hypothesis, and Its Effect on Research at the Mt. Wilson Observatory." In

The Earth, the Heavens, and the Carnegie Institution of Washington, ed. G. A. Good, pp. 113–23. Washington, D.C.: American Geophysical Union. Published as a special edition of *History of Geophysics* 5 (1993).

———. 1995. Review of *Alexander A. Friedmann: The Man Who Made the Universe Expand,* by E. A. Troop et al. *Historical Studies in the Physical and Biological Sciences* 25:387–88.

———. 1996. *Hubble's Cosmology: A Guided Study of Selected Texts.* Tucson: Pachart Publishing.

———. 1997. Review of *Lemaître, Big Bang, and the Quantum Universe,* by M. Heller. *Historical Studies in the Physical and Biological Sciences* 27:363.

Hubble, E. 1929. "A Relation between Distance and Radial Velocity among Extra-Galactic Nebulae." *Proceedings of the National Academy of Sciences* 15:168–73.

———. 1931. Letter to Willem de Sitter, 23 September 1931. Edwin Hubble Collection, Henry Huntington Library, San Marino, California. Quoted in N. Hetherington, "Philosophical Values and Observation in Edwin Hubble's Choice of a Model of the Universe," *Historical Studies in the Physical Sciences* 13 (1982): 41–67.

———. 1938. "Explorations in the Realm of the Nebulae." In *Cooperation in Research,* pp. 91–102. Publication no. 501. Washington, D.C.: Carnegie Institution.

Hubble, E., and R. C. Tolman. 1935. "Two Methods of Investigating the Nature of the Nebular Red-Shift." *Astrophysical Journal* 82:302–37.

Hull, D. 1979. "In Defense of Presentism." *History and Theory* 18:1–15.

Jastrow, R. 1978. *God and the Astronomers.* New York: W. W. Norton.

Kahn, C., and F. Kahn. 1975. "Letters from Einstein to de Sitter on the Nature of the Universe." *Nature* 257:451–58.

Kerszberg, P. 1989. *The Invented Universe: The Einstein–De Sitter Controversy (1916–17) and the Rise of Relativistic Cosmology.* Oxford: Oxford University Press.

———. 1993. "Relativistic Cosmology." In *Encyclopedia of Cosmology: Historical, Philosophical, and Scientific Foundations of Modern Cosmology,* ed. N. Hetherington, pp. 566–79. New York: Garland Publishing.

Kuhn, T. S. 1968. "The History of Science." In *International Encyclopedia of the Social Sciences,* pp. 74–83. Vol. 14. New York: Crowell Collier and Macmillan.

———. 1977. *The Essential Tension: Selected Studies in Scientific Tradition and Change.* Chicago: University of Chicago Press.

Lemaître, G. 1927. "Un univers homogène de masse constante et de rayon croissant, rendant compte de la vitesse radials des nebuleuses éxtra-galactiques." *Annals de la Société Scientifique de Bruxelles* 47:49–56. Trans. and reprinted in *Monthly Notices of the Royal Astronomical Society* 91 (1931): 483–90.

Lundmark, K. 1924. "The Determination of the Curvature of Space-Time in de Sitter's World." *Monthly Notices of the Royal Astronomical Society* 84:747–70.

Newall, H. F. 1917. "Stellar Spectroscopy in 1916." *Monthly Notices of the Royal Astronomical Society* 77:382–87.

North, J. D. 1965. *The Measure of the Universe.* Oxford: Oxford University Press.

Oort, J. H. 1927. "Investigations Concerning the Rotational Motion of the Galactic System, Together with New Determinations of Secular Parallaxes, Precession, and Motion of the Equinox." *Bulletin of the Astronomical Institutes of the Netherlands* 4:79–89.

Robertson, H. P. 1928. "On Relativistic Cosmology." *Philosophical Magazine* 15:835–48.

Russell, H. N. 1916. "Percival Lowell and His Work." *Outlook* 114:781–83.

Silberstein, L. 1924. "The Curvature of de Sitter's Space-Time Derived from Globular Clusters." *Monthly Notices of the Royal Astronomical Society* 84:363–66.

Sitter, W. de. 1917a. "On Einstein's Theory of Gravitation and Its Astronomical Consequences. Third Paper." *Monthly Notices of the Royal Astronomical Society* 78:3–28.

———. 1917b. "On the Relativity of Inertia: Remarks Concerning Einstein's Latest Hypothesis." *Proceedings of the Royal Academy of Amsterdam* 19:1217–22.

———. 1921. "On the Possibility of Statistical Equilibrium of the Universe." *Proceedings of the Royal Academy of Amsterdam* 23:866–68.

———. 1930. "On the Magnitudes, Diameters, and Distances of the Extragalactic Nebulae, and Their Apparent Radial Velocities." *Bulletin of the Astronomical Institutes of the Netherlands* 5:157–71.

———. 1932. *Kosmos*. Cambridge: Harvard University Press.

Slipher, V. M. 1915. "Spectrographic Observations of Nebulae." *Popular Astronomy* 23:21–24.

———. 1917. "Nebulae." *Proceedings of the American Philosophical Society*. 56:403–9.

———. 1922. "Further Notes on Spectrographic Observations of Nebulae and Clusters." *Popular Astronomy* 30:9–11.

Smith, R. W. 1982. *The Expanding Universe: Astronomy's 'Great Debate,' 1900–1931*. Cambridge: Cambridge University Press.

Stent, G. S. 1972a. "Prematurity and Uniqueness in Scientific Discovery." *Advances in the Biosciences* 8:433–49.

———. 1972b. "Prematurity and Uniqueness in Scientific Discovery." *Scientific American* 227 (December): 84–93.

Strömberg, G. 1925. "Analysis of Radial Velocities of Globular Clusters and Non-Galactic Nebulae." *Astrophysics Journal* 61:353–62.

Thomson, K. S. 1983. "The Sense of Discovery and Vice Versa." *American Scientist* 71:522–24.

Troop, E. A., V. Y. Frenkel, and A. D. Chernin. 1993. *Alexander A. Friedmann: The Man Who Made the Universe Expand*. Trans. A. Dron and M. Burov. Cambridge: Cambridge University Press.

CHAPTER TEN

Interdisciplinary Dissonance and Prematurity

Ida Noddack's Suggestion of Nuclear Fission

Ernest B. Hook

INTERDISCIPLINARY DISSONANCE

Enrico Fermi and his colleagues observed products of artificially induced nuclear splitting in 1934. But they did not recognize them for what they were. Not until 1939 did the combined work of Otto Hahn, Fritz Strassmann, Lise Meitner, and Otto Robert Frisch lead to the realization that what Frisch termed "nuclear fission" explained the reported observations of Fermi's group.[1] This interval provokes some pertinent questions bearing on Gunther Stent's definition of the prematurity of scientific discovery. For if a premature discovery (implicitly, a premature claim or a hypothesis) is one that cannot be connected to generally accepted knowledge by a series of simple logical steps, then, for work at disciplinary boundaries—which the nuclear fission example illustrates—one may inquire: a connection to the canon of which discipline?

As someone who, in a rather old-fashioned manner, still believes more or less in the possibility, indeed in the general albeit not inevitable historical trend to progress in scientific understanding, I note that the canon (understood here as generally accepted knowledge)[2] may be in a greater or lesser state of development in various fields. The stage that different fields have reached may result in apparent irreconcilable perspectives at their boundaries. As the following account of the recognition of nuclear fission illustrates, a hypothesis that violates a canon in one discipline may cohere with or appear innocent to that of another, indeed perhaps just because the latter is less developed theoretically. And, even for fields not comparable in this developmental manner, doctrines canonical in one may be inert or hereti-

---

1. Fermi 1934; Fermi et al. 1934; Hahn and Strassmann 1939a; Meitner and Frisch 1939a; Frisch 1939. See also Seaborg, chapter 3 in this volume, on which this chapter is an expansion.
2. For an alternative formulation of canonical knowledge, see Hook, chapter 1 in this volume.

cal in another. Moreover, individuals even in the same or closely related fields may disagree about precisely what is canonical, especially at the periphery.

Nuclear fission illustrates another phenomenon pertaining to but distinct from prematurity, one that I define as "interdisciplinary dissonance": any tendency of observations and theoretical considerations in one discipline to inhibit and obstruct discovery in another, usually at the boundaries of overlapping, disparate foci of study. As we shall see, interdisciplinary dissonance may arise precisely because a claim, hypothesis, or proposal may be premature in one field but not in another.

OTTO HAHN, FRITZ STRASSMANN, AND ENRICO FERMI

The most striking evidence of interdisciplinary dissonance that I can find to date appears in the article by Hahn and Strassmann in January 1939. This report led shortly afterward to the recognition of nuclear fission. Bombarding uranium with neutrons, they expected to find products near uranium ($Z = 92$) in atomic number. But what they presumed to be assays for radium (Ra, $Z = 88$), actinium (Ac, $Z = 89$), and thorium (Th, $Z = 90$) on further checking appeared to provide startling evidence of elements much lower in the periodic table: barium (Ba, $Z = 56$), lanthanum (La, $Z = 57$), and cesium (Ce, $Z = 55$). In the conclusion of their report, they wrote a now well-known paragraph: "As chemists, we really ought to revise the decay scheme given above [in their paper] and insert the symbols Ba, La, Ce in place of Ra, Ac, Th. However, as 'nuclear chemists' working very close to the field of physics, we cannot bring ourselves yet to take such a drastic step which goes against all previous experience in nuclear physics. There could be perhaps a series of unusual coincidences which has given us false indications."[3]

Shortly before, Hahn had written in a letter to Meitner: "Continually we come to the frightful conclusion [*schrecklichen Schluss*] our Ra-isotopes do not behave like Ra but like Ba. Perhaps you can suggest some fantastic solution. We know ourselves that into Ba, it *can* not really split" (emphasis in the original).[4]

Within a month, following the appearance of the theoretical rationale by Meitner and Frisch, and experimental work of a completely different nature by Frisch and others,[5] Hahn and Strassmann took the step that their analytical chemistry clearly indicated but which nuclear chemistry and nuclear physics had appeared to forbid. They concluded unequivocally that they had split the atomic nucleus.[6]

Of course, as Glenn Seaborg notes, others had had similar conceptual difficulties.[7] From 1934 to 1938 these had contributed to the false identification of some

---

3. Hahn and Strassmann 1939a, as translated by Graetzer and Anderson 1971.
4. Hahn to Meitner, 12 December 1938; my translation from Krafft 1981, pp. 263–64.
5. Meitner and Frisch 1939a; Frisch 1939; references in Turner 1940.
6. Hahn and Strassmann 1939b.
7. See Seaborg, chapter 3 in this volume.

neutron products of uranium bombardment as new transuranic elements with atomic numbers greater than 92. Indeed on 12 December 1938, Enrico Fermi received the Nobel Prize in physics "for his demonstration of the existence of *new radioactive elements produced by neutron irradiation,* and for his related discovery of nuclear reactions brought about by slow neutrons"[8] (emphasis added). Ironically, precisely that week Hahn and Strassmann were finishing the work that would lead directly to the recognition of nuclear fission and the rejection of Fermi's "demonstration" of these "new" elements.[9]

In their first reports on transuranium elements in 1934, Fermi and his colleagues had been very cautious. Fermi titled his *Nature* paper "Possible Production of Elements of Atomic Number Higher Than 92."[10] He and his colleagues tested the 13-minute half-life product of neutron bombardment of uranium and found no analytical chemical evidence of either uranium (Z = 92) or nearby elements with the lower atomic numbers 91, 90, 89, 88, 83, or 82. (They excluded those with numbers 87 and 86 on the grounds of their unknown chemical behavior.) Based on these observations, Fermi and his colleagues inferred that "it appears we have excluded the elements in question," which thus "suggests the possibility that the atomic number of the element may be greater than 92."[11] Later, with similar evidence and caution, they inferred that another product, one with about a 90-minute half-life, was possibly element 94. In their last paper published on the matter, in which they reported further work that excluded elements between 82 and 92, they concluded cautiously only that "the simplest interpretation consistent with the known facts is to assume that the . . . [observed activities] . . . are chain products, probably with atomic number[s] 92, 93, 94 respectively."[12]

But others, in commenting on their initial findings and their report, soon became less cautious. Franco Rasetti published a textbook in 1936 on nuclear physics that included the following as a historical claim regarding the decay products investigated: "Fermi and his co-workers reported . . . neither [product 93 nor 94] was isotopic with any element of atomic number between 82 and 92, and that, *therefore,*

---

8. Nobel Foundation 1965, p. 407. Despite this explicit citation of "new radioactive elements," which could only have been 93 and 94, Graetzer and Anderson (1971, p. 16) claim that the award was in essence *only* for Fermi's fundamental work, "not for discovering element 93 *[sic]*"! Segrè (1970, p. 99) alleges the same, inferring this from the published conclusion of the presentation speech by H . Pleijel (1965, p. 413). Pleijel's text refers to (in addition to the slow neutron work) discovery not of new elements but rather of "new radioactive substances belonging to the entire field of the elements." I regard this as being of much less evidential value than Segrè does. Being on the scene, Pleijel likely had an opportunity before publication, after the discovery of fission, to alter the wording of his statement. I find this to be the simplest explanation for the discrepancy. The original wording of the actual award, unlike his address, would, however, have been already widely circulated, and not alterable.

9. The letter by Hahn to Meitner first telling her that the Ra isotopes act like Ba was written the evening of 19 December 1938 (Sime 1996, p. 233).

10. Fermi 1934.

11. Fermi et al. 1934.

12. Amaldi et al. 1935, p. 553.

these *must* be transuranic elements"[13] (emphasis added). If nothing else, this illustrates the treachery of memory, as, extraordinarily, Rasetti was a member of Fermi's team, a coauthor of the Italian version of the original report, and a coauthor of subsequent papers, including the paper by Amaldi and colleagues cited in the previous paragraph, all of which had been cautious.

Rasetti's recollection of what they had actually concluded a year or so earlier may have been clouded by the appearance of the subsequent work of others, for he continues, "More extensive work was carried out later by Hahn and Meitner, who confirmed the existence of transuranic elements."

Certainly, by 1937, further work by the Berlin group appeared to substantiate the discovery of both 93 and 94. According to Ruth Sime's translation, they wrote, "In general, the chemical behavior of the transuranes . . . is such that their position in the periodic system is no longer in doubt. *Above all, their chemical distinction from all previously known elements needs no further discussion*"[14] (emphasis in the original). Orso Corbino, the "elder" physicist in Rome, had written shortly before his death in early 1937, in referring to the transuranium elements, that "the two greatest experts in radioactive chemistry, Lise Meitner and Otto Hahn of Berlin, have fully confirmed Fermi's discovery."[15] A history of the chemical elements (appearing in 1939 but probably written in 1938) notes without reservation the existence of elements 93 and 94.[16] One may appreciate how Fermi, who had left this aspect of the field to go on to other investigations, could in his Nobel Prize speech of December 1938 give his first unqualified endorsement of the transuranics in his brief account of their history: "We concluded that the carriers were one or more elements of atomic number larger than 92: we in Rome use *[sic]* to call the elements 93 and 94 Ausperium and Hesperium respectively. It is known that O. Hahn and L. Meitner have investigated very carefully and extensively the decay products of irradiated uranium, and were able to trace among them elements up to number 96."[17] If the outstanding radiochemists Hahn and Meitner had confirmed his initial tentative conclusions, and the Nobel committee had endorsed them, one can understand his willingness to regard the matter as concluded.

An asterisk inserted at the end of the cited passage refers to a footnote added later, probably in January or February 1939. This mentions the discovery by Hahn and Strassmann of barium among the disintegration products of bombarded uranium. Such, Fermi wrote, "makes it necessary to reexamine all the problems of the transuranic elements[,] as many of them might be found to be products of a splitting uranium." The cautious expression in Fermi's footnote about the need only

---

13. Rasetti 1936, p. 271. The 1934 reports also had not excluded all of the elements down to lead (Pb, $Z = 82$), although subsequent work by the Rome and the Berlin groups did so (Amaldi et al. 1935).

14. Sime 1996, p. 174, translating Hahn et al. 1937.

15. Corbino, cited from *Nuova Antologia* (no other reference) by Laura Fermi 1954, p. 93.

16. Weeks 1939, pp. 439–43.

17. Fermi 1965.

to "reexamine" the evidence may stem from the fact that Hahn and Strassmann had *not* reported on investigation of a putative transuranium substance among the products of uranium bombardment. They had examined only products with a presumed atomic number below uranium, one of which they had thought to be radium ($Z = 88$).[18] They found that, in fact, barium and not radium had resulted, and that, although it provided evidence of fission, the finding still did not directly undermine the evidence of any claimed transuranic. For Fermi and others, however, it of course immediately posed this question.[19]

Glenn Seaborg notes that, for many, including himself, announcement of the first Hahn-Strassmann results pulled the rug out from under all the claimed transuranics at once.[20] Yet Hahn and Strassmann still insisted on the existence of the transuranics they had reported. In a subsequent note in which they give their first unequivocal endorsement of fission and without inhibition write "barium" instead of "radium," they still insist on them. Point 5 in their summary states that, despite a demonstration of fission, "it is our belief that the 'transuranium elements' still retain their placement without a change, as previously described."[21] And in response to a note in press in March 1939 mentioning their previous oversights about fission, their polemical answer (which never appeared) stated with regard to the 16 different kinds of elements they reported (including among these the many claimed isomers of transuranics), "We don't take back a single one"![22]

18. This was a 3.5 hour substance reported by Curie and Savitch 1938. The latter had tentatively suggested that it might be a *new* 93, different from those previously reported, and implied the possible need to increment by one unit the atomic numbers of the transuranics previously claimed by the Berlin group. Hahn and Strassmann believed they had excluded this by showing that the substance was radium, but this also involved some theoretical difficulties, as it implied the implausible ejection of two alpha particles (each with $Z = 2$) from uranium ($Z = 92$) or, only slightly more plausible, two stepwise alpha emissions with thorium ($Z = 90$) as an intermediate. See Herrmann 1990, p. 491.

19. Neutron bombardment of uranium produces, it is now known, results in dozens of different products, some of which are fission products; others, transuranic elements. The many substances produced have different but overlapping half-lives.

20. See Seaborg, chapter 3 in this volume.

21. Summary of Hahn and Strassmann 1939b, translated in Graetzer and Anderson 1971, p. 48.

22. "Wir möchten nur das Eine sagen, . . . Zahl von 16 verschiedenen Atomareten, die wie gefunden haben sollen, keine einzige zurückziehen" (O. Hahn and F. Strassmann 1939, manuscript, in Krafft 1981, pp. 319–20). This was probably written in mid-March 1939 in response to the not-yet-published attack (and priority claim) by Ida Noddack (1939) (see below). The bulk of the reply, written by Strassmann, appears appropriate. The additional material quoted here was added by Hahn to Strassmann's draft. However, no formal response by them was published. The editor, Paul Rosebaud, may well have sought to prevent Hahn and Strassmann from appearing too foolish by still insisting at this late date on all their previously claimed transuranics (but see below). Hopper (1990, p. 82) claims that Rosebaud made the decision himself not to run the reply. Instead, a dignified editor's comment in response appeared immediately beneath Noddack's note, stating, "The gentlemen [Hahn and Strassmann] . . . inform us they have neither the time nor desire to reply." There appears to be a puzzling inconsistency between their never-published note and what Hahn was conceding about the same time to Meitner in a letter written 13 March 1939: "From your findings we must indeed declare that the transuranes are dead"! (Sime 1996, p. 267). Furthermore, in a later note to Meitner on 7 April 1939, Hahn says that

The first report stating that a substance previously thought to be a transuranium element was in fact a fission product (tellurium, $Z = 52$) was published on 15 February 1939, by Philip Abelson, then a graduate student at the University of California, Berkeley.[23] Only further work resulting from a visit by Meitner to Frisch's laboratory in Copenhagen in March 1939 produced evidence that, apparently, was sufficient to eventually convince Hahn and Strassmann that the transuranics they had claimed did not exist.[24]

IDA TACKE NODDACK

By mid-1939, judging from the available literature, there appears to have been general agreement that the putative transuranic elements suspected initially by Fermi and colleagues and "confirmed" by others were instead fission products, and that Fermi and colleagues had unknowingly observed artificially induced nuclear splitting. Every account of the history of nuclear fission must confront the fact that precisely this interpretation of Fermi's results had been published in 1934, shortly after his paper appeared, but that it had been almost universally ignored. Ida Tacke Noddack, a German analytical chemist, made this suggestion in a paper titled (in translation) "On Element 93," which appeared in the German chemical journal *Angewandte Chemie (Applied Chemistry)*. She stated that, rather than conclude that the production of transuranic elements had occurred, "one could assume equally well that[,] when neutrons are used to produce nuclear disintegrations, some new nuclear reactions take place which have not been observed previously with proton or alpha-particle bombardment of atomic nuclei[,] . . . [and that] when nuclei are bombarded by neutrons, it is conceivable that the nucleus breaks up into several *larger* fragments, which would of course be isotopes of known elements but would not be neighbors of the irradiated element"[25] (emphasis in the original). At first impression, this earliest published appearance of what would later be termed nuclear "fission" as a conceivable explanation for any observed experiments appears to have been ignored simply because it was premature in Stent's sense. But before concluding this, it is worth considering the other factors I have summarized elsewhere as possible contributors to early neglect of claims, hypotheses, or proposals later accepted.[26]

Noddack was unsatisfied with their reply and that a sharp clash might be coming, and that "Rosebaud was for giving her a sharp explanation, since he thinks that one must speak one's mind to her once and for all" (Hopper 1990, p. 79). Krafft (1981, p. 319) gives the original text as: "Rosebaud war aber sehr für unsere scharfe Erklärung; denn er meint, man müsse ihr einmal etwas gründlich die Meinung sagen." If correct, this implies that Hahn himself decided not to answer. Perhaps this occurred just after he got news from Meitner and Frisch about the implications of their work.

23. Abelson 1939.

24. Meitner and Frisch 1939b; Sime 1996, p. 267.

25. Noddack 1934b. The quotation is my corrected translation originally by H. G. Graetzer of a statement in Graetzer and Anderson 1971 (pp. 16–19).

26. See Hook, chapter 1 in this volume.

Ignorance

Acceptable evidence exists that individuals in two major groups in nuclear chemistry in the mid-1930s had seen Noddack's paper. Emilio Segrè states that Noddack had sent Fermi's unit in Rome a copy and they had considered it.[27] (For a discussion of why they did not follow up, see below.) Letters exchanged between Hahn and Meitner in 1939 indicate Hahn had seen Noddack's paper earlier and Meitner noted a dim memory of it.[28] But I can find no explicit documentation of prior awareness of Noddack's suggestion by individuals in the Curie group in Paris or in the Lawrence group in Berkeley, although such seems likely.[29] Comments citing her paper and referring at least to her objections to Fermi's evidence—including the need to exclude all elements before inferring the existence of a transuranic, not just those down to number 82—did appear in both textbooks and reviews published before the discovery of fission.[30] (But as the discussion below notes, these may have misled readers.)

Among the key figures in nuclear chemistry at the time, only Strassmann, to my knowledge, later explicitly denied prior knowledge of her paper.[31] I discuss below the possible implications of this.

Prejudice Because of Gender

One physicist and one historian of science initially unfamiliar with the historical episode suggested to me, on first learning of it, that Noddack's hypothesis was ignored on grounds of gender, that is, because it emanated from a woman. Whatever may have been the situation in other fields in the 1930s, I find this a highly implausible explanation for such a reception of a hypothesis in nuclear chemistry or nuclear physics. Awareness of the work and career of Marie Curie, who died in 1934, was very strong. Her daughter Irène had just done work that was to lead to a Nobel Prize in physics the following year. Lise Meitner had international recognition for her own contributions. Certainly, it seems highly unlikely either Meitner or Irène Curie, who took each other's work seriously, would have ignored a suggestion because it emanated from another woman! Moreover, Ida Noddack herself, along with her husband, Walter Noddack, had been nominated for the Nobel Prize in chemistry in 1932, and was also to be nominated in subsequent years. The nominators were prestigious figures, including Walter Nernst, in whose laboratory the Noddacks had worked in the 1920s.[32]

27. Segrè 1955, p. 259; 1970, p. 76.
28. Krafft 1981, p. 318.
29. Glenn Seaborg, personal communication, July 1998; and Seaborg, chapter 3 in this volume.
30. Rasetti 1936, pp. 271; Quill 1938, p. 120.
31. Strassmann, in Krafft 1981, p. 317.
32. Crawford et al. 1987.

While judgment from a present-day perspective is of course subject to its own biases, I find it difficult to believe that a reasonable individual in nuclear chemistry or physics at the time would have been foolish enough to reject or ignore any hypothesis because it emanated from a woman. Certainly, Noddack did encounter professional difficulties because of gender.[33] But after reviewing the available evidence, I find it highly unlikely that gender bias caused her hypothesis to have been ignored.

Prejudice Because of Questionable Scientific Reputation

Existence of another sort of prejudice against Ida Noddack appears more plausible. It accounts for some retroactive explanations, invoked after the discovery of fission, for why her suggestion had been ignored earlier. This prejudice arose from the fact that Ida (then Ida Tacke); her husband-to-be, Walter Noddack; and Otto Berg claimed to have discovered elements 43 and 75 in 1925 while working in the laboratory of Walter Nernst. For these elements, they proposed the names (and symbols) "masurium" (Ma) and "rhenium" (Rh), respectively.[34] Initial skepticism about both claims arose. The group met readily the objections regarding element 75, and so rhenium is enshrined in the current periodic table. However, the evidence they offered for their claim to have found naturally occurring element 43 was much weaker. They could not produce analyzable amounts of the substance as they did eventually for element 75. Moreover, when requested, they could not even produce the original x-ray plates, which would have provided their key evidence of the element. These had "accidentally been broken" and, for unaccountable reasons, the Noddacks had no additional plates they could produce.[35] One must regard the reported description of their behavior as suspicious. To make matters worse, they continued to insist on their putative element 43 long after others dismissed it.

The announcement by Carlo Perrier and Emilio Segrè in 1937 that they had produced element 43 by artificial means heightened skepticism about the Noddacks' original claim.[36] The Perrier and Segrè report and its confirmation implied, to many, that element 43 does not occur naturally, at least not in amounts anything close to those claimed to have been detected by the Noddacks—a view that still appears likely—despite an attempt to reinstate some basis for the Noddacks' claim.[37]

33. Noddack's professional difficulties that resulted from gender bias have been reported by Hopper 1990, p. 23.

34. Noddack et al. 1925; see also Tacke 1925.

35. Segrè 1993, pp. 117–18; see also Hopper 1990, p. 18.

36. Perrier and Segrè 1937.

37. Van Assche 1988. But Herrmann (1989) states in response that, while it is correct that spontaneous fission of uranium in nature does produce minute amounts of element 43, in essence a few unstable molecules, these are too rare for the Noddacks to have found. The Noddacks' x-ray fluorescence spectroscopy would have had to be five orders of magnitude more sensitive than they claimed it to be to detect this amount.

The Italian discovery eventually led to a change in the periodic table: the term "masurium" and its symbol Ma were displaced by "technetium" (reflecting its creation by technology) and the symbol Tc.[38]

The work on rhenium was well accepted and appears to have enhanced the Noddack's standing among many chemists. But to others, this was insufficient to overcome the damage to their reputation from their perceived gaffe over element 43. It was not, as Glenn Seaborg recalled the episode for me, merely that they had made a simple analytical mistake. Rather, by claiming erroneously to have found a substance in nature that does not even exist naturally, they had committed a colossal blunder; moreover, not only did they fail to retract their claim, but they kept insisting on it.[39] Seaborg told me that Ida Noddack had been so discredited after the masurium episode that, by 1938, it would not have been "normal" for her suggestion about fission even to have been remembered, and thus cited, when subsequently confirmed.[40] William Brock, a historian of chemistry, has suggested explicitly that one strong reason, if not the main reason, Noddack's hypothesis was ignored was because "it was felt her expertise was in doubt after the identification of masurium."[41] Yet on consideration, I find these views implausible.

Certainly, prejudice toward the Noddacks may have existed among many of their contemporaries because of the masurium episode, along with active personal dislike for other reasons (see below). But I think that, for two reasons, personal antipathy is insufficient to explain the lack of attention given her paper. First, Noddack proposed her hypothesis in 1934. The work that appeared to disprove that element 43 occurred naturally, did not appear until 1937.[42] Second, whatever single mistake the Noddacks had made, they had of course made some other important discoveries. At that time one error, even a colossal one, would not, at least on logical grounds, justify ignoring Ida Noddack's hypothesis or undermine completely her relevant expertise. Of course, many do not react logically to unappealing hypotheses and may seek irrelevant grounds to justify their view. Years later it was claimed that whenever her husband inquired of Otto Hahn why he did not investigate Ida Noddack's suggestion, Hahn would retort "Ein Fehler reicht!" (One mis-

---

38. Segrè 1993, pp. 113–15. Until the term "technetium" was adopted, "masurium" was still used for element 43, despite the discrediting of the Noddacks' work.

39. Interview with Glenn Seaborg, July 1998.

40. Interview with Glenn Seaborg, 9 December 1997.

41. Brock 1993, p. 343.

42. Glenn Seaborg, however, told me that, even before the 1937 work of Perrier and Segrè, the claims regarding "masurium" were widely doubted because naturally occurring element 43 seemed unlikely on theoretical grounds, and that the Noddacks were already regarded with great skepticism and disdain (interview, July 1998). Segrè (1993, p. 115) notes, "By now [February 1937] . . . I knew the 'masurium' . . . was probably a mistake." And, he adds, in any event, "nuclear systematics raised strong suspicions about its stability," leading to questions about whether there could be enough of it around naturally to have been detected by the Noddacks.

take is enough!) [43] Assuming this anecdote is correct (and it sounds consistent with the impression I have formed of Hahn from the literature), I interpret the cited reply as simply a rhetorical device to justify Hahn's dismissal of the hypothesis he found uncongenial for other reasons.

I conclude, on review of the evidence, that any cited bias and prejudice offered after the discovery of fission, as even a partial explanation for Noddack's suggestion being ignored, resulted simply from the retroactive, irrelevant association of a memory of dislike for or bias against Noddack.[44] But this was not the reason her hypothesis was ignored from 1934 through 1938.

Prejudice Attributable to Personal or Political Dislike

Some evidence suggests that there was widespread dislike for Ida Noddack in the 1930s and 1940s. In March 1939, Lise Meitner wrote to Hahn about Noddack, to whom she referred somewhat disparagingly as "Frau Ida": "Daß sie eine unangenehme Ursche ist, habe ich immer gewußt"[45] (That she is a disagreeable thing, I have always known).[46] The letter was in response to Hahn's private complaints to Meitner, after fission had been discovered, about a letter by Noddack in press. In this, Noddack practically taunted Hahn for having overlooked her earlier suggestion and having persisted in his errors about transuranics.[47] Meitner in her atypical personal comment appears also to have had separate grounds for her unpleasant view of Noddack. And from today's perspective, aspects of Noddack's personality that one can infer from her written comments on already retracted published work of another seem likely to have provoked personal animosity.[48]

43. Pieter Van Assche, personal communication to Ruth Sime, 5 June 1990, cited in Sime 1996, p. 464, n. 66. See also p. 464, n. 61. Jungk (1953, p. 62) quotes Ida Noddack as alleging (in either an undated interview or an uncited letter to him) that Hahn, when asked why he had made no reference to Ida Noddack's suggestion at a talk, replied to Walter Noddack that he had not wanted to embarrass the latter's wife by doing so.

44. See, for example, Alvarez 1987, p. 73; Libby 1979, p. 43; Amaldi 1977, 1989.

45. Krafft 1981, p. 318.

46. "Disagreeable thing" is Sime's translation of "unangenehme Ursche" (1996, p. 272). Hopper (1990, p. 79) translates it pungently as "hag."

47. Noddack 1939.

48. Footnote 9 of Noddack 1934b illustrates this. Her paper on element 93 also discusses the work of a Czech, O. Koblic, who previously claimed to have isolated element 93, which he named "bohemium." She had investigated this and proved it false, noting that she had found tungsten and vanadium in the material Koblic had sent her for assay. At the time of her publication, as a result of communications from Noddack, Koblic had already retracted his claim in print twice. Yet she not only unnecessarily repeated the history of the episode in the text of her paper but also in the footnote indicated that he had insufficiently retracted the claim. In his published retractions, he had admitted only to the presence of tungsten in his "bohemium," although, she continued, "he was informed by letter [from her] concerning . . . both elements [vanadium and tungsten] in his samples." It was not enough for him to admit his error and retract his claim. She had to emphasize (one is tempted to say "rub his nose in it") in print the true depth of his blunder. She appears to imply that, as he was informed by

To make matters worse she and her husband were regarded by many as having been somewhat sympathetic to the Nazis, if not worse.[49] Hahn implies, in a letter to Meitner, that Noddack had allies and supporters among them.[50] But while some may have found her personality or politics repellent, after reviewing the evidence I conclude this cannot account for her 1934 suggestion being ignored. Certainly these factors may have contributed to the lack of expressed sympathy for Noddack when Hahn and Strassmann, Meitner and Frisch, and others failed to cite her in papers written in the flurry of excitement after the discovery of fission.[51] But that is a separate and less important matter.

Prematurity

While not all the other "usual suspects" for the rejection or oversight of work later found correct have been excluded to this point, it appears worthwhile here to consider prematurity, or a related concept, as an explanation. Indeed, several commentators have, in essence, made this suggestion.[52] There appears to be general agreement with the view expressed by Luis Alvarez that, in the early and mid-1930s, "the nucleus we thought . . . was harder than the hardest rock, bound by powerful forces—powerful enough to resist the electrical repulsion of all the protons.

---

letter, he could not evade true knowledge of the depth of his blunder by having forgotten the mention of vanadium. Perhaps Meitner was aware of this or similar episodes.

49. Sime 1996 (p. 465, nn. 67 and 68) mentions some of the evidence for this. Among other things, she cites an interview with Emilio Segrè, 12 May 1985. When Walter Noddack visited him in Palermo, Sicily, in 1937 to discuss element 43, Noddack wore "an irregular military uniform complete with swastikas," Segrè reported.

50. Otto Hahn, letter to Lise Meitner, 7 April 1939, cited by Sime 1996, p. 273.

51. From today's perspective it would have been appropriate for Hahn and Strassmann, as well as Meitner and Frisch, to have acknowledged Noddack's earlier suggestion had they recollected it at the time they wrote their key papers. From a current viewpoint, even after these appeared and they became aware of (or remembered) her earlier work, such retroactive acknowledgment would have been appropriate and gracious. (Hahn did acknowledge it long afterward, for example in Hahn 1958). Perhaps high-stakes research in a tense and competitive atmosphere led (then as often now) to an increase in unfair citation practices. About the same time that Hahn complained to Meitner about Noddack's demand for credit, he also lamented that the French had not been fair to him in their citations about fission. They cited Meitner and Frisch but not Hahn and Strassmann! Noddack certainly did have a point to make in 1939 about her priority and not being cited. But the way in which she made this point was sufficiently ungracious for the editor who ran her letter, and apparently others, to regard her claim with some disdain. Moreover, she overstated her own case—see below. Even if Noddack was, unlike Hahn, a Nazi sympathizer with friends in high places as some believed, I am aware of no evidence that the Nazi political apparatus sought to give Noddack some of the credit, as it might well have done had, say, the Curies in Paris or those of "impure" blood made the discovery.

52. For instance, Leona [Woods] Marshall Libby simply says, "Noddack was ahead of her time" (Libby 1979, p. 43), although of course this does not imply Stent's concept of prematurity *sensu strictu.*

Everyone knew that the alpha particle . . . was the largest chunk of nuclear material that could be clipped out of an atom."[53]

It appeared highly implausible that the neutrons used by Fermi in his bombardment could have a major effect on atomic number. Indeed, even the presumed effect of a simple neutron knocking out just two alpha particles ($Z = 2$) from uranium ($Z = 92$) to get directly or indirectly via thorium ($Z = 90$) to radium ($Z = 88$) appeared implausible. It was only in attempts to nail down and confirm such an unlikely product that Hahn and Strassmann discovered the even more implausible product barium, the consequence of the unimagined nuclear fission. The phenomenon of fission was as unexpected as if, after one had fired projectiles at a succession of increasingly larger buildings, inflicting only minor damage on each, when one reached the ninety-second the building itself collapsed into two parts.

This provides the primary explanation for why Noddack's hypothesis was ignored by Hahn and Meitner, by the Curies and their collaborators in Paris, by Fermi's Italian collaborators, and by others elsewhere who were aware of it.[54] However, this does not appear sufficient to explain why Fermi, in particular, did not undertake or propose any experiments that would follow up on Noddack's paper.

53. Alvarez 1987, p. 73. The omitted material states that this was the view at least before Bohr elaborated the liquid-drop model (in 1936). If indeed a "softening" of views about the *relative* hardness of the nucleus and likely response to neutron bombardment occurred after that time, but before the appearance of the first Meitner-Frisch paper in 1939, I am unaware of any evidence for it. In a more technical explanation than that in Alvarez 1987, Herrmann (1990, p. 482) notes, "All experimental facts known until then [1939] suggest[ed] that the products of nuclear reactions [were] confined to the vicinity of the original nucleus; a differing phenomenon had never been observed. . . . Theory corroborate[d] this view. According to Gamow's 1928 theory of alpha-decay, one of the early triumphs of quantum mechanics, even the alpha particle with its two nuclear charges can hardly tunnel through the electrostatic Coulomb barrier of the nucleus; for larger fragments this barrier increases rapidly and the tunneling becomes extremely unlikely. . . . This reasoning is indeed correct. . . . The crucial point is . . . that the breakup of a heavy nucleus into two fragments of similar size should not be treated as a tunneling process. Quantum mechanics was more of an obstacle than a help for the discovery of fission."

54. While Meitner may have ignored Noddack's proposal in 1934, S. Flügge, who was the "in-house" theorist in Berlin in 1938, recalls discussing with Meitner the possibility that, in essence, Curie and Savitch (1938), working in Paris, had achieved without being aware of it what was later known as fission. They had reported as a product of neutron bombardment a substance with properties behaving almost like lanthanum ($Z = 57$), an element adjacent to barium on the periodic table and thus, if present, a fission product. Flügge comments on the possibility that this really could have been lanthanum rather than some difficult-to-understand transuranic, which was referred to as curiosum in Berlin (Sime 1996, p. 183). Flügge writes, "It would have been a very bold suggestion to say [it was lanthanum,] because it seemed so totally absurd at the time. I remember discussing this point with Miss Meitner and both of us quite agreeing on its impossibility because of the supposedly high energy barrier" (Flügge 1989, p. 27). (It is extraordinary that Meitner had any time at all for work then—May and June 1938—given the intrusive events in her life. See Sime 1996, pp. 184–209). It was determined later, after the discovery of nuclear fission, that the substance found by the Parisian workers was lanthanum contaminated by yttrium ($Z = 39$) (Herrmann 1990).

Fermi Again

Some clear but somewhat contradictory comments in the literature imply Fermi did *not* ignore Noddack's suggestion. Segrè, in his biography of Fermi, writes, "The possibility of fission, however, escaped us although it was called specifically to our attention by Ida Noddack, who sent us an article in which she clearly indicated the possibility.... The reason for our blindness is not clear. *Fermi said, many years later, that the available data on mass defect at that time were misleading and seemed to preclude the possibility of fission*" (emphasis added).[55] The emphasized material in this somewhat contradictory passage implies Fermi himself did in fact consider the possibility suggested by Noddack and did calculations based on available data on mass defect at that time, which led him to conclude fission was not possible. Edward Teller also insists that Fermi did a specific calculation, the results of which excluded Noddack's suggestion. He argues also that it was based on "wrong experimental information. Aston's experiment had at that time introduced a systematic error into calculating the mass and energy of nuclei."[56] Richard Rhodes, in his volume on the atomic bomb, provides some further comment. Rhodes implies that, when he interviewed Segrè in 1983 and raised Teller's comment, Segrè agreed Fermi had "sat down and performed the necessary calculations."[57] However, Rhodes continues, citing the same interview, "Segrè finds Teller's version of the story possible but not persuasive. The helium mass number problem would not necessarily have ruled out breaking up the uranium nucleus." Unfortunately, Rhodes did not go on to ask Segrè what calculations, if any, he thought Fermi might have done on the issue, if not those mentioned by Teller. Certainly Segrè's earlier comment in 1970 is consistent with Teller's later account.[58]

---

55. Segrè 1970, p. 76. In 1955, he noted that he had in fact discussed the subject with Fermi on more than one occasion and the latter had always cited wrong ideas about the mass defects of the nuclei (Segrè 1955, p. 262).

56. Teller 1979, p. 140.

57. Rhodes 1986, pp. 231–32.

58. Segrè's various accounts appear somewhat inconsistent. One wonders if simply a disinclination to concede that Teller's knowledge of Fermi was greater than his own accounts for some of his apparent waffling. In any event I find it remarkable that no one to date has redone the calculations using data available in 1934 on mass defect to determine whether Teller or Segrè is correct. Some, such as Andersen (1996, p. 486) apparently choose to ignore the possibility of such a consideration by Fermi, and Andersen has even gone so far as to claim that "what Noddack proposed was simply meaningless" to others in the scientific community. And a comment attributed to Leona Woods (as she was known at first while a student with Fermi; she was later Leona Woods Marshall and then Leona Marshall Libby), quoted by Rhodes from Libby 1979 in support of Teller's suggestion, appears to imply instead a somewhat different rationale for Fermi's lack of follow-up: "Bohr's liquid-drop model of the nucleus had not yet been formulated[,] and so there was at hand no accepted way to calculate whether breaking up into several large fragments was allowed" (Libby 1979, p. 43). I think what Libby means here is that the liquid-drop model was not "available" to provide that model eventually invoked to explain fission. But while Meitner and Frisch used Bohr's liquid-drop model to explain fission in 1939, this does not preclude there having been a prior method "allowed" or acceptable to attempt calculations pertinent to

Edoardo Amaldi, a collaborator with Fermi and Segrè on the work, wrote in 1977 only that, after the discovery of fission, "we were not able to understand the reason" the possibility had been rejected.[59] Over a decade later he reported that Noddack's "suggestion was hastily set aside because it involved a completely new type of reaction: fission. Enrico Fermi, and all of us grown at his school followed him *[sic]*, was always very reluctant to invoke new phenomena as soon as something new was observed: New phenomena have to be proved!"[60] However, Amaldi was not in as close touch with Fermi as Segrè had been just after the discovery of fission and during the years of World War 2 and so may not have heard of such a calculation from Fermi.[61]

Nevertheless, in a book written while her husband, Enrico, was still alive, and published in 1954, Laura Fermi implies (by omission) a viewpoint similar to that of Amaldi. In explaining the situation to her after the fact, her husband had invoked a simple failure of imagination, and nothing else, for his inability to realize that nuclear fission, not transuranics, accounted for his observations in 1934. According to her book, he cited neither calculations nor misleading data from the literature as having kept him from doing any follow-up. Her charming report of the conversation reads:

---

mass defect. The nuanced history of the model provided in Stuewer 1994 points out that the static aspects of the liquid-drop model had been proposed by Gamow in 1928, published shortly thereafter, and apparently forgotten by Bohr. And by 1938 Bohr's own contribution had been in the literature for about two years. So in that sense it was "available" in 1936 when Bohr published on it. But not even Bohr had written about seeing in the liquid-drop model a mechanism that explained—in terms of a drop that split into parts—the puzzling experiments interpreted as transuranics. Meitner and Frisch did so in 1939 when confronted by the reality of the reported chemical results and Meitner's belief that Hahn was too good a chemist to make an error (see Sime 1996, p. 236, citing Frisch's account). In any event, the mass defect data cited by Teller appears to have been a completely separate obstacle—in addition to lack of an available model, liquid drop or other—to explaining fission. I asked Teller's assistant to inquire of him what he might still remember of the episode. He reiterated his earlier published account that Fermi had undoubtedly calculated accurately the probability of a neutron penetrating the potential barrier, but that the size of the barrier was too high because of erroneous measurements involving mass defect. He recalled that Fermi's error had had the same basis as Szilard's erroneous report of beryllium as being subject to radioactive decay (although he could not recall the exact connection at the time of the inquiry). Judith Shoolery, personal communication, 11 April 2001.

59. Amaldi 1977, p. 304. However, at the time of the discovery of fission Segrè was much more frequently in touch with Fermi, both being then in the United States, while Amaldi remained in Italy. Segrè published with Fermi on fission in 1941 (see Segrè 1981, p. 10).

60. Amaldi 1989, p. 15.

61. In a letter to Fritz Strassmann on 16 March 1978, Amaldi writes explicitly that Fermi considered the implications of her suggestion as "unrealistic or even nonsense." Amaldi adds, "If we had looked at the tables of the masses we would have probably found that fission was energetically possible. But none of us has investigated this point that was not considered by I. Noddack" (Krafft 1981, p. 316). From this and his subsequent writings, I infer that Amaldi was unaware of the comments of Segrè and Teller on mass defect data even sometime after they were made.

[LF]: Then in Rome . . . you must have produced fission without recognizing it.

[EF]: That is exactly what happened. *We did not have enough imagination* to think that a different process of disintegration might occur in uranium than in any other element, and we tried to identify the radioactive products with elements close to uranium on the periodic table. . . . Moreover, we did not know enough chemistry to separate the products of uranium disintegration from one another, and we believed we had about four of them, while actually their number was closer to fifty. [Emphasis added]

[LF]: But then, what has become of your element 93?

[EF]: What at the time we thought might be element 93 has proven to be a mixture of disintegration products. We suspected it for a long time; now we are sure of it.[62]

So perhaps Fermi simply concluded *cif (con intuito formidabile)*—that is, with the "formidable intuition" that he used so frequently, albeit atypically incorrectly in this case—that fission was not possible, as Segrè implies in another interview with Rhodes.[63]

Laura Fermi wrote her book while her husband was still alive, so he would have been in a position to correct any errors. But whereas in her acknowledgments she offers special thanks to "my family, who have endured life with a writing housekeeper and have not complained," perplexingly she thanks, for "reading the parts of the manuscript dealing with scientific matters," not Enrico Fermi but "Dr. Emilio Segrè"![64] One must wonder how much Laura Fermi remembered precisely about a conversation 14 years earlier—recollections that may have been subject to some revision by her scientific reader.

Segrè has written of Fermi that "in scientific matters he was conservative. He hated to conclude from an experiment or calculation more than the results indicated." And, "because [he] loathed being in error, and error is occasionally unavoidable, he wanted to be in error only for having claimed too little."[65] Ironically, the incorrect interpretation of transuranics from his 1934 data, which he himself put forth very reluctantly, was indeed more conservative than the eventual astonishing explanation of fission.

Pre-1939 Interpretations of Noddack's 1934 Paper

In 1940, Louis A. Turner, the earliest to consider Noddack's 1934 suggestion in light of the recognition of fission, wrote: "If this early suggestion of what has turned out

---

62. Fermi 1954, p. 157. Presumably, by a "long time" Fermi meant since the discovery of fission.

63. Rhodes 1986, p. 232. Segrè (1993, p. 151) defined *cif* as a "joking acronym [for *con intuito formidabile*] we used for statements by Fermi that were true, but [which] he could not prove."

64. Fermi 1954.

65. Segrè (1970, pp. 102–3) and Amaldi (1962, pp. 808–9) wrote similarly about Fermi's caution and the need to avoid preconceived interpretations, however plausible, that would make objective appraisal difficult.

to be the correct explanation was anything more than speculation[,] it is regrettable that the reasons for its being considered possible were not more fully developed. *It seems to have been offered more by way of pointing out a lack of rigor in the argument for the existence of element 93 than as a serious explanation of the observations*" (emphasis added).[66] After a close reading of Noddack's paper and some reactions to it, I find his view plausible.

Noddack's suggestion appears in the tenth of 18 paragraphs that comprise the original text of her paper. While Fermi's proposed element 93 gets the greatest attention, she discusses two other suggestions about the discovery of an element 93. Whatever Noddack intended, both her wording ("one could assume equally well … ") and the placement of the comments of the tenth key paragraph between two others that focus on chemical criticisms contribute to the impression Turner had. She reiterates the possibility of split nuclei in no other paragraph of the paper. She points out at some length that Fermi did not even exclude all the elements down to lead ($Z = 82$). She emphasizes the difficulties with the chemical evidence he offered and reports some observations to counter his chemical claims. I am left, after reading it, with the same impression described by Turner, that her suggestion was intended simply to indicate one logical gap among many that she found.

The fact that all elements down to mercury ($Z = 80$), even below lead ($Z = 82$), and in particular polonium ($Z = 84$), later were excluded in the work of the Italian and Hahn-Meitner groups seems likely to have obscured for many the other aspects of Noddack's observations and her suggestion that the nucleus might break up into several large fragments of much lower atomic number.

Others cited Noddack's paper in subsequent reviews or comments on the transuranics published before the recognition of fission. But in the references I can retrieve, discussion tends to obscure the thrust of Noddack's point about the possibility of nuclear splitting and to imply, falsely, that it had been countered. For instance, the 1936 textbook of nuclear physics written by Franco Rasetti cites Noddack's paper in the context of discussion of criticism by A. V. Grosse of the chemical evidence offered for transuranics: "Dr. Ida Noddack also pointed out the need for more convincing proof, for … he [Fermi] did not consider element 84 (polonium) *nor any of the elements proceeding it in the periodic table.* She found[,] moreover, that a number of known elements are carried down with the manganese dioxide.… Both Dr. Grosse and Dr. Noddack, however, regard the artificial production of elements heavier than uranium as entirely within the realm of possibility[,] and, indeed, both independently predicted it"[67] (emphasis added).

The emphasized material at least implies (correctly from today's perspective) that, before one accepts a claimed transuranic, one must exclude the possible ex-

---

66. Turner 1940, p. 2.

67. Rasetti 1936, pp. 272–73. The cited mention by Noddack of the possibility of transuranics is found in an earlier paper (Noddack 1934a). She thought that some, at least 94 and 96, should occur *naturally.* A number of such claims and proposals of the latter type had been made since the 1920s. For references, see Quill 1938, pp. 88–89.

istence of what we would now call a fission product with an atomic number much lower than 84. But from the context of the passage, a reader would infer that her remark addressed only the need for "more convincing proof." And the implication given here and in other accounts of the time is that the chemical objections she offered had been met by others.

Writing in 1938, Lawrence L. Quill, in reviewing the transuranium elements, discusses first the criticism by Grosse and M. S. Agruss that the material thought by Fermi and colleagues to be element 93 was really element 91 (proactinium).[68] He notes that Grosse and Agruss withdrew their objections to "Fermi's conclusions" when Fermi provided further evidence that the material was *not* element 91. Quill continues, "The work of Hahn, Meitner, and Strassmann also emphasized the correctness of Fermi's conclusion." He leaves ambiguity here as to whether the "conclusion" they confirmed was the exclusion of element 91 or confirmation of 93. A reader might well infer falsely the latter.

Quill then discusses Noddack's objections explicitly but obscures the point she suggested—the possibility that neutron bombardment induces elements with a much lower atomic number than lead ($Z = 82$). Yet he does report explicitly that, in checking the chemical methods Fermi used, she found that these could not exclude titanium ($Z = 22$), columbium (now known as niobium [$Z = 41$]), tantalum ($Z = 73$), tungsten ($Z = 74$), iridium ($Z = 77$), platinum ($Z = 78$), gold ($Z = 79$), silicon ($Z = 14$), antimony ($Z = 51$), nickel ($Z = 28$), and cobalt ($Z = 27$), implying that some of these might have accounted for the observations of material interpreted as a transuranic. Quill then notes that Noddack showed also that polonium ($Z = 84$), which Fermi did not consider, might well be found in the product. But he concludes the pertinent paragraph by stating, "Her objections were explained later by Fermi as well as by Hahn, Meitner and Strassmann." For this conclusion (of course, incorrect from today's perspective) he provides no citation. He continues in the next paragraph: "The work of Hahn and Meitner [references] has done much to clarify the positions of elements of 93 and 94." The clear implication given the reader is that all of Noddack's concerns had been addressed.

As Spencer Weart has written, subsequently "scientists who reviewed the uranium research wrote their articles as if all objections, Grosse's and Noddack's together, were answered."[69] Of course, the concern of Noddack's viewed as most serious—that elements near uranium had not been excluded—was addressed experimentally. But I cannot find in the literature of the time a comment on the perceived theoretical obstacles (erroneous as they would have later proved to be) to experimental investigation of her hypothesis for production of elements of lower atomic number as an alternative to the claims of transuranics.

Whatever false inferences others may have drawn, and whatever the logical gaps in the comments on work at the time thought pertinent to her suggestion, I find

68. Quill 1938, pp. 119–20.
69. Weart 1983.

no documented evidence whatsoever that Noddack herself ever pointed out these gaps, reiterated her hypothesis in print, demonstrated that her hypothesis had not been explicitly excluded, published further on the outstanding logical deficiencies in the proof of the transuranics, or made any attempts to refute the logical inadequacies of the statements claiming that her suggestion had been proved false.[70] Nor did she enunciate her suggestion as part of the grand hypothesis explaining some puzzling experimental observations. She did not reiterate it in print even in 1938, when increasing complexities of putative transuranic decay schemes—all of which eventually collapsed—appeared ever more puzzling. Nor does evidence exist that she made any attempt to investigate the matter by herself or in collaboration with others.[71]

While to analytical non-nuclear chemists (like Noddack and her husband) her hypothesis may have seemed possible and even plausible, to nuclear chemists who were aware of it, or who remembered it after the polonium ($Z = 84$) and proactinium ($Z = 91$) questions about Fermi's work had been cleared up—and who even had the equipment and expertise to investigate the possibility—it seemed so far-fetched as to be not worth the effort. These reasons contributed to the lack of any direct response in the literature by radiochemists.

Noddack's comments at a logical and critical empirical level came from a gifted analytical chemist, but one with insufficient theoretical background to appreciate the enormity of what she proposed. There is no written evidence from before the discovery of fission, whatever she said later, that even she took her own proposal very seriously. Indeed, I am left with the impression that she herself, like everyone else prior to 1939 and the recognition of fission, eventually accepted the work of the Hahn and Meitner unit that appeared to confirm the existence of the claimed transuranics.

Strassmann Again

One further episode remains to be examined regarding the lack of reaction to Noddack's suggestion. As noted above, Fritz Strassmann in later years stated that he had been unaware of Noddack's 1934 paper until 1939.[72] He noted that, while he was affiliated with Hahn's unit in 1934—prior to receiving a regular position in 1935—he had been so busy eking out a living that he learned everything in the sci-

---

70. Only after fission was discovered did Noddack write that she and/or her husband had raised the matter verbally with Hahn, and apparently only with him (Noddack 1939).

71. I judge it unlikely that either the Hahn or Curie group would have been receptive to such an overture on her part. I suspect, however, that the Fermi group in Rome would have been delighted to have such a gifted analytical chemist as Ida Noddack join them. Nevertheless, the etiquette and understood formalities at the time conceivably may have been obstacles to her presence in Rome then, because of her gender, although I find this difficult to believe of Fermi. In any event, there is no evidence that she made any attempt to come to Rome to work with Fermi at that time.

72. Krafft 1981, p. 317.

entific literature directly from Hahn. He did not ever recall Hahn mentioning Noddack's paper to him. But he did remember later that, in 1936, he had found evidence of barium ($Z = 56$) as a product of neutron bombardment of the uranium, that is, induced radioactivity in the barium fraction! He had mentioned this to Meitner. She had objected that he had not excluded adsorption effects (that is, the adherence of nonbarium radioactive substances, presumably of much higher atomic number, to the barium). He conceded this, and she discouraged any follow-up. Her negative reaction, he wrote later, was "treundlich, aber energisch" (amiable but energetically firm).[73]

Strassmann deferred to Meitner's intellectual and moral authority, which was considerable. Ruth Sime, Meitner's biographer, notes that, even to the Geheimrat and director Otto Hahn, Meitner could be heard to say on occasion, "Hähnchen, geh'nach oben, von Physik verstehst Du nichts." (Hahn, dear, go upstairs. You understand nothing of physics.) And he would obey.[74] If Hahn, the director of the institute, deferred to Meitner, then even had Strassmann known of Noddack's suggestion, one must question whether he, as a junior assistant working directly under Meitner, would have persisted in attempting to confirm radioactive barium in the face of such views and authority, and thus in 1936 have found such compelling evidence of fission.[75]

Noddack: Coda

While Noddack's suggestion of fission provides an example of prematurity, after reviewing the episode closely I conclude that it is in one sense a relatively minor one. The proposal came from an individual who did not appreciate the theoretical difficulties it provoked, who made her suggestion in one paragraph as only one of many logical criticisms of others' interpretations of data, who not only did not amplify her suggestion but also never even repeated it in print, and who made no documented effort in any way to investigate it. I began this review convinced that she had been done a grave injustice on the matter of her "priority" but conclude that the view implied by Turner in 1940, just after the discovery of fission, is correct.

Nevertheless, while a minor exemplar of prematurity in what one may term an epistemological sense, the failure of the scientific world to take Noddack's sugges-

---

73. Strassmann to Alfred Klemm, July 1969, cited in Krafft 1981, p. 221.

74. Sime (1996, p. 178) at least implies that he complied. Meitner and Strassmann worked on the ground floor, Hahn one flight above them.

75. Sime (ibid., pp. 235–36 and p. 454, n. 26) cites evidence that Hahn later believed that, had Meitner stayed in Berlin, they would not have found fission: "She would have talked us out of barium"! Sime notes that this is unlikely. But Weizsäcker (1996) reported that other members of Hahn's institute told him later that "it is quite possible" Hahn would not have started the experiment that detected barium if Meitner had stayed on.

tion seriously in 1934 likely had enormous social consequences. For had it done otherwise, the Nazis might well have developed an atomic bomb.[76]

The tenth paragraph of Noddack's 1934 paper likely will always be a footnote in any history of fission. But such attention to her comment may be considerably more than merited by any real insight she had on the matter, whatever pause its possible consequences may give to those of us who choose to muse on historical contingency.

INTERDISCIPLINARY DISSONANCE: SOME OTHER ISSUES

The evident conceptual disciplinary divide in the 1930s between (analytical) chemists and nuclear (or "radio") chemists clashed forcefully in the first Hahn and Strassmann paper. Nuclear chemists were, of course, very close to the field of physics and restrained by its theoretical bans. However, (analytical) chemists, judging by the example of Noddack and apparently Strassmann, appeared to have had few constraints from this quarter (at least in the 1930s) but were more concerned with direct "analytical" evidence of claims. Judging from Quill's comments, it appears that such analytical non-nuclear chemists had a much less developed set of canons, and that they depended much less on theoretical considerations and calculations than on experimental results. Located in camps that were becoming increasingly disparate because of the developing implications of their observations and theory, Hahn and—to a lesser extent—Strassmann experienced great strain, from which only Meitner and Frisch rescued them. The fission episode illustrates my point at the beginning of this chapter about the differences between fields in their developmental stages, or in canons. Claims, hypotheses, or proposals that are premature to some fields or individuals may be acceptable to another.

I have searched for other examples or commentary that appears directly pertinent to the concept of interdisciplinary dissonance defined at the beginning of this chapter. In a separate area, which lies at the intersection of the fields of chemistry and physics and appears relevant to the Noddack episode, I have found a possible example in a quotation from Hans Suess. He collaborated with a physicist, A. D. H. Jensen, on work on the "shell model" of the nucleus in the late 1940s, and his observation of "magic numbers" led to his and Jensen's recognition of strong spin-orbit coupling. With regard to the collaboration, he stated,

> Jensen once remarked that if he had known more theoretical nuclear physics, he would never have believed a word of what I had told him. It was really a difficult job for a mere chemist, who uses different methods than a theoretical physicist, to convince him. I used what is generally considered to be "circumstantial evidence."

76. No one familiar with the pre–World War 2 history of nuclear fission with whom I have discussed the matter has disputed this conjecture. Nevertheless, other issues, such as possible unavailability of uranium ore, difficulty in production of a fissionable transuranic, the necessary diversion of technological resources required for success, and so on, might still have been obstacles not overcome.

Chemists are used to considering simultaneously a number of facts and then deriving a conclusion from them, whereas theoretical physicists usually wish to consider the result of one single experiment, or one phenomenon they wish to interpret.[77] (Emphasis added)

Clearly, the Noddack episode is not a unique example of interdisciplinary dissonance. It and the example above suggest, moreover, that the inhibiting influences tend to flow, albeit not necessarily, from the more theoretical field to the more empirical one, as from physics to chemistry in these cases. And the material emphasized in the quotation above suggests one plausible source of interdisciplinary dissonance between individuals with a foot in each field, and between collaborators from different fields working together. Those at the boundaries of fields, of which one has heavy theoretical emphasis and the other is much more experimental, may be particularly susceptible to interdisciplinary dissonance, as indeed was Otto Hahn.[78]

Another likely example of interdisciplinary dissonance, suggested to me by Elihu Gerson, concerns the influence of the physicist Lord Kelvin on the reception of certain of Charles Darwin's proposals involving the evolution of species. Kelvin insisted the earth was much younger than was indicated by some aspects of Darwin's theory. His authority significantly inhibited these proposals' acceptance.[79] Norriss Hetherington has suggested another example involving the age of the earth: the dilemma of the astronomer Edwin Hubble in reconciling the age implied by his model of the expanding universe with an age supported by geological data.[80]

Gerson proposed as another instance the clash of Alfred Wegener's theory of continental drift with objections from other geologists.[81] I think the last case represents a much more typical phenomenon, namely, a dispute within a discipline that inhibits further development of what is later recognized as a major discovery. In analogy to interdisciplinary dissonance, one may define this as *intra*disciplinary dissonance. Questions at issue here may involve which type of evidence is to be granted greater weight, or what *is* the evidence—for example, whether an experimental claim can be replicated. Moreover, personal friction, or fear of it, may be enough to markedly inhibit developments in a field.

Examples of interdisciplinary dissonance in the history of science are, I believe, much less frequent, indeed are rare, although they seem to offer a particularly fascinating perspective on development of scientific fields. Of course, there is not necessarily a sharp boundary between all the types of inter- and intradisciplinary dissonance. For example, a new subfield may emerge in any single discipline and have, compared to existing subfields, a different focus: some new type of

---

77. Cited in Stuewer 1979, p. 37.
78. See also Herrmann (1990, p. 482) for another example pertinent to nuclear fission.
79. Burchfield 1975.
80. Hetherington 1989.
81. For references and further discussion of this episode, see Glen, chapter 8 in this volume.

method and/or strong differences in or emphasis on some aspect of theory. Clashes between these subfields may have important inhibiting effects on further development in the new subfield, which may in fact be emerging as a separate discipline.

Finally, we may consider the relationship of intra- and interdisciplinary dissonance to prematurity. Some proportion of cases of intradisciplinary dissonance, but by no means all, will provide exemplars of prematurity as originally defined by Gunther Stent. If, as I suggest, we expand his notion of prematurity to include the failure of connection across disciplinary boundaries,[82] then cases of interdisciplinary dissonance may also provide exemplars of the expanded notion. Such circumstances may arise especially often for claims, hypotheses, or proposals put forward and unchallenged by, or unthreatening to, workers in a relatively undeveloped field—ideas that cannot be connected to or may even be contradicted by the canon of another that is more fundamental or extensive. The overall explanatory success of the latter field may give greater credibility to the objections of those within it, and be sufficient to inhibit further investigation or follow-up by those in both fields.

While interdisciplinary dissonance does not necessarily imply prematurity in the expanded sense, I suspect almost all cases of the former will be found to be cases of the latter. But this remains only a hypothesis, still to be tested in the light of additional historical evidence.

ACKNOWLEDGMENTS

I am indebted to Glenn Seaborg for his affable cooperation in presenting an interdisciplinary seminar at Berkeley—which first made me aware of Noddack's work—and patiently enduring some of the great puzzlement I subsequently expressed about past events.[83] Indeed, this chapter started as a brief addendum to that of his own in this volume.

BIBLIOGRAPHY

Abelson, P. 1939. "Cleavage of the Uranium Nucleus." *Phys. Rev.* 55:418.

Alvarez, L. 1987. *Alvarez: Adventures of a Physicist.* New York: Basic Books.

Amaldi, E. 1962. "No. 112–119," pp. 808–11. In *Collected Papers* (Note e. Memorie), by Fermi, E. Vol. 1: *Italy 1921–1938.* Chicago: University of Chicago Press; Roma: Academia Nazionale die Lincie.

———. 1977. "Personal Notes on Neutron Work in Rome in the Thirties and Post-war European Collaboration in High-Energy Physics." In *History of Twentieth Century Physics,* ed.

---

82. As I read Stent, despite two of his examples, his notion as a category of *significant* explanation for lack of follow-up of some claims, hypotheses, or proposals, applies or is intended to apply within, not across, fields.

83. Glenn Seaborg, "Patterns of Discovery in the Sciences: A Personal Perspective" (paper presented to seminar no. 25 in the series Patterns of Scientific Discovery since 1800, University of California, Berkeley, 4 November 1996).

C. Weiner, pp. 294–351. *Proceedings of the International School of Physics "Enrico Fermi," Course 57.* New York: Academic Press.

———. 1989. "The Prelude to Fission." In *Fifty Years with Nuclear Fission,* ed. J. W. Behrens and A. D. Carlson, 1:10–19. La Grange Park, Ill.: American Nuclear Society.

Amaldi, E., O. D'Agostino, E. Fermi, B. Pontecorvo, F. Rasetti, and E. Segrè. 1935. "Artificial Radioactivity Produced by Nuclear Bombardment. II." *Proc. Roy. Soc.* (London), ser. A, 146:183–500.

Andersen, H. 1996. "Categorization, Anomalies, and the Discovery of Nuclear Fission." *Stud. Hist. Phil. Mod. Phys.* 27:463–92.

Brock, W. H. 1993. *The Norton History of Chemistry.* New York: W. W. Norton.

Burchfield, J. 1975. *Lord Kelvin and the Age of the Earth.* New York: Science History Publications.

Crawford, E., J. L. Heilbron, and R. Ullrich. 1987. *The Nobel Population, 1901–1937: A Census of the Nominators and Nominees for the Prizes in Physics and Chemistry.* Berkeley: University of California, Office for the History of Science and Technology; Uppsala: Uppsala University, Office for the History of Science.

Curie, I., and P. Savitch. 1938. "Sur les radioéléments formés dans l'uranium irradié par les neutrons II." *J. Phys. Radium* 9:355–59.

Fermi, E. 1934. "Possible Production of Elements of Atomic Number Higher Than 92." *Nature* 133:898–99.

———. 1965. "Artificial Radioactivity Produced by Neutron Bombardment: Nobel Lecture, December 12, 1938." In *Nobel Lectures including Presentation Speeches and Laureates' Biographies: Physics 1922–1941,* ed. Nobel Foundation, pp. 414–21. Amsterdam: Elsevier Publishing.

Fermi, E., F. Rasetti, and O. D'Agostino. 1934. "Sulla possibilità di produrre elementi di numero atomico maggiore di 92." *Ricerca Scientifica* 5:536–37.

Fermi, L. 1954. *Atoms in the Family: My Life with Enrico Fermi.* Chicago: University of Chicago Press.

Flügge, S. 1989. "How Fission Was Discovered." In *Fifty Years with Nuclear Fission,* ed. J. W. Behrens and A. D. Carlson, 1:20–29. La Grange Park, Ill.: American Nuclear Society.

Frisch, O. R. 1939. "Physical Evidence for the Division of Heavy Nuclei under Neutron Bombardment." *Nature* 143:276.

———. 1967. "The Discovery of Fission: How It All Began." *Physics Today* 20, no. 11 (November): 43–48.

Graetzer, H. G., and D. L Anderson. 1971. *The Discovery of Nuclear Fission: A Documentary History.* New York: Van Nostrand Reinhold.

Grosse, A. V., and M. S. Agruss. 1934. "The Chemistry of Element 93 and Fermi's Discovery." *Phys. Rev.* 46:241.

———. 1935. "The Identity of Fermi's Reactions of Element 93 with Element 91." *J. Am. Chem. Soc.* 57:438–39.

Hahn, O. 1958. "The Discovery of Fission." *Scientific American* 198, no. 2 (February): 76–84.

———. 1966. *A Scientific Autobiography.* Trans. and ed. W. Ley. New York: Charles Scribner's Sons.

Hahn, O., L. Meitner, F. Strassmann. 1937. "Über die Trans-Urane und ihr chemisches Verhalten." *Ber. Deutsch. Chemis. Ges.,* ser. B, 70:1374–92.

Hahn, O., and F. Strassmann. 1939a. "Über den Nachweis und das Verhalten der bei der Bestrahlung des Urans mittels Neutronen enstehenden Erdalkalimetalle." *Naturwissenschaften* 27:11–15.

———. 1939b. "Nachweis der Enstehung aktiver Bariumisotope aus Uran und Thorium durch Neutronenbeststrahlung; Nachweis weiterer aktiver Brüchstucke bei der Uranspatung." *Naturwissenschaften* 27:89–95.

———. 1939c. "Zur Frage nach der Existenz der 'Trans-Urane.' I. Die endgültige Streichung von Eka-Platin und Eka-Iridium." *Naturwissenschaften* 27:451–53.

Herrmann, G. 1989. "Technetium or Masurium—a Comment on the History of Element 43." *Nuclear Physics,* ser. A, 505:352–60.

———. 1990. "Five Decades Ago: From the 'Transuranics' to Nuclear Fission." Trans. K. L. Kirchen. *Angew. Chem., Int. Ed. Engl.* 29:481–508. First published in *Angew. Chem.* 102 (1990): 469–96.

Hetherington, N. S. 1989. "Geological Time versus Astronomical Time: Are Scientific Theories Falsifiable?" *Earth Sci. Hist.* 8:167–69.

Hopper, T. 1990. " 'She Was Ignored': Ida Noddack and the Discovery of Nuclear Fission." Master's thesis, Department of History and Philosophy of Science, Stanford University. Submitted to Peter Galison.

Jungk, R. 1958. Brighter Than a Thousand Suns: A Personal History of the Atomic Scientists. Trans. J. Cleugh. New York: Harcourt, Brace, and Company.

Krafft, F. 1981. *Im Schatten der Sensation: Leben und Wirken von Fritz Strassmann.* Weinheim: Verlag Chemie.

———. 1983. "Internal and External Conditions for the Discovery of Nuclear Fission by the Berlin Team." In *Otto Hahn and the Rise of Nuclear Physics,* ed. W. R. Shea, pp. 135–65. Dordrecht: D. Reidel.

Libby, L. M. 1979. *The Uranium People.* New York: Crane Russak and Charles Scribner's.

Meitner, L., and O. R. Frisch. 1939a. "Disintegration of Uranium by Neutrons: A New Type of Nuclear Reaction." *Nature* 143:239–40.

———. 1939b. "Products of the Fission of the Uranium Nucleus." *Nature* 143:471–72.

Nobel Foundation. 1965. "Physics 1938." In *Nobel Lectures, Including Presentation Speeches and Laureates' Biographies: Physics 1922–1941,* p. 407. Amsterdam: Elsevier Publishing.

Noddack, I. 1934a. "Das Periodische System der Elemente und Seine Lücken." *Angew. Chem.* 47:301–5.

———. 1934b. "Über das Element 93." *Z. Angew. Chem.* 47:653–55.

———. 1939. "Bemerkung zu den Untersuchungen von O. Hahn, L. Meitner und F. Strassmann über die Produkte, die bei der Bestrahlung von Uran mit Neutronen enstehen." *Naturwissenschaften* 27:212–13.

Noddack, W., I. Tacke, and O. Berg. 1925. "Die Ekamangane." *Naturwissenschaften* 13:567.

Perrier, C., and E. Segrè. 1937. "Some Chemical Properties of Element 43." *J. Chem. Physics* 5:716.

Pleijel, H. 1965. "Physics 1938." In *Nobel Lectures, Including Presentation Speeches and Laureates' Biographies: Physics 1922–1941,* ed. Nobel Foundation, pp. 409–13. Amsterdam: Elsevier Publishing.

Quill, L. L. 1938. "The Transuranium Elements." *Chemical Reviews* 23:87–155.

Rasetti, F. 1936. *Elements of Nuclear Physics.* New York: Prentice-Hall.

Rhodes, R. 1986. *The Making of the Atomic Bomb.* New York: Simon and Schuster.

Segrè, E. 1955. "Fermi and Neutron Physics." *Rev. Mod. Phys.* 27:257–63.

———. 1970. *Enrico Fermi, Physicist.* Chicago: University of Chicago Press.

———. 1981. "Fifty Years Up and Down a Strenuous and Scenic Trail." *Ann. Rev. Nucl. Part. Sci.* 31:1–18.

———. 1993. *A Mind Always in Motion: The Autobiography of Emilio Segrè.* Berkeley and Los Angeles: University of California Press.

Sime, R. L. 1996. *Lise Meitner: A Life in Physics.* Berkeley and Los Angeles: University of California Press.

Stuewer, R. H. 1994. "The Origin of the Liquid-Drop Model and the Interpretation of Nuclear Fission." *Persp. on Sci.* 2:76–129.

Stuewer, R. H., ed. 1979. *Nuclear Physics in Retrospect: Proceedings of a Symposium on the 1930s.* Minneapolis: University of Minnesota Press.

Tacke, I. 1925. "Zur Auffindung der Ekamangane." *Z. Angew. Chem.* 38:1157–60.

Teller, E. 1979. *Energy from Heaven and Earth.* San Francisco: W. H. Freeman.

Turner, L. A. 1940. "Nuclear Fission." *Rev. Mod. Phys.* 12:1–29.

Van Assche, P. H. M. 1988. "The Ignored Discovery of the Element Z = 43." *Nuclear Physics,* ser. A, 40:205–14.

Weart, S. R. 1983. "The Discovery of Fission and Nuclear Physics Paradigm." In *Otto Hahn and the Rise of Nuclear Physics,* ed. W. R. Shea, pp. 91–133. Dordrecht: D. Reidel.

Weeks, M. E. 1939. *Discovery of the Elements.* 4th ed. Easton, Penn.: Journal of Chemical Education.

Weizsäcker, C. F. 1996. "Hahn's Nobel Was Well Deserved." *Nature* 383:294.

SECTION B

Disputable Cases

CHAPTER ELEVEN

Michael Polanyi's Theory
of Surface Adsorption

How Premature?

Mary Jo Nye

Adsorption is a process whereby gases are attracted and held to the surface of a solid. In his 1972 essays on prematurity and scientific discovery, Gunther S. Stent presented Michael Polanyi's potential theory of adsorption as an example of "delayed appreciation" of a scientific discovery.[1] Drawing upon Polanyi's article "Potential Theory of Adsorption," published in *Science* in 1963, Stent wrote:

> Despite the fact that Polanyi was able to provide strong experimental evidence in favor of his theory, it was generally rejected. Not only was the theory rejected, but it was considered . . . ridiculous by the leading authorities of the time. . . . At the very time he put it forward the role of electrical forces in the architecture of matter had just been discovered[, a] . . . point of view . . . irreconcilable with Polanyi's basic assumption of the mutual independence of individual gas molecules in the adsorption process. . . . It was only in the 1930s, after F. London developed his new theory of cohesive molecular forces . . . that it became conceivable that gas molecules *could* behave in the way which Polanyi's experiments indicated. . . . Meanwhile, Langmuir's theory had become so well-established, and Polanyi's had been consigned so authoritatively to the ash can of crackpot ideas, that Polanyi's theory was rediscovered only in the 1950s.[2]

At the time he published the *Science* article, Polanyi had long since left his chair in physical chemistry at the University of Manchester for a position in social studies especially created for him in 1948. His book *Personal Knowledge: Towards a Post-critical Philosophy* had appeared in 1958, contending, among other things, that scientific knowledge is achieved through the authority of a scientific community in which commitment to a dominant knowledge framework is strong and discipline

---

1. See Stent, chapter 2 in this volume. Yves Gingras refers to the "death and resurrection of Michael Polanyi's potential theory of adsorption" as illustrating "the crucial role of temporality in science." Gingras 1995, p. 145.

2. For Polanyi's article, see Polanyi 1963b; the quotation is from Stent 1972a, p. 437.

against deviants can be severe.[3] That is to say, scientific method is most often not a method of doubt (as Karl Popper was arguing), but of belief. Orthodoxy runs strong in scientific disciplines, in Polanyi's view.

At a symposium organized by Alistair Crombie at Oxford in July 1961, Polanyi, now a research fellow at Merton College, was one of the commentators. Thomas S. Kuhn gave a paper, "The Function of Dogma in Scientific Research," in which he summarized his thesis on paradigms, normal science, and scientific revolutions, which was about to be published by the University of Chicago Press.[4] At the beginning of his paper, Kuhn noted the similarity of his and Polanyi's views on the "importance of quasi-dogmatic commitments as a requisite for productive scientific research."[5]

In the discussion at Oxford, Polanyi remarked that he had tried in vain for many years to call attention to the steadfastness of scientists' commitment to established beliefs, the view that Kuhn now was arguing.[6] Later, talking with Kuhn in Berkeley in February of 1962, Polanyi brought up his old adsorption theory, noting, as he would in his 1963 *Science* article, that the theory had been a casualty in the 1920s to theories of electric forces—dismissed by Albert Einstein and other scientists in favor of the Langmuir theory.[7]

In bringing up his adsorption theory, Polanyi was aware, as he was in the case of work he had done on the structure of cellulose in the 1920s, that an interpretation that he had proposed—and that had been rejected at the time—had later come to be part of mainstream scientific theory.[8] In regard to the adsorption theory, physical chemistry textbooks in the 1950s were teaching students that, while the Langmuir equation was elegant, readily derived, and easily understood, it most often failed to account for heterogeneous surfaces, and that, as Farrington Daniels and Robert Alberty put it in their classic textbook of 1955, "it is now believed that most surfaces are heterogeneous."[9]

But had Polanyi's theory been rejected earlier because it was premature? Was it ridiculed as a crackpot idea? The answer to each question, I think, is negative, and most certainly negative in the second instance. Yet, still, both Polanyi and Stent have been correct in noting the intrinsic interest of the reception of Polanyi's adsorption theory to studies of the nature of scientific practice. A brief look at the history of this theory will suggest what we can learn from that history and from Polanyi's reminiscences.

While undertaking medical studies in Budapest during the years 1908 to 1913, Polanyi carried out experimental work in biochemistry.[10] In 1913, the year that he

---

3. See, for example, Polanyi 1958, pp. 150–51, 163–64.
4. Kuhn 1962.
5. Kuhn 1963, p. 347, n. 1.
6. See Polanyi 1963a, p. 375.
7. See Polanyi 1962a.
8. On x-ray crystallography and cellulose, see Polanyi 1969; also Nye 2000 and Nye 2001.
9. See Daniels and Alberty 1961, p. 610.
10. On Polanyi's life, especially the early years, see Wigner and Hodgkin 1977; and Palló 1998.

completed his medical degree, he entered the Technische Hochschule in Karlsruhe in order to study chemistry with Georg Bredig and Kasimir Fajans. In August 1914 Polanyi entered the Austrian army as a military surgeon, but he spent much of the war period on leave or on light duty, largely for reasons of ill health. In 1917 he completed a doctoral thesis in physical chemistry, which he defended at the University of Budapest. Polanyi worked briefly as an assistant to Georg de Hevesy at the University of Budapest before he, and then Hevesy, was dismissed due to anti-Semitic and antiliberal political views in Admiral Miklós Horthy's new government.[11]

In September of 1920, Polanyi took a position in Berlin at the Kaiser Wilhelm Institut für Faserstoffchemie (Institute for Fiber Chemistry), which was housed in the buildings of the Kaiser Wilhelm Institute for Physical Chemistry and Electrochemistry directed by Fritz Haber.[12] By this time Polanyi had published papers in several areas of thermodynamics, including one on Walther Nernst's heat theorem and another on Einstein's quantum theory for specific heats.[13] Polanyi's doctoral thesis, like some of his early papers, focused on the adsorption of gases on the surface of colloidal droplets and other solids, using data in the published literature on adsorption of CO_2 by charcoal.[14]

In the thesis, Polanyi built on contemporary work by Arnold Eucken in order to explain adsorption data through the derivation of an adsorption isotherm, which expresses the relation between gas pressure and the volume of gas adsorbed on a solid surface at a given temperature.[15] Polanyi assumed that there are attractive forces of some kind between the adsorbing solid and the atoms or molecules of gas that locate themselves in several layers at the surface of the solid. He described the attractive force by a simple functional equation relating the magnitude of the adsorption potential to the volume in which this potential is present. Polanyi assumed that the potential is independent of the temperature of the adsorbing wall, and that the pressure exerted by the adsorbed material on its immediate neighborhood is the same as that which the adsorbed material would exert, at that density and temperature, if it were in the free state.[16] He described the gas density as decreasing continuously outward from the solid surface in the same way that the density of the earth's atmosphere decreases upward as we move away from the solid crust of the earth.[17] Polanyi's approach lay within the framework of late-nineteenth-century classical thermodynamics.

11. See Palló 1998, p. 42.

12. Polanyi 1933a. Of great utility is Cash 1977; see also Wigner and Hodgkin 1977, pp. 413–15.

13. Polanyi 1913a, esp. p. 157; and Polanyi 1913b, discussed in Scott 1983, pp. 282–83. See also Wigner and Hodgkin 1977, p. 416.

14. Polanyi 1917.

15. Polanyi noted that Eucken introduced the term "Adsorptionspotential" in 1914, a few months before Polanyi's first paper on the subject (1963b, p. 1013, n. 2).

16. Polanyi 1916. See Scott 1983, p. 283; Wigner and Hodgkin 1977, p. 417.

17. See the discussion in Söderbaum 1966, p. 283.

About the same time, from 1916 to 1918, Irving Langmuir at the General Electric Laboratory in Schenectady, New York, published experimental results on the adsorption of gases on mica surfaces and on water, arguing that the surface layer is monomolecular in its constitution, with a structure determined entirely by electrostatic forces. Langmuir, who had taken his doctoral degree with Nernst in 1906, framed this work within the newly emerging electron theory of chemical valence, first articulated by G. N. Lewis in 1916, arguing that the force that retains the gas particles on the solid surface results from the chemical field of forces of the surface atoms. Since the unsaturated and residual valences of the surface atoms work from fixed points and at fixed distances, the adsorbed gas particles fit into a latticelike layer as if in a chessboard where each square can be occupied by only a single gas particle. Thus, adsorption ceases when the lattice is fully occupied.[18] Langmuir also devised a surface-film balance that allowed the measurement of the properties of molecular layers on liquid surfaces, like water.[19]

Langmuir's theory constituted a bold break with classical theory, couched as it was in terms of G. N. Lewis's and Langmuir's chemical theories of the electron-pair bond. These were just becoming known in Britain and in Europe around 1920, at the time when Niels Bohr's first quantum theory of the atom was to undergo considerable amendment and further development.

In 1921, shortly after Polanyi joined the staff of the Kaiser Wilhelm Institute, Fritz Haber invited him to give a full account of adsorption theory at Haber's colloquium.[20] Polanyi's paper drew considerable criticism from both Haber and Einstein, who faulted Polanyi for disregarding the new electron theories of the structure of matter. Polanyi later said, "Professionally, I survived the occasion only by the skin of my teeth."[21] He began to doubt that he currently could have published his earliest adsorption work in a German scientific journal.[22]

Among Polanyi's colleagues at the institute was Herbert Freundlich, who headed the colloid department that had grown out of Freundlich's wartime work on gas masks. Freundlich gave a full account of Polanyi's theory in the 1922 edition of his textbook *Kapillarchemie*.[23] Polanyi remembers, however, that Freundlich was ambivalent and, at the time, had said, "I am heavily committed now to your theory myself; I hope it is correct."[24] Hermann F. Mark, a colleague in Polanyi's Institute for Fiber Chemistry, interested himself in the work on adsorption and its implica-

18. Ibid., pp. 283–84.

19. Langmuir 1916, 1917. See Gaines 1993.

20. Cited as *Festschrift der Kaiser-Wilhelm-Gesellschaft* 1921, p. 171, in Polanyi 1929c, p. 431.

21. Polanyi 1963b, p. 1011.

22. Ibid., p. 1012. Polanyi also resented the fact that Eucken criticized his adsorption theory while adopting some of Polanyi's assumptions. See Eucken 1922.

23. Mentioned in Polanyi 1963b, p. 1010.

24. Ibid.

tions for catalysis, later recalling that most chemists on the whole found Polanyi's theory perfectly satisfactory.[25]

From 1914 to 1922 Polanyi wrote twelve papers on adsorption, the best received of which was a 1921 paper published in the *Zeitschrift für Elektrochemie*. Here he discussed energy levels for activation in catalysis and concluded that the adsorption layer is in an activated energy state in which atoms or radicals are at a lower energy level than in the gas.[26] Yet, for six years following that period, Polanyi did little work on adsorption, focusing instead on x-ray diffraction studies of fibers, crystals, and metals, as well as chemical reaction rates and activation energies.

His interest in adsorption seems to have revived following the arrival of Fritz London in Berlin in 1927.[27] Polanyi recognized that a theoretical justification for his potential function might be found in Walter Heitler and London's approach to chemical binding forces and to Van der Waals forces through quantum-mechanical exchange forces between electrons.

London frequented Haber's seminars at the Kaiser Wilhelm Institute, as well as the weekly Wednesday physics colloquia at the University of Berlin, collaborating with H. Kallmann at the institute on theoretical work in 1929 and 1930.[28] London's interest in Van der Waals forces, which originally inspired his study of electron binding, attracted Polanyi's attention because Van der Waals forces between two molecules, unlike valence forces between atoms, are additive and relatively unaffected when a third molecule is brought into the vicinity of the two molecules. In 1930 London gave the name "dispersion forces," by analogy to dispersion of light, to these long-range forces between molecules. The dispersion forces account for the phenomena of capillarity and adsorption, as well as sublimation. These forces are independent of temperature.[29]

With Polanyi's interest refocused on adsorption theory, he began new investigations with colleagues in 1928 and undertook the organization of an all-day colloquium on heterogeneous catalysis during a four-day meeting of the Deutsche Bunsen Gesellschaft (the national organization for physical chemistry) in Berlin in May 1929. The seven colloquium speakers included Max Bodenstein, Hugh S. Taylor, London, and Polanyi, with Haber giving an introductory speech in which he extolled the importance of heterogeneous catalysis for German industry and noted the exciting possibilities in the new wave mechanics for clarifying theories of catalysis and adsorption.[30]

25. Scott 1983, p. 284; also see Mark 1962, p. 603.

26. Polanyi 1921, noted in Schwab 1937, pp. 241–43.

27. See Wigner and Hodgkin 1977, p. 418.

28. Gavroglu 1995, pp. 50, 59.

29. Ibid., pp. 67–68. See London 1930.

30. Haber 1929. The January 1929 issue of *Zeitschrift für Elektrochemie* carried the initial announcement of the colloquium, on p. 1. Also see the April 1929 issue, pp. 161–62, for the program.

Polanyi's paper on activation processes at surfaces was one that he committed to memory the evening before the Friday colloquium.[31] In this paper he proposed a general equation from which Langmuir's adsorption potential could be derived as a special case, while attempting to account for some observational differences between Langmuir's and his theories, allowing that there might be a monomolecular layer adsorbed on some solid surfaces. Polanyi later said that Haber concluded from Polanyi's remarks that he had given up his theory. Eugene Wigner, who joined Polanyi's research group in Berlin in 1923, recalled that he, too, thought that Polanyi was leaning toward Langmuir's view that electrical or valence forces could not attract several layers of gas.[32] In fact, Polanyi was insisting on a third, molecular force to explain multiple layers and arguing that Langmuir's formula represented an idealization that was not obeyed in all cases.[33]

This was a theme that Polanyi repeated in a paper sent in June to the *Zeitschrift für Elektrochemie* responding to some criticisms of his theory and of his recent experimental work with F. Goldmann, K. Welke, and W. Heyne.[34] Polanyi demonstrated that Langmuir's adsorption isotherm (which predicts that adsorption is independent of increasing pressure and stops once the solid's single-layered surface is covered) is not obeyed in many cases.[35] But many researchers thought that Polanyi's objection was not fatal to Langmuir's theory, since it appeared that the most active sites of adsorption on a surface, if not all sites, did follow the Langmuir equation.[36]

If we look, then, at the status of the potential theory of adsorption that Polanyi had developed about 1930, we find that his theory was part of the general discourse and central discussion of surface adsorption and catalysis. It was recognized that adsorption was no simple matter, as simple as neither Polanyi's isotherm nor the Langmuir isotherm.

Polanyi's theory had seemed old-fashioned and insufficiently rooted in contemporary electron theory in the 1920s, but in a paper they coauthored in 1930, London and Polanyi used quantum mechanical resonance between electronic systems to demonstrate that the adsorption potential of a solid wall decreases with the distance from the wall ("inverse third power law") just as Polanyi had first argued in 1914.[37] With this publication, Polanyi thought he would be on his way to winning the argument.

In January of 1932 the Faraday Society sponsored a symposium at Oxford on the subject of adsorption of gases. The three invited keynote speakers were Eric Rideal of Cambridge and Freundlich and Polanyi of Berlin. In fact, none of the

31. See Polanyi's diary, 9 May 1929, in Polanyi 1929b.
32. Wigner 1992, pp. 78–80; Wigner and Hodgkin 1977, p. 417.
33. Polanyi 1929a, pp. 431–32. This is in part a response to Zeise 1929.
34. Polanyi 1929c.
35. See discussion in Schwab 1937, pp. 187, 193–94.
36. Ibid., 194.
37. London and Polanyi 1930.

Berlin participants made the meeting, having been "prevented," it was said, "at the last moment from coming to England,"[38] presumably for financial reasons having to do with the terrible depression in Germany during the winter of 1931–32.

Hugh Taylor, who was on leave from Princeton at Manchester University and who had participated in the Berlin colloquium in 1929, introduced the agenda for the Oxford symposium by saying that two rival theories about surfaces had been struggling for supremacy, one assuming thick compressed films and long-range forces of attraction extending outward from solid surfaces, and the other emphasizing extremely short-range interatomic and molecular forces resulting in a unimolecular adsorption layer. Now, Taylor claimed, "as to adsorption, one can summarize the situation by saying that the thick compressed film has during the last decade become progressively thinner until now the tendency is to reinterpret the ideas of the compressed film in terms of the unimolecular layer."[39] Of Polanyi's work, Taylor highlighted the most recent work with Henry Eyring on activation energies and the London resonance energies between atoms of the adsorbent and adsorbate.[40]

In Freundlich's paper, circulated at the symposium, he suggested, as he had done in concluding remarks at the Berlin colloquium in 1929, that current experimental results did not permit a clear decision between the rival theories. "Monomolecular layers," stated Freundlich, "are perhaps the rule at low pressures, polymolecular ones at higher pressures, especially near to the saturation pressure.... Theories using an adsorption potential lead to other formulae correlating the amount adsorbed with the equilibrium pressure."[41] He offered an equation relating the volume of adsorbed material to the equilibrium pressure, which still is known as the Freundlich equation.

Polanyi's 1932 Oxford paper reiterated the success of the resonance approach to cohesion or dispersion forces as an explanation of adsorption. It also reiterated his claims that (1) the potential is not essentially dependent on temperature, (2) the potential is independent of the constitution of neighboring space, and (3) molecules in the adsorbed state exert approximately the same force on one another as they do when they are free.[42] There are different kinds of adsorption forces: electrical, valence, and dispersion forces.[43] Langmuir was partly right, and so was Polanyi. Later, the different forces would come to be distinguished as chemisorption and physical adsorption.

38. As noted by Mond 1932, p. 130.

39. Taylor 1932, p. 132.

40. Ibid., p. 138, citing Eyring and Polanyi 1931; and Eyring 1931.

41. Freundlich 1932, p. 198; also Freundlich 1929.

42. The same points are emphasized in *Zeitschrift für Elektrochemie* 35 (1929): 431. One of the points at issue in Polanyi's critique of Langmuir's equation was the exact way in which constants in the equation are temperature dependent.

43. Polanyi 1932, pp. 321–22.

Although Polanyi did not make it to Oxford in January of 1932, he visited England on a couple of occasions during the course of the year, negotiating with Manchester chemists and administrators who were offering him an appointment in physical chemistry. On 13 January 1933, Polanyi ended ten months of discussions by declining a chair in physical chemistry at Manchester. He gave as his reason a bout of rheumatism triggered by his last visit to damp and smoggy Manchester.[44] What he declined was an annual salary of £1,500.[45] The offer also included a newly constructed laboratory of physical chemistry estimated to cost £40,000, along with apparatus and funds for some 20 research assistants and coworkers. He was to have light teaching duties and absolute independence in running his laboratory.[46] This was hardly an offer to an unappreciated or undistinguished chemist. Polanyi decided to stay in Berlin, a decision which had to be abandoned after elections at the end of January resulted in Adolf Hitler's appointment to the German chancellorship. In the autumn of 1933, Polanyi and his family moved to Manchester, where he became a professor of physical chemistry.

The 1963 article in *Science* had the unintended effect of leading many of its readers to think that Polanyi had been a failed scientist, a physical chemist whose work was ignored, refuted, and rejected, leaving him an outsider in the chemical community. We have seen that this was not the case. Nor was his theory premature, in the ordinary meaning of the word. Indeed, as initially posed, Polanyi's adsorption theory was old-fashioned. As *re*fashioned in collaboration with London, the theory entered into the new framework of atomic theory and quantum mechanics that dominated cutting-edge theoretical physics and theoretical chemistry in the 1920s and 1930s. Yet Langmuir's theory received the laurels, emphasizing as it did chemical adsorption in terms of electrical forces and chemical valency, even though physical chemists recognized that Langmuir's equation, Langmuir's isotherm, and Langmuir's theory were not universal in coverage and that the adsorbate was not monomolecular in the majority of cases.

Langmuir received the Nobel Prize in chemistry in the autumn of 1932 "for his discoveries and investigations in surface chemistry." His Nobel lecture made specific mention of a few colleagues, among them Hugh Taylor, with passing reference to A. Sherman and Eyring.[47] Although it was not publicly known, of course, Langmuir had been nominated ever since 1916 for a Nobel award in both chemistry and physics, with nominations in physics in 1928 and 1929 from Niels Bohr and in chemistry in 1929 from Polanyi's Berlin colleague Freundlich. Among nominations, an especially influential one, no doubt, was the 1931 nomination in chemistry by the Swedish physical chemist Theodor Svedberg, whose work on molecular dimensions

44. Polanyi 1933c. On 30 January 1933 Adolf Hitler was appointed chancellor of Germany; also see Polanyi 1933b.

45. Lapworth 1932.

46. Allemand 1932a,b.

47. Langmuir 1966.

helped establish what came to be called "molecular reality."[48] Langmuir's approach fit perfectly with Svedberg's.

Polanyi later said that he chose not to introduce students at Manchester to his adsorption theory because they would not then be able to pass the required examinations: "I could not undertake to force on them views totally opposed to generally accepted opinion."[49] He wrote only one more article on adsorption after leaving Germany, an article on adsorption and catalysis for the *Journal of the Society of Chemistry and Industry.*[50]

In explaining the continued sway of Langmuir's theory from the mid-1930s to the mid-1950s, Polanyi mused in his *Science* article of 1963:

> It is . . . difficult to understand why more than 15 years passed after the presentation of my paper of 1932, in which the original objections had been proved unfounded, before the rediscovery and gradual rehabilitation of the theory set in. I suppose so much confusion was left over from the previous period that it took some time for scientists to take cognizance of the new situation, and that meanwhile my own work, which had been so long discredited, remained suspect. If the problem had been more important, this period of latency would have, no doubt, been shorter.[51]

Polanyi's disappointment in the resistance to his adsorption theory led him to reflect on the nature of scientific discovery and the awarding of priority and recognition within the scientific community. There is no doubt that he was concerned with priority issues throughout his scientific career. In the matter of the adsorption theory, he laid out in print in 1929 the originality of his 1916 theory in comparison to both Eucken's work and Langmuir's work. His own contribution, as Polanyi assessed it, lay in establishing the postulate that the adsorption forces include cohesion forces that act between adsorbed molecules, a point neglected by Langmuir.[52]

Priorities help establish the reputation of a scientist as a leader in his or her field. "The example of great scientists is the light which guides all workers in science," wrote Polanyi in 1962, adding, "but we must guard against being blinded by it. There has been too much talk about the flash of discovery[,] and this has tended to obscure the fact that discoveries, however great, can only give effect to some intrinsic

48. On the nominations, see Crawford et al. 1987.

49. Polanyi 1963, p. 1013.

50. Polanyi 1935. In 1938 Stephen Brunauer, Paul H. Emmett, and Edward Teller published a paper on adsorption that became known as the "BET theory." It modified Langmuir's theory in favor of multilayer adsorption, with reference to, among others, Goldman and Polanyi's paper of 1928 focusing on adsorption isotherms for charcoal (Brunauer et al. 1938).

51. Polanyi 1963b, p. 1012.

52. "Die Voraussetzungen 1. und 2. hat bereits vor mir A. Eucken eingeführt und verwertet; mein Beitrag bestand darin, die in der Voraussetzung 3. gelegene Approximation zu prüfen. In ihrer Verwendung liegt der grundsätzliche Unterschied gegenüber dem Ansatz von Langmuir, der die Wirkung der Kohäsionskräfte zwischen den adsorbierten Molekülen vernachlässigt." Polanyi 1929c, p. 431. Polanyi's three postulates are mentioned earlier in the chapter.

potentiality of the intellectual situation in which scientists find themselves. It is easier to see this for the kind of work that I have done than it is for major discoveries."[53]

Thus Polanyi came to characterize his work on the adsorption theory and much of his other work as *typical,* rather than atypical of the scientific process. The work did not fit within the "predominantly accepted scientific view of the nature of things" at the time.[54] It is inaccurate, then, to describe Polanyi's adsorption work as "crackpot" or "premature," even in Stent's sense of prematurity as work that fails to connect to contemporary scientific discourse and canonical knowledge. Polanyi's work on adsorption did connect to contemporary knowledge and it was part of a lively and long-term scientific discussion. This was a debate that Polanyi lost in the short run. Yet, even so, Polanyi's work on adsorption was part of a general body of scientific investigation that earned him considerable esteem in the scientific community of the 1920s and 1930s, even though he did not receive the accolades that he perhaps most coveted as author of a great discovery.

SOURCES

This essay makes use of the Michael Polanyi Papers, which are held in the Special Collections of the Regenstein Library at the University of Chicago. I am grateful to have had permission to consult these papers. Research for this project was supported by the National Science Foundation grant no. SBR-9321305 and by the Thomas Hart and Mary Jones Horning Endowment. I thank Mary Singleton and Gunther Stent for helpful comments on an earlier version of this chapter. I also have discussed Polanyi's work on adsorption in the essay "At the Boundaries: Michael Polanyi's Work on Surfaces and the Solid State," in *Chemical Sciences in the 20th Century: Bridging Boundaries,* ed. Carsten Reinhardt, pp. 246–57 (Weinheim: Wiley-VCH, 2001).

Bibliographies of Michael Polanyi's publications may be found in E. P. Wigner and R. A. Hodgkin, "Michael Polanyi, 12 March 1891–22 February 1976," *Biographical Memoirs of Fellows of the Royal Society* 23 (1977): 413–48; Marjorie Grene, ed., *The Logic of Personal Knowledge: Essays Presented to Michael Polanyi on His Seventieth Birthday, 11th March 1961* (London: Routledge and Kegan Paul, 1961); and Harry Prosch, *Michael Polanyi: A Critical Exposition* (Albany: State University of New York Press, 1986).

BIBLIOGRAPHY

Allemand, A. J. 1932a. Letter to M. Polanyi, 15 May. Michael Polanyi Papers, University of Chicago, Regenstein Library Special Collections, Box 2, Folder 8.

———. 1932b. Letter to M. Polanyi, 29 November. Michael Polanyi Papers, University of Chicago, Regenstein Library Special Collections, Box 2, Folder 10.

53. Polanyi 1969, p. 97.
54. Polanyi 1963b, p. 1012.

Brunauer, S., P. H. Emmett, and E. Teller. 1938. "Adsorption Gases in Multimolecular Layers." *Journal of the American Chemical Society (J. Amer. Chem. Soc.)* 60:309–19.

Cash, J. M. 1977. *Guide to the Papers of Michael Polanyi.* Chicago: Joseph Regenstein Library.

Crawford, E., et al. 1987. *The Nobel Population, 1901–1937.* Berkeley: Office for History of Science and Technology.

Daniels, F., and R. A. Alberty. 1961. *Physical Chemistry.* 2d ed. New York: John Wiley.

Eucken, A. 1922. Letter to M. Polanyi, 31 March. Michael Polanyi Papers, University of Chicago, Regenstein Library Special Collections, Box 1, Folder 17.

Ewald, P. P., ed. 1962. *Fifty Years of X-Ray Diffraction.* Utrecht: Oosthoek.

Eyring, H. 1931. "The Energy of Activation for Bimolecular Reactions Involving Hydrogen and the Halogens, According to the Quantum Mechanics." *J. Amer. Chem. Soc.* 53:2537–49.

Eyring, H., and M. Polanyi. 1931. "Über einfache Gasreaktionen." *Zeitschrift für physikalische Chemie (Z. Physik. Chem.),* ser. B, 12:279–311.

Freundlich, H. 1929. "Diskussion." *Zeitschrift für Elektrochemie und angewandte physikalische Chemie (Z. Elektrochemie.)* 35:585.

———. 1932. "Introductory Paper to Section II." *Transactions of the Faraday Society (Trans. Far. Soc.)* 28:195–201.

Gaines, G. 1993. "Irving Langmuir (1881–1957)." In *Nobel Laureates in Chemistry 1901–1993,* ed. L. K. James, pp. 205–10. Washington, D.C.: American Chemical Society.

Gavroglu, K. 1995. *Fritz London: A Scientific Biography.* Cambridge: Cambridge University Press.

Gingras, Y. 1995. "Following Scientists through Society? Yes, but at Arm's Length." In *Scientific Practice: Theories and Stories of Doing Physics,* ed. J. Z. Buchwald, pp. 123–150. Chicago: University of Chicago Press.

Goldmann, F., and M. Polanyi. 1928. "Adsorption von Dämpfen an Kohle und die Wärmeausdehnung der Benetzungsschicht." *Z. Physik. Chem.* 132:321–70.

Grene, M., ed. 1961. *The Logic of Personal Knowledge: Essays Presented to Michael Polanyi on His Seventieth Birthday, 11th March 1961.* London: Routledge and Kegan Paul.

———. 1969. *Knowing and Being: Essays by Michael Polanyi.* London: Routledge and Kegan Paul.

Haber, F. 1929. "Einleitung." *Zeitschrift für Elektrochemie und Angewandte Physikalische Chemie* 35:533–34.

Heyne, W., and M. Polanyi. 1928. "Adsorption aus Lösungen." *Z. Physik. Chem.* 35:384–98.

Kuhn, T. S. 1962. *The Structure of Scientific Revolutions.* Chicago: University of Chicago Press.

———. 1963. "The Function of Dogma in Scientific Research." In *Scientific Change,* ed. A. C. Crombie, pp. 347–69. New York: Basic Books.

Langmuir, I. 1916. "The Constitution and Fundamental Properties of Solids and Liquids. I. Solids." *J. Amer. Chem. Soc.* 38:2221–95.

———. 1917. "The Constitution and Fundamental Properties of Solids and Liquids. II. Liquids." *J. Amer. Chem. Soc.* 39:1848–1906.

———. 1966. "Surface Chemistry: Nobel Lecture in Chemistry in 1932." In *Nobel Lectures: Chemistry, 1922–1941,* pp. 287–325. Amsterdam: Elsevier Publishing.

Lapworth, A. 1932. Letter to M. Polanyi, 1 March. Michael Polanyi Papers, University of Chicago, Regenstein Library Special Collections, Box 2, Folder 8.

London, F., with R. Eisenschitz. 1930. "Über das Verhältnis der Van der Waalsschen Kräfte zu den homöopolaren Bindungskräften." *Zeitschrift für Physik* 60:491–527.

London, F., and M. Polanyi. 1930. "Über die atomtheoretische Deutung der Adsorptionskräfte." *Die Naturwissenschaften* 18:1099–1100.

Mark, H. 1962. "Recollections of Dahlem and Ludwigshafen." In *Fifty Years of X-Ray Diffraction*, ed. P. P. Ewald, pp. 603–7. Utrecht: Oosthoek, 1962.

Mond, R. 1932. Introductory remarks to the colloquium "The Adsorption of Gases: A General Discussion" (12–13 January 1932). *Trans. Far. Soc.* 28:129–447.

Nye, M. J. 2000. "Laboratory Practice and the Physical Chemistry of Michael Polanyi." In *Instruments and Experimentation in the History of Chemistry*, ed. F. L. Holmes and Trevor Levere, pp. 367–400. Cambridge: MIT Press.

———. 2001. "At the Boundaries: Michael Polanyi's Work on Surfaces and the Solid State." In *Chemical Sciences in the 20th Century: Bridging Boundaries*, ed. Carsten Reinhardt, pp. 246–57. Weinheim: Wiley-VCH.

Palló, G. 1998. "Michael Polányi's Early Years in Science." *Bulletin for the History of Chemistry* 21:39–43.

Polanyi, M. 1913a. "Eine neue thermodynamische Folgerung aus der Quantenhypothese." *Verhandlungen der deutschen physikalischen Gesellschaft (V. deut. physik. Gesell.)* 15:156–61.

———. 1913b. "Neue thermodynamische Folgerungen aus der Quantenhypothese." *Z. Physik. Chem.* 83:339–69.

———. 1916. "Adsorption von Gasen (Dämpfen) durch ein festes nichtpflüchtiges Adsorbens." *V. deut. physik. Gesell.* 18:55–80.

———. 1917. "Gázok absorptiója szilárd, nem illanó adszorbensen" [Absorption of gases by a solid non-volatile adsorbent]. Ph.D. diss., University of Budapest.

———. 1921. "Über Adsorptionskatalyse." *Z. Electrochemie* 27:143–50.

———. 1922. Letter to A. Eucken, 4 April. Michael Polanyi Papers, University of Chicago, Regenstein Library Special Collections, Box 1, Folder 19.

———. 1929a. "Betrachtungen über den Aktivierungsvorgang an Grenzflächen." *Z. Elektrochemie* 35:561–67.

———. 1929b. Diary. Michael Polanyi Papers, University of Chicago, Regenstein Library Special Collections, Box 44, Folder 4.

———. 1929c. "Grundlagen der Potentialtheorie der Adsorption." *Z. Elektrochemie* 35:431–32.

———. 1932. "Introductory Paper to Section III." *Trans. Far. Soc.* 28:316–33.

———. 1933a. "Curriculum Vitae," June. Michael Polanyi Papers, University of Chicago, Regenstein Library Special Collections, Box 2, Folder 12.

———. 1933b. Letter to F. G. Donnan, 17 January. Michael Polanyi Papers, University of Chicago, Regenstein Library Special Collections, Box 44, Folder 4.

———. 1933c. Letter to Arthur Lapworth, 13 January. Michael Polanyi Papers, University of Chicago, Regenstein Library Special Collections, Box 2, Folder 11.

———. 1935. "Adsorption and Catalysis." *J. Soc. Chem. Ind.* 54:123.

———. 1958. *Personal Knowledge: Towards a Post-critical Philosophy.* Chicago: University of Chicago Press.

———. 1962. Interview by Thomas Kuhn at Berkeley, February 15. Archives for History of Quantum Physics, American Institute of Physics, Niels Bohr Library, Transcript, pp. 9–10.

———. 1963a. "Commentary by Michael Polanyi." In *Scientific Change*, ed. A. C. Crombie, pp. 375–80. New York: Basic Books.

———. 1963b. "Potential Theory of Adsorption." *Science* 14:1010–13. Reprinted in *Knowing and Being: Essays by Michael Polanyi*, ed. M. Grene (London: Routledge and Kegan Paul, 1969), pp. 87–96.

———. 1969. "My Time with X-Rays and Crystals." In *Fifty Years of X-Ray Diffraction*, ed. P. P. Ewald, pp. 629–36. Utrecht: Oosthoek, 1962. Reprinted in *Knowing and Being: Essays by Michael Polanyi*, ed. M. Grene (London: Routledge and Kegan Paul, 1969), pp. 97–104.

Polanyi, M., and K. Welke. 1928. "Adsorption, Adsorptionswärme und Bindungscharakter von Schwefeldioxyd an Kohle bei geringen Belegungen." *Z. Physik. Chem.* 132:371–83.

Schwab, G. M. 1937. *Catalysis from the Standpoint of Chemical Catalysis*. Translated by H. S. Taylor and R. Spence. New York: Van Nostrand.

Scott, W. T. 1983. "Michael Polanyi's Creativity in Chemistry." In *Springs of Scientific Creativity*, ed. R. Aris et al., pp. 279–307. Minneapolis: University of Minnesota Press.

Söderbaum, H. G. 1966. "Presentation Speech: Nobel Prize in Chemistry in 1932." In *Nobel Lectures: Chemistry, 1922–1941*, pp. 283–86. Amsterdam: Elsevier Publishing.

Stent, G. S. 1972a. "Prematurity and Uniqueness in Scientific Discovery." *Advances in the Biosciences* 8:433–49.

———. 1972b. "Prematurity and Uniqueness in Scientific Discovery." *Scientific American* 227 (December): 84–93.

Taylor, H. 1932. "The Adsorption of Gases: A General Discussion (12–13 January 1932)." *Trans. Far. Soc.* 28:132–38.

Wigner, E. P. 1992. *The Recollections of Eugene P. Wigner, as told to Andrew Szanton*. New York: Plenum.

Wigner, E. P., and R. A. Hodgkin. 1977. "Michael Polanyi, 12 March 1891–22 February 1976." *Biographical Memoirs of Fellows of the Royal Society* 23:413–48.

Zeise, H. 1929. "Die Adsorption von Gasen und Dämpfen und die Langmuirsche Theorie." *Z. Elektrochemie* 35:426–31.

CHAPTER TWELVE

Prematurity and the Dynamics of Scientific Change

Frederic L. Holmes

On the "author's page" of the issue of *Scientific American* in which one version of Gunther Stent's article "Prematurity and Uniqueness in Scientific Discovery" appeared in 1972, he explained that he had first presented these ideas in a brief commentary at a meeting in 1970 at the American Academy of Arts and Sciences. The vigorous discussion that had ensued, he said, persuaded him that he should focus his ideas more sharply.[1] I was at this discussion, and I remember the lively response that Stent's ideas elicited. Robert Merton particularly was impressed that his view of premature discoveries was a novel and revealing insight. Those of us who took part (I rather passively as a fledgling historian) felt that Stent himself might have made a discovery.

Ernest Hook comments that Stent's concept of prematurity has received "only modest attention in the literature of the history, philosophy, or sociology of science."[2] To the reasons he has adduced for what he clearly feels is a disappointing response to an idea that has a "powerful heuristic attraction," I add that, during the decades since 1972, the nature of scientific discovery itself has received less attention from historians and sociologists of science than one might expect if one regards discovery as the central aim of science. Philosophers are said to have "rediscovered" discovery as a subject for philosophical analysis by questioning the dictum of Karl Popper that discovery is a psychological, not a philosophical, problem. But historians and sociologists have been more interested during this period in various "contextual" dimensions of science than in the process previously seen as the heart of the scientific enterprise. Some historians and sociologists even regard discovery as an inappropriate category for demarcating the steps in what they treat as the "construction" of scientific knowledge.

---

1. His note appears on p. 11 in the same issue as Stent 1972b.
2. See Hook, chapter 1 in this volume.

Without delving into complex questions about the ultimate status of scientific knowledge, I believe that scientists obviously discover things in their systematic pursuit of specialized knowledge in the same way that ordinary people discover things in the course of their lives. It is not so obvious, however, what a "unit" of scientific discovery is.

THE IMPORT OF SOCIAL RECOGNITION

As Robert Merton said in the 1960s, the reward system that maintains the norms through which science functions successfully requires appropriate recognition of the individual discoveries that scientists make. Such recognition not only has consequences for careers and offers the opportunity to continue work, but, as Merton puts it, assures the individual that "one's work really matters, that one has measured up to the hard standards maintained by a community of scientists."[3] In 1965 Warren Hagstrom introduced a discussion of "competition for recognition" with the statement: "Recognition is normally given for the first formal presentation of an innovation or discovery to the scientific community." He defined a professional scientist as "one to whom discoveries are attributed and to whom the recognition for them is awarded."[4] Events ranging from the award of Nobel Prizes to the pervasiveness of priority disputes can be explained as actions intended to sustain the appropriate recognition of degrees of scientific achievement measured in discoveries.

The sociological importance of distributing such packages of recognition fairly and accurately has contributed much to maintaining the tacit assumption that scientific discoveries occur at discrete times and places in a form that can be readily and permanently identified. In his *Structure of Scientific Revolutions*, which, unlike his ubiquitous "paradigm," has received only moderate attention, Thomas Kuhn showed how deceptive this view can be. Disagreements about when and by whom a discovery such as that of oxygen was made inevitably are not resolved, because "the sentence 'Oxygen was discovered' misleads by suggesting that discovering something is a single act assimilable to our usual (and also questionable) concept of seeing." Summarizing the several events to which this particular discovery has variously been attached by one historian or another, Kuhn concluded that "we can safely say that oxygen had not been discovered before 1774, and we would probably also say that it had been discovered by 1777 or shortly thereafter. But within those limits or others like them, any attempt to date the discovery must inevitably be arbitrary because discovering a new sort of phenomenon is necessarily a complex event.... Discovery is a process and must take time."[5] By the same argument, one cannot definitively attribute this discovery to Joseph Priestley, or Antoine-Laurent Lavoisier, or Carl Wilhelm Scheele but can only say that each of them,

3. Merton 1973, p. 400.
4. Hagstrom 1965, p. 69.
5. Kuhn 1970, pp. 53–55.

and some others as well, had a part in it. Many of the developments in science that have been treated as single discoveries were far more complex processes than the discovery of oxygen, and extended over far longer periods of time.

Even when a discovery can be pronounced complete, its meaning is not fixed. By 1777 oxygen had not yet acquired even its permanent name. Lavoisier called it at that point "the purest part of the air," and a little afterward "eminently respirable air," terms that clearly associated the new air with properties different from that implied by the term "acidifying principle," which he adopted a year later. The concept embedded in the latter phrase he hid within the permanent vocabulary of chemistry in 1780 by rendering it in Greek as "oxygen principle." Did the discovery of oxygen end at that point, or must it be extended in time to cover all of the subsequent changes in the meaning of what we understand by the word "oxygen"?

Discovery is, therefore, a more complex process than it is commonly taken to be, and the identities of discoveries are more fluid than the packages we commonly wrap around their historical advent suppose. These considerations are pertinent to the present assessment of the validity and value of the concept of premature discoveries. Retrospective recognition and classification of a discovery as premature implies that it is seen as equivalent to a subsequent discovery recognized in its own time. But how can we be sure that the earlier discovery was, in its own time, the same discovery that it later appeared to be? I illustrate this problem by discussing briefly the two cases on which Stent concentrated his own analysis.

MENDEL AND MENDELISM

The most famous case of discovery not appreciated in its own time is, as Stent points out, that of Gregor Mendel, whose paper "Experiments on Plant Hybrids," published in 1865, went almost unnoticed until its rediscovery in 1900. The most exceptional feature of Mendel's case is that, after so long an interval for canonical knowledge to catch up, Mendel's work was not only cited retrospectively, but it exerted the direct influence on subsequent investigation that eluded it in its own time. But what was Mendel's discovery? That there is some uncertainty about this is hinted at in Stent's own treatment. In the shorter version of his paper, published in *Scientific American,* he wrote that Mendel's "discovery of the gene in 1865 had to wait 35 years before it was 'rediscovered' at the turn of the century."[6] In the expanded version of the paper he changed this passage and referred to Mendel's "discovery of the particulate nature of heredity."[7]

In his paper Mendel mentioned neither genes nor the particulate nature of heredity, and it requires considerable hindsight to infer these conceptions from discussions in Mendel's paper that did not invoke these terms. In 1865 it would not have been possible for him or his contemporaries to describe in this manner what-

6. Stent 1972b, p. 86.
7. Stent 1972a, p. 437.

ever he did discover. In his own language his "task" was to "observe the development of . . . hybrids in their progeny."[8] Although he did not summarize the results of his experiments in any specific set of conclusions, he referred several times to "laws" of development formulated for the case of *Pisum,* whose applicability to other plants he intended to study further.[9]

Robert Olby has noted that the rediscoverers of Mendel's papers rephrased these laws in ways that attached to them meanings one cannot find in Mendel's own writing. In brief, they extended the applicability of the laws from the formation of hybrids to general laws of inheritance and provided these laws with language that implied the existence of paired "factors" in somatic cells whether or not the factors were associated with identical or contrasted characters (which they expressed in the terms "heterozygote" and "homozygote" that were coined at that time). They did this, as Olby shows, because they associated the factors with the cytological evidence for paired chromosomes. Mendel, who worked before chromosomes had even been observed, let alone identified as the potential material substrates for the factors, was not concerned with the nature of this substrate. The reason he wrote A + 2Aa + a, whereas the post-1900 Mendelians wrote AA + 2Aa + aa, is that he was basing his laws on the conformity of his results with a mathematical formalism, whereas they were connecting the same results to the concept of discrete units that entered into the composition of the germ cells and the zygotes. Even those who, like William Bateson, hesitated to accept the complete identification of these units—which Bateson called "factors," with chromosomes or portions thereof—conceived of them as particulate. Mendel had written that "constant characteristics that occur in different forms of a tribe of plants can, by means of repeated artificial fertilization, appear in all the unions that are possible according to the rules of combination." Bateson described the *process* through which all these associations result as "segregation"—a term that implies the particulate nature of what is being separated in the process. By these changes in language, Bateson and his contemporary Mendelians transformed Mendel's conclusions into the "discovery of particulate inheritance" that Stent's expanded paper attributes to Mendel himself.[10]

That Stent made no distinction between Mendel's discovery and later interpretations of his results was not because he blurred what had previously been clear but because the Mendelians themselves merged Mendel's laws with the further inferences they had drawn from them. Here is how Bateson put it:

> The fact of *segregation* was the essential discovery which Mendel made. As we now know, such segregation is one of the normal phenomena of nature. It is segregation which determines the regularity perceptible in the hereditary transmission of differences, and the definiteness or discontinuity so often conspicuous in the variation of animals and plants is a consequence of the same phenomenon. Segregation thus

8. Mendel 1865, p. 3.
9. Ibid., pp. 3, 32, 35, 42.
10. Ibid., p. 23; Olby 1985, pp. 234–58.

defines the units concerned in the constitution of organisms and provides the clue by which an analysis of the complex heterogeneity of living forms may be begun.[11]

In this passage Bateson has at once redefined one of the central laws Mendel proposed, and veiled the change by calling his own version of the law "the essential discovery which Mendel made." Is this a sleight of hand that Bateson made deliberately? My suggestion is that it was instead an unconscious distortion explainable by the general norm that scientists must be appropriately recognized for contributions counted as discoveries. Bateson was particularly concerned that Mendel receive such credit because it had so long been denied him. Because discoveries are seen in science as stable once they have been made, Bateson was probably unaware that, in fitting Mendel's discovery with a new language and associating it with concepts that had emerged since Mendel's time, he was turning it into a somewhat different discovery than the one Mendel had originally made.

Olby concluded that "the problem of the neglect of the *Versuche* is to a large extent a pseudo-problem. *When viewed within the context of the period,* Mendel's discussion of hybrids, though rigorous, brilliant and systematic, would not appear to have broken entirely fresh ground."[12] I do not agree that the problem is so easily solved. Placing Mendel in the context of the 1860s shows that he was not, without a great deal of later mediation, the founder of twentieth-century genetics, but it leaves untouched the question of why the experiments themselves, which all commentators agree were astonishingly original in their conception and execution, received so little attention. In fact, to show that Mendel was concerned with problems of plant breeding and the nature of hybrids that were characteristic of the mid–nineteenth century rather than the early twentieth only sharpens the problem, because it removes the explanation—that the requisite "canonical knowledge" with which to connect Mendel's work did not become available until later.

One way to elude the conclusion that Mendel did not discover the same laws that formed the foundation for twentieth-century Mendelian genetics would be to adopt the view that a discovery may include implications that are not apparent to its discoverer. If readers of Mendel's paper in 1900 and afterward found meanings in it that Mendel did not explicitly state, can we not say that these meanings were all along inherent in his discovery and were only "brought to light" by the availability of later canonical knowledge? That approach would suggest a further articulation of Stent's concept of prematurity: a discovery may be premature to such an extent that its full significance escapes even its own author.

I see no objection in principle to such a strategy, but it is a dangerous one, because it is very hard to know when we reach the limit of what we may attribute in hindsight to scientists by the light of knowledge they did not have. That is a recipe for what was called a few decades ago "precursoritis." To me, it seems more real-

---

11. Bateson 1913, p. 13.
12. Olby 1985, p. 253. Emphasis in original.

istic to apply Kuhn's view of discovery. The discovery of the gene, or of the particulate nature of heredity—if we accept Stent's formulation of what the discovery was—was a process extending over a little more than forty years, in which Mendel's experiments on plant hybridization, and the constant laws he inferred from them, played a large role.

Do these considerations cast doubt on Stent's general conception of prematurity in scientific discovery, do they eliminate Mendel's work as an example, or do they merely lead us to redefine the elements of prematurity in this case? Even in his own time, Mendel obviously discovered *something* that was little noticed immediately but was seen later as very important. The discovery closest to an immediate result of the experiments was that, in the offspring of the first hybrid generation the "constant forms" appeared in fixed ratios, and that these forms could be further segregated according to whether or not they gave rise to constant forms in the next generation, again in fixed ratios. These ratios being statistical, Stent has suggested that their significance was missed because statistical methods were not used in botany or biology in that era. That Mendel was familiar with them was due to his training in mathematics and physics. But it is not clear whether the contemporary lack of attention was a problem in timing or in disciplinary specialization. Crossing disciplines has often led in the history of science to innovative solutions to problems that were then not fully understood or appreciated by those in the field into which the interloper had crossed. If the canonical knowledge needed to connect the discovery with the current concerns of practitioners in a field was available, but in another field unfamiliar to them, prematurity may not be the best category to which to assign the reasons for their inattention.

AVERY AND TRANSFORMATION

The case that gave rise to Stent's conception of prematurity was, of course, that of Oswald Avery's place in the discovery that DNA is the genetic material. In attempting to answer the question "*Why* was Avery's discovery not appreciated in its day?" Stent came to the conclusion that its implications could not "be connected by a series of simple logical steps to contemporary canonical knowledge." Though the discovery did not go unnoticed, "no one seemed to be able to do much with it, or build upon it, except for the students of the transformation phenomenon. That is to say, Avery's discovery had virtually no effect on general genetic discourse."[13]

Stent's assertions were quickly challenged. Olby's *The Path to the Double Helix*, published in 1974, presented Avery's discovery as one whose significance was quickly grasped by such leading figures as Theodosius Dobzhansky, Hermann Muller, Sir Henry Dale, and Macfarlane Burnet; that was widely discussed; and that led to experiments in other laboratories that confirmed his results.[14] In a biography of

13. Stent 1972a, p. 434–35. (This paper is reprinted in part as chapter 2 in this volume.)
14. Olby 1974, pp. 181–206.

Avery published in 1976, René Dubos invoked the same events to argue that "contrary to Stent's assertion, the 'general discourse of genetics' was immediately affected by the view that DNA is involved in genetic phenomena."[15]

In 1979 Rollin Hotchkiss, Avery's last research assistant, presented the paper "The Identification of Nucleic Acids as Genetic Determinants," which also countered Stent's view. It was not Avery's discovery that was premature, according to Hotchkiss, but only some of the early generalizations drawn from it. The fact that it required "5–8 years" to go from Avery's DNA-transformation to the Watson-Crick replication model was not due to lack of appreciation or to inability to build on Avery's discovery, but simply to the "slow process . . . [of] imagining and designing the new kinds of experiments that could give force and generality to those ideas."[16] Erwin Chargaff has weighed in against Stent's interpretation by stressing that "Avery's discovery made" a "profound impression" on him, inducing him to take up the studies of DNA variability that led to the Chargaff ratios and, hence, to "the double helix proposal of Crick and Watson."[17]

In a rather heated reference to the "flamboyant theoretical declarations of the phage group," which included Stent, Dubos charged that *their* lack of appreciation for Avery's work on DNA was due to their preference for "cosmic riddles" over "living organisms," for their deprecation and avoidance of the kind of patient, disciplined biochemical research that Avery had employed to reach his proof that the transforming principle is DNA.[18] In a refreshingly candid admission of the subjectivity of his view, Stent had stated that he presented Avery's case as an example of premature discovery "mainly because of my own failure to appreciate it when I joined *Delbrück's* phage group and took the Cold Spring Harbor Phage course in 1948."[19] In an unpublished autobiography, he has further elucidated the nature of this experience. It was not merely his personal failure, but the fact that Avery's work was not discussed at Cold Spring Harbor then, that accounted for his belief that the work had "no effect on general genetic discourse." Leaving aside the polemical tone of Dubos's remarks, it does seem clear that the members of the phage group were, because of their own predilections, left out of a discourse that was taking place elsewhere.

As in the case of Mendel, therefore, the difficulty in connecting Avery's discovery to canonical knowledge appears to be more a disciplinary than a temporal problem. The two cases are also linked by problems in the retrospective definitions of what the discovery was. In a letter in which Carl Lamanna complained that in an earlier essay on the origins of molecular biology Stent had omitted to mention Avery or "DNA-mediated bacterial transformation," Lamanna described the 1944 paper by Avery, C. M. MacLeod, and M. McCarty as "the definitive proof of DNA

15. Dubos 1976, pp. 155–59.
16. Hotchkiss 1979, pp. 321–42.
17. Chargaff 1979, pp. 348, 354.
18. Dubos 1976, pp. 155–58.
19. Stent 1972a, p. 436.

as the basic hereditary substance."[20] Accepting both the criticism and this definition of what was at issue, Stent agreed that he "should have really mentioned ... Avery's proof in 1944 that DNA is the hereditary substance."

Readers of the original paper will know that Avery and his coauthors made no such claim. Their conclusion was: "The evidence supports the belief that a nucleic acid of the deoxyribose type is the fundamental unit of the transforming principle of Pneumococcus Type III." They explicitly anticipated the possibility that their proof could be interpreted as implying that DNA is the genetic material. "The inducing substance has been likened to a gene," they wrote, "and the capsular antigen which is produced in response to it has been regarded as a gene product." They did not, however, identify themselves with this interpretation, which they ascribed to Dobzhansky, and they discussed as an alternative possibility the analogy drawn by Wendell Stanley between the "activity of the transforming agent and that of a virus."[21]

Avery's supporters have argued that it was only his "scientific puritanism," his reticence about speculating beyond what his evidence proved, that prevented Avery from expressing publicly his private opinion that DNA was the hereditary material. But private opinions and insightful anticipations are not discoveries. It was only in the light of subsequent events that Avery's "discovery" could be redefined as the discovery that DNA is the hereditary material. As in the case of Mendel, it seems to be concern for the appropriate recognition of individual contributions to the advance of science that has induced scientists and historians alike to broaden Avery's demonstration of the nature of a bacterial transforming principle to the discovery that DNA is the "genetic material." This concern about individual recognition is vividly illustrated in Dubos's avid defense of Avery:

> During the late 1930's, Avery had been nominated for the Nobel Prize in recognition of his immunochemical studies. After the 1944 paper, the Nobel committee was immediately alerted to the fact that he had once more made a fundamental contribution to biological science. But the 1944 paper was ineffective from the public relations point of view; ... it failed to extrapolate from the role of DNA in a single bacterial species to the role of DNA in other living things. In other words, it did not make it obvious that the findings opened the door to a new era in biology. ... Yet, the very phenomenon of transformation, representing as it did the first example of directed change in hereditary characteristics, was in itself a biological landmark worthy of the Nobel Prize.[22]

Just as 35 years after publication of Mendel's work his laws of hybrid formation appeared to be general laws of heredity and to imply segregating particles, so Avery's identification of the transforming principle appeared a decade later to be the identification of the material that controls hereditary characteristics in

---

20. Lamanna 1968, 1397–98.
21. Avery et al. 1944, pp. 155.
22. Dubos 1976, p. 159.

general. Here again we may ask whether it is justified to read back into an earlier discovery implications that depend on later developments, or whether it is better to question whether the generalization that DNA is the genetic material can be attributed to any single discovery, if a discovery is taken to occur at a specific time and place. Rollin Hotchkiss took the longer view: "The DNA Revolution," he declared, "concerned the operational identification of the DNA molecule as the gene material itself. It took place in the quarter century of 1930 to 1956."[23]

If we compare this case to Kuhn's discussion of the discovery of oxygen, we may say that before 1930 it had not been discovered that DNA is the gene material, and that by 1956 it had been, but that to locate the discovery at any point in time between these boundaries would be arbitrary.

IDENTIFICATION OF CRITERIA FOR PREMATURITY

The cases of Mendel and of Avery are similar in that to each man has been attributed retrospectively a "discovery" that appears instead the outcome of more complex processes in which he played a part. The two discoveries differ in that one appears genuinely to have gone unappreciated in its own time, whereas the other was neglected only by a particular group within a wider scientific community. Stent's applications of his concept of prematurity require, therefore, revision. The question remains: If we are careful to define discoveries as their authors defined them, and if we examine all the available evidence concerning the contemporary attention they received, will we still be able to identify historical contributions to science that fit the criterion "A discovery is premature if its implications cannot be connected by a series of simple logical steps to contemporary canonical knowledge"?

This question should not be answered a priori. If there are few other examples as spectacular as that of Mendel of the contemporary neglect of work later seen to be of fundamental importance, historians should not dismiss the widespread feeling among scientists that some potentially important discoveries are overlooked, or resisted, or accepted only after abnormally long periods of delay. Before we can seek the reasons for individual instances of the kind, however, we need to examine some of the tacit assumptions underlying both the concept of prematurity and the intuition that the acceptance of some discoveries are "delayed." All of these ideas presume some "normal" rate of scientific advance and a normal lapse of time between the first presentation of a discovery and its assimilation into a field. Both the idea of premature discoveries and the converse proposal of Harriet Zuckerman and Joshua Lederberg that there are postmature discoveries imply deviations from a normal rate of progress, response, further examination, confirmation, and

23. Hotchkiss 1979, p. 321.

integration into canonical knowledge.[24] There are, however, no controls against which to measure the deviations. We may use citation indices to construct an average half-life for a scientific paper, and conclude that most discoveries are accepted within a certain time interval or forgotten. Any paper whose citation curve begins to rise only after a longer interval following its publication might be a candidate for prematurity. Aside from the fact that such information is available only for the very recent past, such a procedure may be the least efficient means to capture those events of large historical meaning that seem to be what Stent aimed to explain.

Whether the acceptance of a discovery appears to be rapid or slow, accelerated or delayed, or whether it is met with enthusiasm or resistance, depends not only on the interval measured in months, years, or scientific generations but also on the subjective perspectives of those involved and of those who interpret such events historically. It was for a long time customary among historians who described William Harvey's discovery of the circulation to emphasize the opposition it received and the long time that elapsed between the publication of *De Motu Cordis* in 1628 and the general acceptance of the circulation by the 1650s. That viewpoint fit the image Harvey himself promulgated, one sustained until recently, that, as the first modern, he confronted the dead weight of a canonical tradition into which the circulation did not fit. Looking at the same events today, and taking into account the pace of investigation in the seventeenth century and the extent of revision necessary to adapt older views to the discovery, we may in fact be astonished at how quickly others came to support, confirm, and extend his discovery.

Similarly, accounts of the reception of Lavoisier's oxygen theory have often highlighted the long campaign that he and his followers had to carry on before his discoveries were accepted into the canonical framework of chemistry. If, however, we compare the length of time between Lavoisier's initial announcement of a very imperfect theory of combustion in 1773 and the discovery of the composition of water in the early 1780s—which overcame the last remaining deep difficulties in his theoretical structure—with the additional time necessary to win over the majority of chemists to the new view, then the latter process also appears relatively rapid.

As these examples suggest, it is not an easy matter to decide when the acceptance of a discovery is sufficiently "delayed" to make it a candidate for "prematurity." Nevertheless, it is still a worthwhile task to search for unequivocal historical examples of discoveries unappreciated in their own time but later recognized to be important, and to test them against Gunther Stent's stimulating contention that they will prove to lack simple connections to contemporary canonical knowledge. Whether the outcome of such searches proves to be positive or negative, the effort can still illuminate our understanding of the dynamics of scientific change.

24. Zuckerman and Lederberg 1986.

BIBLIOGRAPHY

Avery, O. T., C. M. MacLeod, and M. McCarty. 1944. "Studies on the Chemical Nature of the Substance Inducing Transformation of Pneumococcal Types," *Journal of Experimental Medicine* 79:137–57.

Bateson, W. 1913. *Mendel's Principles of Heredity.* Cambridge: Cambridge University Press.

Chargaff, E. 1979. "How Genetics Got a Chemical Education." In *The Origins of Modern Biochemistry: A Retrospect on Proteins,* ed. P. R. Srinivasan, J. S. Fruton, and J. T. Edsall. New York: New York Academy of Sciences.

Dubos, R. J. 1976. *The Professor the Institute and DNA.* New York: Rockefeller University Press.

Hagstrom, W. O. 1965. *The Scientific Community.* New York: Basic Books.

Hotchkiss, R. D. 1979. "The Identification of Nucleic Acids as Genetic Determinants." In *The Origins of Modern Biochemistry: A Retrospect on Proteins,* ed. P. R. Srinivasan, J. S. Fruton, and J. T. Edsall. New York: New York Academy of Sciences.

Kuhn, T. S. 1970. *The Structure of Scientific Revolutions.* 2d ed. Chicago: University of Chicago Press.

Lamanna, Carl. 1968. "DNA Discovery in Perspective." *Science* 160:1397–98.

Mendel, G. 1865. "Versuche über Pflanzen-Hybriden." *Verhandlungen des naturforschenden Vereines in Brünn 4: Abhandlungen,* 1–47.

Merton, R. K. 1973. *The Sociology of Science: Theoretical and Empirical Investigations.* Ed. N. W Storer. Chicago: University of Chicago Press.

Olby, R. 1974. *The Path to the Double Helix.* Seattle: University of Washington Press.

———. 1985. *Mendelism.* 2d ed. Chicago: University of Chicago Press.

Stent, G. S. 1972a. "Prematurity and Uniqueness in Scientific Discovery." *Advances in the Biosciences* 8:433–49.

———. 1972b. "Prematurity and Uniqueness in Scientific Discovery." *Scientific American* 227 (December): 84–93.

Zuckerman, H. A., and J. Lederberg. 1986. "Postmature Scientific Discovery?" *Nature* 324:629–31.

CHAPTER THIRTEEN

Barbara McClintock's Controlling Elements

Premature Discovery or Stillborn Theory?

Nathaniel C. Comfort

Barbara McClintock's discovery of movable genetic elements seems to provide a case study in prematurity.[1] The standard version of the story was first articulated in Evelyn Fox Keller's widely read biography.[2] In this version, McClintock's discovery in the late 1940s that genetic elements in maize (Indian corn) could transpose, or move, was met with an initial burst of skepticism and derision. The discovery was greeted with "stony silence"; it "fell like a lead balloon"; "with one or two exceptions, no one understood."[3] Scientists reacted to her findings with "puzzlement, frustration, even hostility."[4] This skepticism soon settled into a humiliating silence. Transposition, it has been said, was too shocking for most scientists to accept; it challenged the "beads-on-a-string" theory of the gene, which held that genes must be static, independent, autonomous units. Geneticists then allegedly ignored McClintock and "marginalized" her in the scientific community.[5]

In the years following Keller's book, the standard version was elaborated on, romanticizing McClintock as a loner and a martyr to truth, toiling away unrecognized. She "stood for a long time on the periphery of her field"[6]—an unproductive vista for a corn geneticist—working "virtually in isolation for over thirty years, holding to her vision of . . . genetics, before her work was recognized. Over the decades, she persisted with minimal funding."[7]

---

1. See Stent, chapter 2 in this volume.
2. Keller 1983.
3. The first and third quotations are from Keller 1983, p. 139; the second is from McGrayne 1993, p. 169.
4. McGrayne 1993, p. 169.
5. Keller 1985, pp. 160, 173.
6. Shteir 1987, p. 31.
7. Shepherd 1993, p. 88.

In this last passage, eventual—and overdue—recognition is implied. In the late 1970s, the standard version continues, molecular biologists discovered transposition in bacteria. The genetics community could no longer ignore McClintock's findings. They at last admitted she had been right all along and soon began awarding her prestigious scientific prizes, culminating in the 1983 Nobel Prize in Physiology or Medicine. In the standard version, McClintock has become the "belatedly named Nobel laureate" who was "not adequately recognized in science."[8] This standard version has been extended by numerous biographers, feminist authors critiquing "macho" science or celebrating "feminine" science, a gamut of scientists from the hard-nosed to the soft-hearted, and even children's writers.[9]

The concept of prematurity is implicit in the standard version. The language of the McClintock story is that of prematurity: "ignored"; "ahead of her time"; "overdue recognition"; "belatedly named Nobel laureate." The diagnosis of prematurity in McClintock's case rests on two assumptions: first, that McClintock was ignored and her work was not accepted in the 1950s and 1960s; second, that her work was accepted in the late 1970s. These in turn rest on the assumption that the concept or evidence that was rejected in the 1950s was the same that was later accepted.

None of these assumptions is valid. Examination of newly available archival collections, especially McClintock's research notes but also the notes and correspondence of her closest associates, forces a reinterpretation of the standard narrative of the discovery of movable genetic elements. This interpretation is supported by interviews with McClintock and her friends and colleagues and by close reading of her published work. McClintock's evidence for transposition was accepted immediately. Her interpretation of its wider significance, however, was and remains doubtful in most scientists' minds. The prematurity argument about McClintock founders on two points: She was neither ignored at the time nor "proven right" later.

A SYSTEM OF CONTROL

McClintock discovered transposition while engaged in a systematic research program that she had been pursuing since the 1930s and which was well-known to her colleagues. The discovery was not disembodied from canonical knowledge or prevailing theories. For McClintock, movement of genetic elements was never especially fascinating in and of itself. It always had a wider meaning. Initially, transposition was a means of explaining the long-standing genetic problem of spontaneous mutations.

---

8. Shteir 1987, p. 31; Rose 1994, p. 162.

9. For the biographies, see Dash 1991; Felder 1996; McGrayne 1993, pp. 144–75; Opfell 1986; Shiels 1985. For the feminist authors, see Arianrhod 1992, p. 43; Keller 1983, p. 139; 1985, p. 154; Morse 1995, p. 12; Rose 1994, pp. 157, 163; Shepherd 1993, p. 88; Shteir 1987, p. 31; Schiebinger 1987, p. 16. For the hard-nosed scientists, see Fedoroff and Botstein 1992; Fincham 1992; Spradling 1993. For the soft-hearted scientists, see Hofstetter 1992; MacColl 1989. For the children's writers, see Heiligman 1994; Fine 1998; Kittredge 1991.

In 1942 McClintock joined the permanent staff of the Carnegie Institution of Washington's Department of Genetics at Cold Spring Harbor, New York. She was by this time one of the nation's leading geneticists and one of the top two or three among those working on maize. McClintock had been the first to distinguish among the ten chromosomes of maize and had developed histological and microscopic techniques for observing them. By inaugurating the microscopic study of maize chromosomes, she accomplished for maize what Thomas Hunt Morgan's group had done for *Drosophila*, enabling cytology to be wedded to genetics, resulting in the powerful set of tools known as cytogenetics. Using these techniques, she and colleague Harriet Creighton had provided a critical, if long-expected, confirmation that the exchange of genes between genetic linkage groups corresponded to physical exchange of chromosomal regions.[10] Genetic linkage in fact represented physical proximity of genes. Subsequent research papers strengthened her reputation as a remarkably perceptive observer and a reliable, imaginative experimentalist.

Her work at Cold Spring Harbor extended research she had begun at the University of Missouri, where she had examined the cytology and genetics of maize. Her most striking discovery at Missouri was the breakage-fusion-bridge (BFB) cycle. (See fig. 13.1.) In the BFB cycle, specially constructed chromosomes undergo chromosome breakage and refusion in successive cell divisions. Repeated cycles of BFB can result in a welter of unusual chromosomal phenomena. Although McClintock first found BFB in strains of corn that had been x-rayed, she soon developed special stocks that underwent the BFB cycle spontaneously. Since 1938, she had been studying the BFB cycle and its attendant effects.[11] Her work on it was recognized as a tour de force and strengthened her reputation as one of the leading cytogeneticists in the country.

In the summer of 1944, McClintock performed an experiment designed to exploit the BFB cycle as a substitute for x rays or chemicals in producing mutations. The resulting progeny had such an abundance of novel and bizarre mutations that even McClintock was surprised. "It had gone wild," she recalled years later. "The genome had gone wild."[12]

Among the swarm of new mutations was a cluster of so-called mutable genes. These variant forms of a gene appear to switch on and off as the plant grows, resulting in streaks, spots, or other variegations. Though well-known in *Drosophila*, mutable genes were rare in maize and so were especially noteworthy. Among these, however, was an even odder gene: rather than produce a visible trait, such as leaf color, this gene caused the chromosome to break. McClintock soon realized that, in order to act, it required a second gene that caused it to "go into action."[13] She

10. Creighton and McClintock 1931.

11. McClintock 1938, 1941, 1942.

12. McClintock 1980. For a detailed analysis of McClintock's discovery of mutable genes and transposition, see Comfort 2001, esp. chaps. 4–6.

13. Folder "Chromosome loss—unclassified" and Folder "Staff meeting, 10/28/46," Box 8, Series V, Barbara McClintock Collection, American Philosophical Society Library, Philadelphia (hereafter MC).

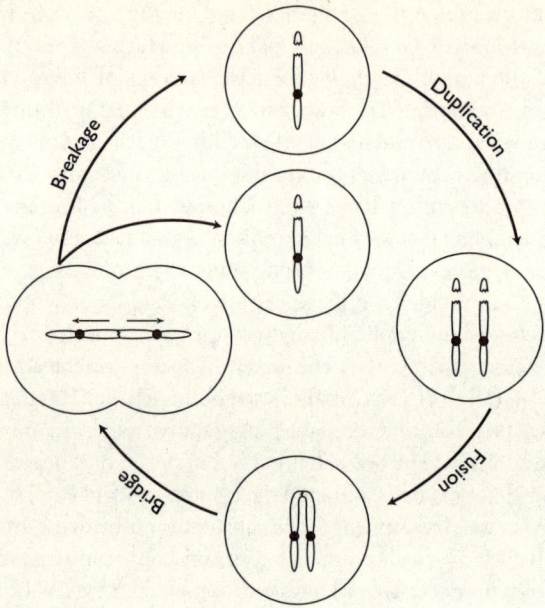

Figure 13.1. Breakage-fusion-bridge cycle.

named the chromosome breakage gene *Ds,* for "dissociation," and the auxiliary gene *Ac,* for "activation."[14]

Ds was easy to map; by the summer of 1947 she had it well-located.[15] *Ac,* however, proved puzzling. By early 1948, all she could say was that it was unlinked to *Ds,* meaning it lay either far away on the same chromosome or on another chromosome altogether. Then even that certainty eroded: she found plants in which *Ac* appeared to lie very close to *Ds.* Where was it? Far or near? At the same time, *Ds* began to show strange behavior. Another new mutable gene had appeared, one that seemed to interact with *Ds:* when it was present, *Ds* seemed to disappear or inactivate, and vice versa. Did the two mutable genes inhibit one another? Was *Ds* in two places at once?[16]

In the spring of 1948, McClintock recognized that one solution solved both sets of problems. The odd behavior of both *Ds* and *Ac* could be explained if the two genes

14. In October 1946, McClintock named the elements *D* for "dissociation" and *V* for "variegation," according to a convention established in the 1920s by Rollins Emerson. By fall 1947 she had changed the notation to *Ds* and *Ac.*

15. Folder "Summer—mid-summer conclusions," Box 8, Series V, MC.

16. Handwritten notes, 20 January 1948, Folder "Modifiers of *Ds* action," Box 8, Series V, MC.

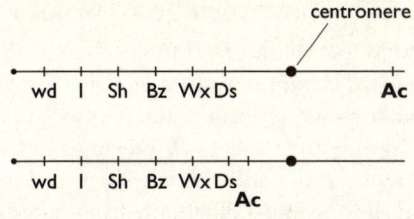

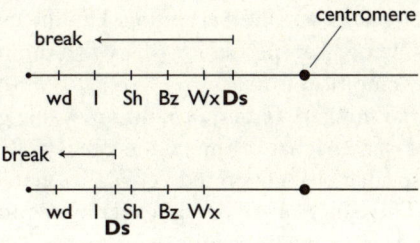

Figure 13.2. Transposition.

physically moved on the chromosomes from one site to another.[17] (See fig. 13.2.) Mc-Clintock adopted for this behavior the genetic term "transposition," used earlier to describe large-scale chromosomal alterations such as translocation and inversion. Transposition per se does not appear to have come particularly as a shock to Mc-Clintock. Her research notes indicate no great surprise at the fact of movable genetic elements. In writing about and discussing her work, she described the mechanism of transposition in conventional terms, as a breakage and refusion of chromosome ends, familiar by that time to all geneticists.[18] It was transposition's implications that evoked her enthusiasm.

McClintock recognized immediately that transposition explained the continued appearance of new mutable genes. Transposition of *Ac* or *Ds* in or next to a normal gene could inhibit that gene, causing an appearance in the plant indistinguishable from a mutation. But if the *Ac* or *Ds* jumped back out again, the affected gene would be disinhibited and would revert to its dominant condition. A single transposition would result in a single mutation. Repeated transpositions in and out of the same gene would result in a mutable form of the gene.[19]

17. First description of transposition of *Ac:* "Main problems 1948," Box 9, Series V, MC; transposition of *Ds:* untitled folder, 28 April 1948, Box 15, Series V, MC.

18. McClintock 1950a, pp. 161–63.

19. "June 1948 Essay," Box 9, Series V, MC.

A THEORY BY WHICH TO WORK

McClintock's recognition that *Ac* and *Ds* transposed immediately began to break her loose from the standard beads-on-a-string model of the genes. When she realized that *Ds* had moved from the *Ds* locus to the *C* locus, the very term "locus" became less meaningful. She began to refer to *Ds* as being in "standard position" when it occurred at the location to which she had first mapped it.[20] She soon became convinced that *Ds* and *Ac* functioned differently from other genes. Between 1948 and 1950, McClintock used movable genetic elements to explain an increasingly broad spectrum of biological phenomena: first, mutable genes; then, all mutations; finally, biological development and even evolution. Her discovery of transposition set her on a course of theoretical speculation that was a dramatic departure from her early, meticulously empirical work.

McClintock began to think of *Ds* and *Ac* not as movable genes but as another class of chromosomal element that regulated the genes.[21] By transposing in and out of genes, these movable elements could "create" a mutable gene out of any normal, stable gene. This, she realized, could explain the production of mutable genes not merely in maize but in all organisms.

Other geneticists welcomed these results. McClintock's theory of mutable genes seemed to fill a gap in the accepted theory of mutation. Mutations were generally believed to be produced by physicochemical changes to the gene molecule. Most mutations are one-way losses of function, which could be explained by damage to the gene. Hermann J. Muller, who in 1946 won a Nobel Prize for his work demonstrating x-ray-induced genetic mutations, championed the chemical-change-in-the-gene model. Mutable genes, however, posed a problem. They lose and regain function repeatedly. To account for mutable genes under the chemical-change model, "lightning"—x rays, ultraviolet radiation, chemical mutagens, and so forth—would have to strike many times at the same minute chromosomal spot, damaging and repairing the gene. Probability militated against that. How then to account for spontaneously mutating genes under the chemical-change model of mutation?

When Muller heard about McClintock's new findings on mutable genes, he believed she had solved the problem. He learned of her results in late 1948, when the phage geneticist Salvador Luria, Muller's colleague at Indiana University, gave a seminar on McClintock's work based on a letter McClintock wrote to the cotton geneticist S. G. Stephens. Stephens was affiliated with the University of Texas, but had spent a sabbatical at Cold Spring Harbor and befriended McClintock. His work on pseudoalleles—families of genes acting with related functions—fascinated her. Luria got hold of McClintock's long letter to Stephens, in which she not only

20. For example, "The Ds locus—Report in detail [outline]," undated, but probably 1949, Box 7, Series V, MC.

21. McClintock to S. G. Stephens, 28 June 1948, Box 12, Marcus Rhoades Collection, Lilly Library, Indiana University, Bloomington (hereafter RC); "Outlines, June 1948—essay," Box 9, Series V, MC. See Comfort 2001, chap. 6, for details.

summarized her recent findings but also elaborated her theory of mutable genes. When Muller heard Luria's seminar, he was so excited he bolted out of the room and headed straight for Western Union: "I tried to telegraph you immediately afterwards," he wrote her, "but the telegraph office was apparently closed so I am sending you this now."[22] He congratulated her on her "magnificent achievement" in solving the old problem of mutable genes—"at least in maize." McClintock responded with an annotated copy of her forthcoming Carnegie report, written in August, and photographs illustrating some of the kernel and leaf patterns produced by the controlling units.[23]

By 1950, she had extended her theory even further. Adopting an idea then in fashion among geneticists, she argued that her movable elements were made of heterochromatin, a dark-staining type of chromosomal material associated with the well-known phenomenon called position effect.[24] Since 1923 it had been known that genes located near heterochromatin were silenced.[25] This effect was noticed when a large-scale movement, such as a translocation or inversion, brought a gene into proximity with heterochromatin. McClintock inverted the mechanism and brought the heterochromatin to the genes instead of the other way around. Her movable elements, she reasoned, must be tiny bits of heterochromatin somehow invested with the power of movement. By jumping in and out of genes, *Ds* and *Ac* could inhibit and disinhibit gene action. She thus made two leaps from accepted heterochromatin theory: her submicroscopic pieces of heterochromatin moved, going to the genes rather than having the genes come to them; and this movement was systematic, not random.

Thinking about these elements as being controlled led McClintock to a solution to a problem outside the domain of mainstream 1940s genetics. Embryologists had long puzzled over the relationship between genetics and biological development. Conventional genetics treated genes as autonomous units with binary, on/off action. Yet embryologists had shown that all cells in the body come from a single progenitor and must therefore share the same set of genes. If all cells have the same genes and genes are always "on," how does the organism produce different tissues? The embryologists looked outside the nucleus for the answer: they argued that the cytoplasm must affect gene action. Different cellular environments would produce different patterns of gene activity, resulting in the differentiation of various tissues during development.[26]

McClintock was familiar with the problems of embryology. Her earliest interest in biology had been in zoology and embryology, and in 1932 she had gone to

22. Muller to McClintock, 27 October 1948, Muller Correspondence, Lilly Library, Indiana University, Bloomington (hereafter LL).

23. McClintock to Muller, 12 November 1948, Muller Correspondence, LL.

24. For the internal history of the position effect, see Carlson 1966, chaps. 13–15; Dunn 1965, pp. 162–63.

25. Sturtevant and Morgan 1923; Sturtevant 1925.

26. See Sapp 1987, chaps. 2 and 3; 1994; Gilbert 1978.

Berlin to work with the iconoclastic Richard Goldschmidt, whose 1938 *Physiological Genetics* would attempt to reconcile embryology and genetics.[27] Although McClintock's investigations since the thirties had led her away from such questions, she found herself once again confronting the problem of gene action during development. It was a problem most geneticists felt was, for the time being, insoluble. The great geneticist Thomas Hunt Morgan was quoted as saying, "Except for the rare case of plastid [i.e., chloroplast, known then to have its own genome] inheritance[,] all known characters can be sufficiently accounted for by the presence of genes in chromosomes. In a word, the cytoplasm may be ignored genetically."[28]

McClintock borrowed a problem from the embryologists, but unlike them she looked within the nucleus for the seat of gene regulation. Her bits of movable heterochromatin could be a means for controlling gene action. For this to occur, the transpositions would have to be nonrandom; something had to control them. The manner in which normal gene action in her plants seemed to have been destabilized suggested to her that such a control system did indeed exist, and that she had inadvertently disrupted it during the 1944 experiment. In a manic burst of letters to Marcus Rhoades in the spring of 1950, McClintock outlined a theory of what she began referring to as "controlling units" and later "controlling elements."[29]

RECEPTION

It was this theory of genetic control by transposition, rather than transposition per se, that met resistance from the genetics community. Transposition seems to have been accepted almost immediately, and largely on faith. It was some time before McClintock made any of her data public. She deliberately withheld her data in her first journal article on the phenomenon, in the *Proceedings of the National Academy of Sciences (PNAS)* in 1950. While preparing this paper, she explained to Marcus Rhoades, "I should like very much to get this information distributed about as soon as it is in the state to do so without going through a detailed discussion with all the data."[30] The 1950 paper was indeed strikingly data-free. It contained no tables, no photographs, no micrographs, no diagrams. She summarized her findings in narrative form, trusting her reader to trust her. She planned to present a full account with all the data, and she even wrote several chapters of a large monograph, but never completed it; only Rhoades and his students ever saw them.[31]

McClintock's paper at the Cold Spring Harbor Symposium the next year convinced her colleagues of transposition. That paper is a cornerstone of the standard

27. Goldschmidt 1938. On Goldschmidt, see Dietrich 1995, pp. 436–38; 1996; Gilbert 1988.

28. Morgan (1926), quoted in Sapp 1994, p. 231.

29. McClintock to Rhoades, 3 March 1950, 4 March 1950, 3 April 1950, 12 April 1950, Folder 21, Box 12, RC.

30. McClintock to Marcus Rhoades, 3 April 1950, Folder 21, Box 12, RC.

31. These chapters may be found in the Rhoades Collection at the Lilly Library of Indiana University and the McClintock Collection at the American Philosophical Society Library.

version of the McClintock story, crucial to accepting the idea that transposition was premature. The standard narrative holds that McClintock was surprised by and disappointed in the reception to her work on movable elements. This was the talk, Keller writes, that was received with "stony silence. With one or two exceptions, no one understood."[32] McClintock's "disappointment must have been enormous," Keller continues.[33] Keller's interpretation was drawn from the source: she quotes McClintock in a 1979 interview as saying, "It was just a surprise that I couldn't communicate; it was a surprise that I was being ridiculed, or being told that I was really mad."[34]

In 1951, however, McClintock had not expected her theory to be understood.[35] Only a year before, she had told Marcus Rhoades she was uncomfortable about making this speculative "dive into the deep,"[36] but she felt strongly enough that genetics needed a shake-up that she went ahead with it. In McClintock's mind, the Cold Spring Harbor paper, too, was an argument for genetic control. It was, she said later, "a very important paper in the way that I tried to bring out that genes had to be controlled."[37] Though McClintock recalled the 1951 meeting as a severe disappointment, she clearly developed this chagrin in hindsight.

In the standard version, no one in the audience at the 1951 Cold Spring Harbor meeting had as thorough a knowledge of maize as McClintock.[38] Perhaps so, but the room was filled with dozens of scientists whose understanding was sufficient to appreciate her arguments. Many had heard her mutable-gene story before, had talked with her extensively, and had seen her slides and kernels. Although the corn and fly geneticists were fewer in number than in previous years, they were represented by some of the leaders in the field, among them L. F. Randolph, Lewis Stadler, Jack Schultz, and Milislav Demerec. Most drosophilists at the time had at least a passing familiarity with the maize literature and researchers. Maize and fruitfly researchers formed sister communities that had grown up together and that, by the 1950s, were growing old together: their organisms were the core of what was rapidly becoming classical genetics.

Genetics was still a small enough field that even many bacterial and viral geneticists were familiar with maize research. Joshua Lederberg and his student Norton Zinder, for example, came to Cold Spring Harbor from Madison, Wisconsin, where, Zinder remembered years later, not only were he and the other bacterial geneticists regularly called out into the cornfields to help during harvest or planting, but they were familiar with the analyses and arguments of the maize geneti-

32. Keller 1983, p. 139.

33. Ibid.

34. Ibid., p. 140.

35. For example, McClintock to Rhoades, 3 April 1950; McClintock to Rhoades, 6 July 1950, Folder 21, Box 12, RC.

36. McClintock to Rhoades, 6 July 1950, Folder 21, Box 12, RC.

37. McClintock 1980.

38. For example, Keller 1983, p. 145; Dash 1991, p. 90.

cists.[39] Milislav Demerec, director of the laboratories at Cold Spring Harbor, was at that time working primarily in bacterial genetics but had worked on *Drosophila* for years and had trained in maize genetics. Cross-talk among the subdisciplines was widespread. Many others in McClintock's audience had other connections to her or to maize genetics and, at least in general terms, knew of her new findings on mutable genes. In all, more than 30 attendees at the 1951 symposium are known, or can be safely assumed, to have been familiar with at least the outlines of Mc-Clintock's recent work.[40]

One of these, Royal Alexander Brink, had already seen transposition himself, in his own maize plants. In 1950, Brink's graduate student Robert A. Nilan had shown that a movable element like McClintock's *Ac* was responsible for the variegations at another, well-known locus, the *P* (pericarp) locus. Brink and Nilan called their new element "modulator," a more purely descriptive term, they felt, than Mc-Clintock's "controller." Nilan defended his thesis in the spring of 1951 and they published the work in 1952.[41] In the 1952 paper, they noted that modulator conformed "in certain respects also to the genetic behavior of the mutable loci . . . ingeniously analyzed by McClintock (1950)."[42]

Though Brink clearly did not doubt the reality of transposition, neither did he accept McClintock's interpretation of movable elements as controllers of development. Though his admiration for McClintock never flagged, he debated with her throughout the 1950s over her interpretation. In a valedictory review in 1960, he argued that development controlled transposition, rather than the other way around. One could, he wrote, "turn McClintock's controller argument around, without compromising the facts, and assume that what she considers primary causes are, themselves, effects."[43] According to one of Brink's former students, Brink questioned what he saw as McClintock's overly speculative interpretation of her "controlling elements." He adopted the term "transposable elements" to contrast his views with hers.[44] Brink's view, the more conservative interpretation, was widely adopted.

In 1953 transposition was confirmed again. The young Iowa maize geneticist Peter Peterson, working with maize strains irradiated by the atomic blasts at Bikini atoll, discovered another transposable element, which he called *En*, for "enhancer."[45] The *En* element operated at the *pale-green* locus and was clearly distinct

39. Zinder 1996.

40. The list includes Edgar Altenburg, David Bonner, Royal A. Brink, Vernon Bryson, Ernst Caspari, Harriet Creighton, Milislav Demerec, E. H. Dollinger, Margaret Emmerling, Boris Ephrussi, Harriet Ephrussi-Taylor, Bentley Glass, Richard Goldschmidt, Alfred Hershey, Rollin Hotchkiss, Esther Lederberg, Joshua Lederberg, Ed Lewis, Salvador Luria, Hermann J. Muller, Kenneth Paigen, L. F. Randolph, Ruth Sager, Tracy Sonneborn, Lewis Stadler, S. G. Stephens, Patricia St. Lawrence, Waclaw Szybalski, Martha Taylor, Bruce Wallace, Evelyn Witkin, and Norton Zinder.

41. Nilan 1951; Brink and Nilan 1952.

42. Brink and Nilan 1952.

43. Brink 1960.

44. Kermicle 1996; Brink 1960.

45. Peterson 1953.

from McClintock's elements. Yet Peterson, like Brink, avoided use of McClintock's "controller" terminology.

In interviews in the 1990s, a number of scientists expressed puzzlement at McClintock's theory of developmental control, but wholesale acceptance of transposition. "She was talking about regulation and development," Zinder said in 1996. "That's what she was [annoyed with us about], and nobody would take that seriously. *Everybody* took seriously the transposition. What she wanted was that transposition was the regulatory control of development. And nobody there, or at least the very bright geneticists in the room, could understand how stochastic processes could really be involved in something as organized as developmental regulatory stages."[46] This reaction, which recurred in many interviews, came unprompted and was unexpected. It is not part of the standard story about McClintock.

In 1953, McClintock demonstrated that a transposable or controlling element could cause mutability at Rhoades's *Dotted* locus,[47] and the next year she discovered a new controlling element, which she called *Suppressor-mutator* (this was later found to be identical with Peterson's *En*). Though transposition had not been confirmed outside maize, it was by the 1950s a well-established phenomenon with ample experimental evidence. Whether it was general to all organisms—and whether it was adaptive—remained open an question.

McClintock was not ignored during the 1950s and 1960s. She was invited to give numerous talks and lectures at prestigious universities, among them an entire semester's worth to the biology graduate students at the California Institute of Technology.[48] She lectured at major universities through the 1950s and 1960s, including the big agricultural schools of the Midwest and many Ivy League schools of the East. Every one of these talks concerned genetic control, gene action, or developmental regulation; none focused solely on the fact and mechanism of transposition.[49] Her mutable-gene work was cited in major review articles, such as the valedictory address by the great maize geneticist Lewis Stadler in 1954, and in at least four textbooks on genetics.[50] J. A. Peters included her 1950 *PNAS* article in his 1959 anthology of *Classic Papers in Genetics*.[51] A number of scientists, especially younger researchers, began using her controlling elements in their own experiments.[52] In short, her argument was heard, her data were accepted, but her interpretation was doubted.

46. Zinder 1996.

47. McClintock 1953.

48. "Caltech lectures 1954," Box Ac-Chromosoma, Series III, MC.

49. Lecture notes are located in Series III, MC.

50. Stadler 1954, p. 818. The textbooks on genetics include Serra 1965, p. 454; Sager and Ryan 1961; King 1962, pp. 187–88; Srb et al. 1965, p. 341.

51. Peters 1959.

52. Some examples from the maize literature employing or considering controlling elements include Nuffer 1955, 1961; Emmerling 1958; Laughnan 1955; Schwartz 1960a,b, 1962; Fabergé 1958; Manglesdorf 1958, p. 419; Coe 1962; Nelson 1959.

GOING MOLECULAR

The argument for prematurity requires that a discovery ultimately "mature" or become accepted among scientists. The standard version of the McClintock story holds that after a long period of being ignored, molecular biologists at last discovered transposition in their own organisms and realized she had been right all along. The evidence for this apparent turnaround seems considerable. A 1976 meeting at Cold Spring Harbor was organized on the topic of insertion elements, plasmids, and episomes and featured many references to and acknowledgments of McClintock. Citations of her work began to increase. (See table 13.1.) Her four major publications of the 1950s were cited a total of 70 times between 1970 and 1974, 166 times between 1975 and 1979, and 308 times between 1980 and 1984. Her two reviews from the Cold Spring Harbor symposia of 1951 and 1956 represent 85 percent of the 308 citations during the years 1980 to 1984 and account for most of the increase since 1970.[53] Tellingly, citations of her 1953 *Genetics* article, the paper that she believed conclusively demonstrated her controller theory, showed no increase between 1970 and 1984 and accounted for just 2 percent of the 1980 to 1984 total. Five years later, in 1981, she won a host of major awards, followed two years later by the most prestigious award in science: an unshared Nobel Prize, in the category Physiology or Medicine. After decades during which, by her own admission, she declined to publish because she felt no one was listening to her, this string of awards and honors indeed marked a remarkable change in fortune.[54]

Underlying the surface events, however, is a story both simpler and more complex. It is simpler in that the consensus view of McClintock's work did not change substantially. It is more complex in that the later molecular biologists redefined transposition so that it became consonant with contemporary theory, rejecting McClintock's interpretation and adopting a new one. Thus redefined, transposition—but not developmental control by transposition—entered the canon of accepted biological fact.

Genetic control became canonical with Francois Jacob and Jacques Monod's operon model of the gene. The operon, proposed in 1961, was part of a large body of work on so-called temperate bacterial viruses, or bacteriophages, by Jacob, Monod, André Lwoff, Elie Wollman, and others at Paris's Pasteur Institute. In contrast to the virulent phages studied by the famous American phage group, temperate phages lysogenize, or integrate into the host chromosome and remain dormant. The operon model offers a simple, linear mechanism for genetic control. (See fig. 13.3.) It posits a structural gene, responsible for making a protein, flanked by controlling regions. Special proteins bind to controlling regions to switch the gene on and off, regulating its production of protein. The impact of the operon's discovery was immediate and immense. It sparked a flurry of new research and earned

---

53. Numbers tabulated from the *Science Citation Index*. The four papers mentioned are McClintock 1950b, 1951, 1953, and 1956.

54. McClintock n.d.

TABLE 13.1 Number of Citations of McClintock's Major
Transposition Papers, 1970–1984, by Five-Year Intervals

	Source				
	Proceedings of the National Academy of Sciences (1950)	Cold Spring Harbor Symposia (1951)	Genetics (1953)	Cold Spring Harbor Symposia (1956)	Total
1970–1974	0	32	6	32	70
1975–1979	23	72	2	69	166
1980–1984	38	141	6	123	308

its discoverers a Nobel Prize in 1965, just four years after publication. By any measure, the operon was rapidly integrated into the canon of accepted fact.

When Jacob and Monod published the operon model, McClintock "went through the ceiling" with delight.[55] She felt the bacterial community had at last recognized the fact of gene regulation. The operon model had no transposition in it, but to McClintock this was unimportant. "There wasn't any transposition" in the parallels between controlling elements and the operon, she said. "The main thing was control. . . . They showed that there was a control."[56] McClintock published an article in the *American Naturalist* in 1961 describing these parallels.[57] To gain acceptance for her theory, she for a time called her elements "operators" and "regulators."[58] Jacob and Monod acknowledged McClintock, but argued that her elements were more likely examples of "episomes"—nonchromosomal elements that bound to chromosomes—than parts of the chromosomes themselves.[59]

Ironically, it was the operon and related bacterial genetics, not McClintock's model, that gave rise to the modern conception of transposition. One of the central problems that emerged from the Pasteur Institute corpus was that of the integration of temperate viruses such as bacteriophage λ into the host chromosome. Did the virus insert into the host genome, or did it remain a distinct episome? If it inserted, were insertion and excision the same molecular reaction run in opposite directions, or were they distinct processes? In 1959 and possibly later, Jacob argued

55. McClintock 1980.
56. Ibid.
57. McClintock 1961.
58. Staff meeting, 28 November 1960, Folder "Cold Spring Harbor Lectures," Box "Chromosome Const. of R. of M.—Ms. #5—Dis. & Char. of Trans. El. #1, Series III, MC; McClintock to Oliver Nelson, 15 April 1965; McClintock to Oliver Nelson, 28 March 1968, Folder "Nelson, O. E. #1," Box "Mo—Ne," Series I, MC.
59. Jacob 1959; Jacob and Monod 1961a,c; Horace Freeland Judson (1996, p. 445) says Monod told him that their failure to cite McClintock in their widely read 1961 summary (Jacob and Monod 1961b) was an unhappy oversight.

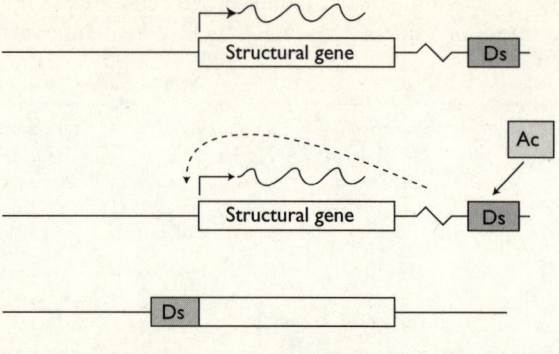

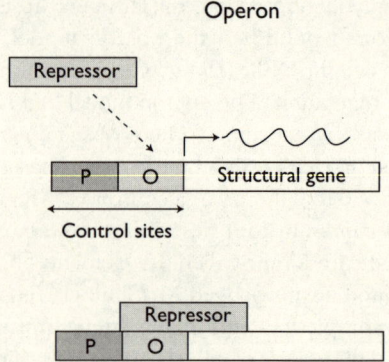

Figure 13.3. Controlling elements.

that λ acted like an episome, "hooking on" to the chromosome rather than insert-ing into it.[60] In 1962, however, Allan Campbell argued that it inserted, as McClin-tock's controlling elements seemed to do.[61] This became known as the "Campbell model" of episome integration.

Between 1967 and 1969, two groups of scientists studying the operon for the *gal* (galactose) gene in the bacterium *E. coli* discovered mutations they could explain only by postulating that segments of DNA were inserted into the genome upstream (as the polymerase flows) from the mutated gene.[62] The Campbell model was in-voked to explain how these elements might integrate into the bacterial chromo-

60. Jacob 1959, p. 14.
61. Campbell 1962.
62. Adhya and Shapiro 1969; Shapiro 1969; Jordan et al. 1967; Jordan and Saedler 1967.

some. These became known as insertion sequence (IS) elements, or simply insertion elements. Insertion elements were later identified using electron microscopy, and, in the early 1970s, were isolated using the new techniques of recombinant DNA technology. Only then, in 1972, were insertion elements recognized as being transposable and superficially similar to McClintock's elements.[63]

In contrast to McClintock's interest in layers of regulation and execution of developmental program, the insertion-element researchers focused on DNA sequences. Their operational definition of transposition was more mechanistic and less interpretive than McClintock's controlling element model. It was easily applied to other problems and observations, and it generated many new experiments. What sequences did the insertion elements recognize? What binding sites were used by enzymes that cut and pasted the elements? By the mid-1970s, insertion elements became recognized as a standard "tool" that cells used to move pieces of DNA around the genome.

Researchers working on various organisms soon found that this tool was surprisingly adaptable. In 1974 and 1975, five research groups published papers linking insertion elements to the transfer of antibiotic resistance genes between bacterial chromosomes and extrachromosomal circlets of DNA called plasmids.[64] The resistance genes were found to be sandwiched between insertion elements. The insertion elements excised from the plasmid and inserted into the chromosome, taking the resistance genes with them. None of the five groups cited McClintock in their first transposition papers.

Viruses could also be transposable. Phage μ, a bacteriophage discovered in 1963 by A. L. Taylor, differed from most temperate phages in that it caused mutations in its host and could integrate at multiple sites in the bacterial chromosome.[65] By the late 1960s a small cadre of "muologists" had coalesced and had begun to explore the use of μ in addressing some of the problems of DNA excision and insertion.[66] While Taylor had known McClintock and seen parallels between the broad outlines of her controlling elements and his new virus, the muologists were immersed in the molecular mechanisms of DNA cutting and pasting. The connection between μ and controlling elements dissolved in the mid-1960s, and by the 1970s a new connection emerged between μ and insertion elements. Phage μ used insertion-element sequences to integrate into the host chromosome. In 1976, muologist Ahmad Bukhari called μ a "transposition element."[67]

Yeast geneticists used insertion elements to explain the hoary problem of mating-type switching. It had been known since the 1960s that yeast could "change sex"

63. Starlinger and Saedler 1972; Malamy et al. 1972. Szybalski's group was collaborating with Shapiro's. This year, then, both of the original discoverers of insertion elements acknowledged McClintock.

64. Kleckner et al. 1975; Berg et al. 1975; Heffron et al. 1975; Hedges and Jacob 1974; Gottesman and Rosner 1975.

65. Taylor 1963; Toussaint 1987, p. 3.

66. Toussaint 1987, p. 3.

67. Bukhari 1976.

from (+) to (−). In 1977, principal investigator Ira Herskowitz, along with Jeffrey Strathern and James Hicks, at the University of Oregon proposed a "cassette" model of mating-type switching.[68] The cassette model contained transposition but not controlling elements. In their model, the (+) and (−) genes were distinct, transposable DNA cassettes that inserted in front of a regulatory gene. Depending on which cassette was inserted into the regulatory site, the yeast cell would assume a (+) or a (−) mating type. Flanking the cassettes was an insertion-element-like DNA sequence that seemed to mediate the insertion and excision events. Hicks, Strathern, and Herskowitz cited earlier proposals that had mentioned McClintock, as well as the bacterial geneticists working on insertion and excision, but not McClintock herself. One of their early papers did cite McClintock,[69] but only to recognize Yashi Oshima and Isamu Takano's use of her concept of controlling elements; Hicks and Herskowitz argued that the controlling-element model did not account for all available data. Once again, the transient connection to McClintock's elements was present in the foundations of this model but had dissolved by the time the model itself was elaborated.

In 1976, McClintock was nominated for a Nobel Prize for the first time. By this time, several molecular biology laboratories had made a link between insertion element–mediated transposition and McClintock's controlling elements. Her work, however, was still understood mainly in the framework of developmental control, in which McClintock herself had framed it. As a precursor to Jacob and Monod's molecular model of bacterial gene regulation, her model still seemed insufficiently convincing. A reviewer declined her Nobel nomination on the grounds that her work did not explain gene control in higher organisms with the same clarity that had been achieved in bacteria.[70]

That same year, McClintock's work began to be recast. The independent lines of molecular transposition research were united in 1976 at a meeting at Cold Spring Harbor, the proceedings of which were published in 1977.[71] A result of conversations at a meeting on plasmids at Squaw Valley, California, the previous year, this meeting was organized by Jim Shapiro and Sankhar Adhya, one of the original insertion-element teams, and the muologist Ahmad Bukhari.[72] Since the meeting was to be held at Cold Spring Harbor, McClintock was called in to advise. By this time, the bacterial, viral, and yeast geneticists had recognized parallels between their work and hers. At least since the 1960s, she had been arguing that a molecular interpretation of transposition would be coming soon and was needed before any detailed understanding of the mechanism of the process would be possible.[73]

68. Hicks et al. 1977.
69. Hicks and Herskowitz 1977.
70. Wettstein 1999.
71. Shapiro et al. 1977.
72. Shapiro 1992, p. 214.
73. Barbara McClintock to Mel Green, 13 February 1967, Folder "Green, MM," Box "Cat—G"; McClintock to Oliver Nelson, 20 March 1968, Folder "Nelson, O. E. #1," Box "Mo—Ne," Series I, MC.

Yet it was probably she who suggested inviting several classical geneticists who worked with movable elements.[74] McClintock presented a paper at the meeting and was thanked in the preface of the published volume, but she declined to contribute a paper to the volume.[75]

The meeting was an explicit recognition of the connections among these different areas and was a deliberate effort to build a new subdiscipline around transposable elements. A team of bacterial geneticists led by Allan Campbell defined transposable elements as "DNA segments which can insert into several sites in a genome."[76] The term "transposon" was adopted as a synonym, giving the elements a stylish, bacterial ring. The definition is striking for both its generality and its mechanism: a transposable element is simply a movable piece of DNA. It says nothing about what transposable elements do, or why; and while it included McClintock's controlling elements, it incorporated many phenomena besides those elements: viruses, episomes, yeast cassettes, and insertion elements. Further, transposition itself was a concept broader than before. Transposition could include not only the physical excision and reinsertion of a chromosome segment, as in McClintock's original conception, but also "replicative" transposition—in which the transposon remains in position but deposits a copy of itself at the new site—and even insertion of foreign DNA not originally part of the chromosome.

The new definition stripped transposable elements of the functional connotations that for McClintock were of primary importance. Sentimentally, the report credited her with the phrase "transposable elements"[77]—the term Brink had used to distinguish his interpretation from hers. That McClintock was credited with the term no doubt reflects the facts that McClintock was the more famous and brilliant scientist, that in the 1970s many of the workers in the new field of transposable elements were friends or acolytes of McClintock, and perhaps also that she, not Brink, was the forerunner-in-residence of the Cold Spring Harbor meeting.

Over the next three years, transposable-element research broadened even further. The cassette model of yeast mating-type switching was elaborated in more detail.[78] In 1979 and 1980, the Cornell University yeast geneticist Gerald Fink published his discovery of a transposon affecting the activity of a gene for a histone protein, calling the unit Ty1, for "yeast transposon one."[79] According to Fink, his search for transposable elements in yeast was suggested by McClintock.[80] Other work from further afield also tied in: retroviruses, discovered in the 1970s, were found to integrate into host DNA through an RNA intermediate. Because they fit

---

74. Green 1967, 1969; Peterson 1953.

75. Program of meeting, "DNA Insertions," 18–21 May 1976, Cold Spring Harbor Laboratory, N.Y. Courtesy David Stewart, Cold Spring Harbor Laboratory, N.Y.; Shapiro, 1997.

76. Campbell et al. 1977, p. 16.

77. Ibid., n. 1.

78. Hicks et al. 1979; Klar and Fogel 1979; Nasmyth and Tatchell 1980.

79. Greer and Fink 1979; Roeder and Fink 1980.

80. Fink 1992, p. 283.

the broad 1977 definition, they were brought into the fold. Also, immunologists had discovered a complex system of gene rearrangements that led to the antigen specificity of immunoglobulins. These too were considered transposable elements.[81]

By 1980, transposable elements had become a major new field of biology. As Cold Spring Harbor Laboratory director James Watson wrote, a Cold Spring Harbor symposium on movable genetic elements was "virtually unavoidable."[82] Transposition now pervaded many fields of basic research, including yeast, bacterial, and viral genetics. It touched on important medical considerations, including antibiotic resistance, cancer (via retroviruses), and immunology. It even connected to the commercially and politically explosive area of recombinant DNA technology: man-made transposons were now ferrying genes from one species into another, making genetic chimeras and creating the possibility of biological factories for insulin and other medically important chemicals.

In 1981, McClintock won seven prestigious awards and prizes, among them the T. H. Morgan Award, a lifetime MacArthur Fellowship, the Wolf Prize in Medicine, and a Lasker Award. She was also renominated for a Nobel Prize, once again without success (since many eventual Nobelists are nominated repeatedly before they win, this indicates the strength, not the weakness, of her reputation—she was, after all, a plant geneticist contending for the Physiology or Medicine prize!). She was nominated once again the following year. In a 17-page review of the nomination, a reviewer wrote, "Whether or not transpositions of the chromatin elements discovered by McClintock play a role in the physiology of an organism such as differentiation as suggested by her cannot be decided at this point."[83] In 1983, McClintock won an unshared Nobel in Physiology or Medicine for "her discovery of mobile genetic elements."[84]

While colleagues tried to connect McClintock to gene regulation, she failed to win major honors; only when transposition was divorced from gene control did she win accolades for this work. In celebrating McClintock and transposition, scientists discarded the theory that was so important to her. Transposition, universally accepted since 1950 as a peculiar event restricted to maize, was now recognized as a phenomenon of widespread and fundamental importance. Developmental regulation by controlling elements, in contrast, was treated skeptically at first and eventually was rejected outright.

---

81. See "Retroviruses as Insertion Elements" and "Rearrangements in Antibody Genes," *Cold Spring Harbor Symposia on Quantitative Biology* 45 (1981). Howard Temin, codiscoverer of reverse transcriptase, mentioned the parallels between reverse transcriptases and controlling elements as early as 1970, though he interpreted them loosely. See Temin 1970.

82. Watson 1981, p. xiii.

83. Wettstein 1999.

84. For a fuller account of McClintock and the Nobel Prize, see Comfort 2001a, chap. 9; Comfort 201b.

CONCLUSION

In the 1940s and 1950s, Barbara McClintock was thoroughly connected to the canonical knowledge of her period. She was well-read in all the genetic literature, was aware of the current problems in the field, and had integrated that work into her own interpretations and theories. Transposition's rapid confirmation and citation in textbooks show that it was incorporated immediately into the working knowledge of maize genetics. Her interpretation of transposition as a mechanism of genetic regulation, however, was not integrated in that way. McClintock never did persuade more than a handful of her colleagues that transposition was orchestrated in normal cells to determine the fate of developing tissues. Even 14 years after her Nobel Prize, few scientists accept this interpretation of the role of transposition.[85]

For McClintock, controlling elements and transposition were one. She did not distinguish between evidence and interpretation. Thus, if any part of her work was doubted, she would have been inclined to take it as wholesale rejection. Her colleagues, however, did distinguish between evidence and interpretation. In their eyes, she was a brilliant scientist who made a major discovery. Transposition indisputably does occur and is biologically important—in ways McClintock could not have dreamed. But in her colleagues' minds, she overextended her evidence, placing her findings in the context of a speculative theory. The theory seemed unsupported in the 1950s; subsequent evidence has failed to corroborate it. In short, transposition was not premature because it was not doubted then; her theory of genetic control was not premature because it is not believed today.

A CODA ON PREMATURITY

I approached this study skeptical of the concept of prematurity. It struck me then, and still strikes me, as a Whiggish approach to history. I set out for this conference with an assumption that no discoveries are premature; every one is, by its nature, made in and of its time. Norton Zinder came to the opposite conclusion, deciding that *all* discoveries are premature, otherwise they would not be discoveries.

It seems to me that Zinder and I might both be right, and herein lies what seems the real problem with prematurity. Although the standard narrative about McClintock is false, there are ways in which controlling elements could be described as premature. Scientists outside maize genetics did not discover transposition until 20 years after McClintock's 1948 discovery. Her vision of integrating genetics with development was beyond the scope of most geneticists in the 1950s; this integration did not begin until the late 1960s.[86]

85. Few, but not none: see for example Martienssen 1996; Martienssen and Richards 1995.
86. See Keller 1995.

But so what? My difficulty with prematurity is methodological, not theoretical. While my colleagues at this symposium have convinced me that at least some discoveries can be viewed as premature, the benefit of such a view to the historian remains, to me, obscure. A bias toward prematurity reinforces the canonical interpretation of history. It makes historical narratives seem unproblematic. It inclines one toward a restoration of heroic history of science that makes most contemporary historians head for the trenches. The assumption of prematurity renders unnecessary, or at least nonobvious, the contextualization that makes good history.

There is, however, a way in which prematurity could be useful to historians: as a sort of null hypothesis. Apparent instances of prematurity serve as blazing red flags calling historians to look deeper at accepted interpretations. What documents can we uncover, what biases can we detect, that allow us to reject the hypothesis of prematurity and place a discovery in the context of its time, place, and contemporary ideas? By erecting signposts to subjects deserving closer study, prematurity could promote good, empirical, revisionist history. Will this approach catch on? It is too soon to say.

ACKNOWLEDGMENTS

This work was supported by the National Science Foundation, Carnegie Institution of Washington, American Philosophical Society, and Lilly Library of Indiana University. The author wishes to thank Carol Greider, Horace Freeland Judson, Mark Lesney, Steve Weiss, and the participants in the symposium on prematurity and scientific discovery for valuable comments on this manuscript and the talk from whence it came.

BIBLIOGRAPHY

Adhya, S. L., and J. A. Shapiro. 1969. "The Galactose Operon of *E. coli* K-12. I. Structural and Pleiotropic Mutations of the Operon." *Genetics* 62:231–47.

Arianrhod, R. 1992. "Physics and Mathematics, Reality and Language." In *The Knowledge Explosion: Generations of Feminist Scholarship,* ed. C. Kramarae and D. Spender. New York: Teachers College Press.

Berg, D. E., J. Davies, B. Allet, and J. D. Rochaix. 1975. "Transposition of R Factor Genes to Bacteriophage Lambda." *Proceedings of the National Academy of Sciences USA* 72:3628–32.

Brink, R. A. 1960. "Paramutation and Chromosome Organization." *Quarterly Review of Biology* 35:120–37.

Brink, R. A., and R. A. Nilan. 1952. "The Relation between Light Variegated and Medium Variegated Pericarp in Maize." *Genetics* 37:518–44.

Bukhari, A. I. 1976. "Bacteriophage Mu as a Transposition Element." *Annual Review of Genetics* 10:389–412.

Campbell, A. 1962. "Episomes." *Advances in Genetics* 11:101–45.

Campbell, A., D. E. Berg, D. Botstein, E. M. Lederberg, R. P. Novick, P. Starlinger, and W. Szybalski. 1977. "Nomenclature of Transposable Elements in Prokaryotes." In *DNA Insertion Elements, Plasmids, and Episomes*, ed. J. A. Shapiro, A. I. Bukhari, and S. L. Adhya, pp. 15–22. Cold Spring Harbor, N.Y.: Cold Spring Harbor Laboratory Press.

Carlson, E. A. 1966. *The Gene: A Critical History.* Philadelphia: Saunders.

Coe, E. H. 1962. "Spontaneous Mutation of the Aleurone Color Inhibitor in Maize." *Genetics* 47:779–83.

Comfort, N. 2001a. *The Tangled Field: Barbara McClintock's Search for the Patterns of Genetic Control.* Cambridge: Harvard University Press.

———. 2001b. "From Controlling Elements to Transposons: Barbara McClintock and the Nobel Prize." *Trends in Biochemical Sciences* 26:454–57. Simultaneously published in *Trends in Genetics* 17:475–78.

Creighton, H., and McClintock, B. 1931. "A Correlation of Cytological and Genetical Crossing-Over in *Zea mays*." *Proceedings of the National Academy of Sciences USA* 17:492–97.

Dash, J. 1991. *The Triumph of Discovery: Women Scientists Who Won the Nobel Prize.* Englewood Cliffs, N.J.: Julian Messner.

Dietrich, M. R. 1995. "Richard Goldschmidt's 'Heresies' and the Evolutionary Synthesis." *Journal of the History of Biology* 28:431–61.

———. 1996. "On the Mutability of Genes and Geneticists: The 'Americanization' of Richard Goldschmidt and Victor Jollos." *Perspectives on Science: Historical, Philosophical, Social* 4:321–45.

Dunn, L. C. 1965. *A Short History of Genetics.* New York: McGraw-Hill.

Emmerling, M. H. 1958. "An Analysis of Intragenic and Extragenic Mutations of the Plant Color Component of the Rr Gene Complex in *Zea mays*." *Cold Spring Harbor Symposia on Quantitative Biology* 23:393–407.

Fabergé, A. C. 1958. "Relation between Chromatid-Type and Chromosome-Type Breakage-Fusion-Bridge Cycles in Maize Endosperm." *Genetics* 43:737–49.

Fedoroff, N., and D. Botstein, eds. 1992. *The Dynamic Genome: Barbara McClintock's Ideas in the Century of Genetics.* Cold Spring Harbor, N.Y.: Cold Spring Harbor Laboratory Press.

Felder, D. 1996. *The 100 Most Influential Women of All Time.* New York: Citadel Press.

Fincham, J. R. S. 1992. "Moving with the Times." *Nature* 358:631–32.

Fine, E. H. 1998. *Barbara McClintock, Nobel Prize Geneticist.* Springfield, N.J.: Enslow Publishers.

Fink, G. 1992. "Transposable Elements (Ty) in Yeast." In *The Dynamic Genome: Barbara McClintock's Ideas in the Century of Genetics,* ed. N. Fedoroff and D. Botstein, pp. 281–87. Cold Spring Harbor, N.Y.: Cold Spring Harbor Laboratory Press.

Gilbert, S. F. 1978. "The Embryological Origins of the Gene Theory." *Journal of the History of Biology* 11:307–51.

———. 1988. "Cellular Politics: Ernest Everett Just, Richard B. Goldschmidt, and the Attempt to Reconcile Embryology and Genetics." In *The American Development of Biology,* ed. R. Rainger, K. Benson, and J. Maienschein, pp. 311–46. New Brunswick, N.J.: Rutgers University Press.

Goldschmidt, R. B. 1988. *Physiological Genetics.* New York: McGraw-Hill.

Gottesman, M. M., and J. L. Rosner. 1975. "Acquisition of a Determinant for Chloramphenicol Resistance By Coliphage Lambda." *Proceedings of the National Academy of Sciences USA* 72:5041–45.

Green, M. M. 1967. "The Genetics of a Mutable Gene at the White Locus of *Drosophila melanogaster*." *Genetics* 56:467–82.

————. 1969. "Controlling Element Mediated Transpositions of the White Gene in *Drosophila melanogaster.*" *Genetics* 61:429–41.

Greer, H., and G. R. Fink. 1979. "Unstable Transpositions of His4 in Yeast." *Proceedings of the National Academy of Sciences USA* 76:4006–10.

Hedges, R. W., and A. E. Jacob. 1974. "Transposition of Ampicillin Resistance from RP4 to Other Replicons." *Molecular and General Genetics* 132:31–40.

Heffron, F., C. Rubens, and S. Falkow. 1975. "Translocation of a Plasmid DNA Sequence Which Mediates Ampicillin Resistance: Molecular Nature and Specificity of Insertion." *Proceedings of the National Academy of Sciences USA* 72:3623–27.

Heiligman, D. 1994. *Barbara McClintock: Alone in Her Field.* New York: W. H. Freeman.

Hicks, J. B., and I. Herskowitz. 1977. "Interconversion of Yeast Mating Types. II. Restoration of Mating Ability to Sterile Mutants in Homothallic and Heterothallic Strains." *Genetics* 85:387–88.

Hicks, J. B., J. N. Strathern, and I. Herskowitz. 1977. "The Cassette Model of Mating-Type Interconversion." In *DNA Insertion Elements, Plasmids, and Episomes,* ed. J. A. Shapiro, A. I. Bukhari, and S. L. Adhya, pp. 457–62. Cold Spring Harbor, N.Y.: Cold Spring Harbor Laboratory Press.

Hicks, J. B., J. N. Strathern, and A. J. S. Klar. 1979. "Transposable Mating Type Genes in *Saccharomyces cerevisiae.*" *Nature* 282:478.

Hofstetter, A. M. 1992. "The New Biology: Barbara McClintock and an Emerging Holistic Science." *Teilhard Studies* 26:1–15.

Jacob, F. 1959. "Genetic Control of Viral Functions." *Harvey Lectures 1959* 1:1–39.

Jacob, F., and J. Monod. 1961a. "Genetic Regulatory Mechanisms in the Synthesis of Proteins." *Journal Molecular Biology* 3:318–56.

————. 1961b. "Regulation of Gene Activity." *Cold Spring Harbor Symposia on Quantitative Biology* 26:193–211.

Jordan, E., and H. Saedler. 1967. "Polarity of Amber Mutations and Suppressed Amber Mutations in the Galactose Operon of *E. coli.*" *Molecular and General Genetics* 100:283–95.

Jordan, E., H. Saedler, and P. Starlinger. 1967. "Strong-Polar Mutations in the Transferase Gene of the Galactose Operon in *E. coli.*" *Molecular and General Genetics* 100:296–6.

Judson, H. F. 1996. *The Eighth Day of Creation: Makers of the Revolution in Biology.* Expanded edition. Cold Spring Harbor, N.Y.: Cold Spring Harbor Laboratory Press.

Keller, E. F. 1983. *A Feeling for the Organism.* New York: W. H. Freeman.

————. 1985. *Reflections on Gender and Science.* New Haven: Yale University Press.

————. 1995. *Refiguring Life: Metaphors of Twentieth Century Biology.* New York: Columbia University Press.

Kermicle, J. 1996. Interview by author. Tape recording. Madison, Wis., 15 October. American Philosophical Society Library, Philadelphia.

King, R. 1962. *Genetics.* New York: Oxford University Press.

Kittredge, M. 1991. *Barbara McClintock.* New York: Chelsea House.

Klar, A. J. S., and S. Fogel. 1979. "Activation of Mating Type Genes by Transposition in *Saccharomyces cerevisiae.*" *Proceedings of the National Academy of Sciences USA* 76:4539.

Kleckner, N. R., K. Chan, B. K. Tye, and D. Botstein. 1975. "Mutagenesis by Insertion of a Drug-Resistance Element Carrying an Inverted Repetition." *Journal of Molecular Biology* 97:561.

Kohler, R. 1994. *Lords of the Fly: Drosophila Genetics and the Experimental Life.* Chicago: University of Chicago Press.

Laughnan, J. 1955. "Structural and Functional Basis for the Action of the *A* Alleles in Maize." *American Naturalist* 89:91–103.

MacColl, S. 1989. "Intimate Observation." *Metascience* 7:90–98.

Malamy, M. H., M. Fiandt, and W. Szybalski. 1972. "Electron Microscopy of Polar Insertions in the Lac Operon of *Escherichia coli.*" *Molecular and General Genetics* 119:207–22.

Mangelsdorf, P. C. 1958. "The Mutagenizing Effect of Hybridizing Maize and Teosinte." *Cold Spring Harbor Symposia on Quantitative Biology* 23:409–21.

Martienssen, R. A. 1996. "Epigenetic Phenomena: Paramutation and Gene Silencing in Plants." *Current Biology* 6:810–13.

Martienssen, R. A., and E. J. Richards. 1995. "DNA Methylation in Eukaryotes." *Current Opinions in Genetics and Development* 5:234–42.

McClintock, B. 1938. "The Fusion of Broken Ends of Sister Half-Chromatids Following Chromatid Breakage at Meiotic Anaphases." *Missouri Agricultural Experiment Station Research Bulletin* 290:1–48.

———. 1941. "The Stability of Broken Ends of Chromosomes in *Zea mays.*" *Genetics* 26:234–82.

———. 1942. "The Fusion of Broken Ends of Chromosomes Following Nuclear Fusion." *Proceedings of the National Academy of Sciences USA* 28:458–63.

———. 1950a. "Mutable Loci in Maize." *Carnegie Institution of Washington Year Book* 49:157–67.

———. 1950b. "The Origin and Behavior of Mutable Loci in Maize." *Proceedings of the National Academy of Sciences USA* 36:344–55.

———. 1951. "Chromosome Organization and Gene Expression." *Cold Spring Harbor Symposia on Quantitative Biology* 16:13–47.

———. 1953. "Induction of Instability at Selected Loci in Maize." *Genetics* 38:579–99.

———. 1956. "Controlling Elements and the Gene." *Cold Spring Harbor Symposia on Quantitative Biology* 21:197–216.

———. 1961. "Some Parallels between Gene Control Systems in Maize and in Bacteria." *American Naturalist* 95:265–77.

———. 1980. Interview by W. B. Provine and P. Sysco. Transcript. Ithaca, N.Y., and Cold Spring Harbor, N.Y. Cornell University Archives, Ithaca, N.Y.

———. (n.d.) Interview by N. Symonds. Tape recording. Cold Spring Harbor, N.Y. Transcribed by the author, in the author's possession.

McGrayne, S. B. 1993. *Nobel Women in Science.* New York: Birch Lane.

Monod, J., and F. Jacob. 1961. "General Conclusions." *Cold Spring Harbor Symposia on Quantitative Biology* 26:389–401.

Morgan, T. H. 1926. "Genetics and the Physiology of Development." *American Naturalist* 60:489–515.

Morse, M. 1995. *Women Changing Science: Voices from a Field in Transition.* New York: Plenum.

Nasmyth, K. A., and K. Tatchell. 1980. "The Structure of Transposable Yeast Mating Type Loci." *Cell* 19:753.

Nelson, O. E. 1959. "Intracistron Recombination in the Wx/wx Region in Maize." *Science* 130:794–95.

Nilan, R. A. 1951. "Genic Control of Mutation Frequency of the Variegated Pericarp Gene in Maize." Ph.D. diss., University of Wisconsin.

Nuffer, M. G. 1955. "Dosage Effect of Multiple *Dt* Loci on the Mutation of *a* in the Maize Endosperm." *Science* 121:399–400.

————. 1961. "Mutation Studies at the A. Locus in Maize I. A Mutable Allele Controlled by Dt." *Genetics* 46:625–40.

Opfell, O. 1986. *The Lady Laureates: Women Who Have Won the Nobel Prize.* 2nd ed. Metuchen, N.J.: Scarecrow Press.

Peters, J. A., ed. 1959. *Classic Papers in Genetics.* Englewood Cliffs, N.J.: Prentice-Hall.

Peterson, P. A. 1953. "A Mutable Pale Green Locus in Maize." *Genetics* 38:682–83.

Roeder, G. S., and G. R. Fink. 1980. "DNA Rearrangements Associated with a Transposable Element in Yeast." *Cell* 21:239–49.

Rose, H. 1994. *Love, Power, and Knowledge: Towards a Feminist Transformation of the Sciences.* Bloomington: Indiana University Press.

Sager, R., and F. Ryan. 1961. *Cell Heredity.* New York: John Wiley and Sons.

Sapp, J. 1987. *Beyond the Gene: Cytoplasmic Inheritance and the Struggle for Authority in Genetics.* New York: Oxford University Press.

————. 1994. "Concepts of Organization: The Leverage of Ciliate Protozoa." In *A Conceptual History of Modern Embryology,* ed. S. Gilbert, pp. 229–58. Baltimore: Johns Hopkins University Press.

Schiebinger, L. 1987. "The History and Philosophy of Women in Science: A Review Essay." In *Sex and Scientific Inquiry,* ed. S. Harding and J. O'Barr. Chicago: University of Chicago Press.

Schwartz, D. 1960a. "Analysis of a Highly Mutable Gene in Maize: A Molecular Model for Gene Instability." *Genetics* 45:1141–52.

————. 1960b. "Electrophoretic and Immunochemical Studies with Endosperm Proteins of Maize Mutants." *Genetics* 45:1419–27.

————. 1962. "Genetic Studies on Mutant Enzymes in Maize. III. Control of Gene Action in the Synthesis of pH 7.5 Esterase." *Genetics* 47:1609–15.

Serra, J. A. 1965. *Modern Genetics.* Vol. 1. London: Academic Press.

Shapiro, J. A. 1969. "Mutations Caused by the Insertion of Genetic Material into the Galactose Operon of *Escherichia coli.*" *Journal of Molecular Biology* 40:93–105.

————. 1992. "Kernels and Colonies: The Challenge of Pattern." In *The Dynamic Genome: Barbara McClintock's Ideas in the Century of Genetics,* ed. N. Fedoroff and D. Botstein, pp. 213–21. Cold Spring Harbor, N.Y.: Cold Spring Harbor Laboratory Press.

————. 1997. Interview by the author. Tape recording. Chicago, 27–28 January. American Philosophical Society Library, Philadelphia.

Shapiro, J. A., A. I. Bukhari, and S. L. Adhya, eds. 1977. *DNA Insertion Elements, Plasmids, and Episomes.* Cold Spring Harbor, N.Y.: Cold Spring Harbor Laboratory Press.

Shepherd, L. J. 1993. *Lifting the Veil: The Feminine Face of Science.* Boston: Shambhala.

Shiels, B. 1985. *Winners: Women and the Nobel Prize.* Minneapolis: Dillon Press.

Shteir, A. B. 1987. "Botany in the Breakfast Room: Women in Early Nineteenth Century British Plant Study." In *Uneasy Careers and Intimate Lives: Women in Science, 1789–1979,* ed. P. Abir-Am and D. Outram. New Brunswick, N.J.: Rutgers University Press.

Spradling, A. 1993. "McClintock Myths: Review of *The Dynamic Genome: Barbara McClintock's Ideas in the Century of Genetics.*" *Science* 259:1206–8.

Srb, A., R. Owen, and R. Edgar. 1965. *General Genetics.* San Francisco: W. H. Freeman.

Stadler, L. J. 1954. "The Gene." *Science* 120:811–19.

Starlinger, P., and H. Saedler. 1972. "Insertion Mutations in Microorganisms." *Biochimie* 54:177.

Stent, G. 1972. "Prematurity and Uniqueness in Scientific Discovery." *Scientific American* 227 (December): 84–93.

Sturtevant, A. H. 1925. "The Effects of Unequal Crossing Over at the Bar Locus in *Drosophila*." *Genetics* 10:117–47.

Sturtevant, A. H., and T. H. Morgan. 1923. "Reverse Mutation of the Bar Gene Correlated with Crossing Over." *Science* 57:746–47.

Taylor, A. L. 1963. "Bacteriophage-Induced Mutation in *Escherichia coli*." *Proceedings of the National Academy of Sciences USA* 50:1043–51.

Temin, H. M. 1970. "Malignant Transformation of Cells by Viruses." *Perspectives in Biology and Medicine* 14:11–26.

Toussaint, A. 1987. "A History of Mu." In *Phage Mu*, ed. N. Symonds, A. Toussaint, P. van de Putte, and M. Howe, pp. 1–23. Cold Spring Harbor, N.Y.: Cold Spring Harbor Laboratory Press.

Watson, J. D. 1980. Foreword to *Cold Spring Harbor Symposia on Quantitative Biology* 45:xiii.

Wettstein, D. von. 1999. Telephone interview by author. 1 July.

Zinder, N. 1996. Interview by author. Tape recording. New York, 12 March. American Philosophical Society Library, Philadelphia.

CHAPTER FOURTEEN

The Work of Joseph Adams
and Archibald Garrod

Possible Examples of Prematurity in Human Genetics

Arno G. Motulsky

Two British physicians, Joseph Adams (1756–1818) and Archibald Garrod (1857–1936), set out principles and facts pertinent to modern human and medical genetics, but their insights were not recognized until more recent times. The contributions of these two men may be examined within the framework of Gunther Stent's concept of prematurity in scientific discovery.

As a physician-scientist I have had the advantage of doing both research and clinical work in human and medical genetics, and I continue to work in both areas. In one sense, understanding the science and practical background of a technical field is an advantage when exploring the history of science and medicine. But approaching history as a nonhistorian who may be less knowledgeable about the intellectual, social, and economic milieu against which the historical work was done may lead to distortions in assessing its ultimate significance. Having worked with patients in genetic diagnosis and counseling, I feel intellectually close to the work that Adams and Garrod did. Some observers might point out that physicians who analyze the work of pioneers in their field may be less than objective because of hero worship and might read too much into the earlier writings. I agree with such criticism to some extent, but often an individual with a perspective from his or her own field may add valid interpretations that are less apparent to historians without the technical background. The two approaches complement one another.

JOSEPH ADAMS (1756–1818)

No one recording the history of genetics referred to Joseph Adams before I wrote my article in 1959. Moreover, I did not publish in a journal dealing with the history of science or medicine. As a young medical school faculty member in internal medicine, I selected the *Archives of Internal Medicine*.[1] Few readers of that journal were interested in the new field of medical genetics. Later, the article was mentioned in textbooks in our field, and Adams eventually entered the literature of the history of human genetics.

In my article, I designated Joseph Adams "a forgotten founder of medical genetics." In 1956–57, on a sabbatical leave at the Galton Laboratory at University College in London, I had spent time searching the shelves of the library of University College. There, I found an old book that had never been taken out on loan from the library. This book by Adams, published in 1814, was titled *A Treatise on the Supposed Hereditary Properties of Disease containing Remarks on the unfounded terrors and ill-judged cautions consequent on such erroneous opinions; with notes illustrative of the subject, particularly in madness and scrofula*.[2] It included 41 pages of "principles"—that is, generalities—and was followed by extensive commentary and documentation. I was amazed by Adams's recognition of concepts of the genetic characteristics of diseases not fully explicated until more than a century later. I am not aware of relevant archives, personal papers, or other unpublished documents that might cast additional light on the development of his ideas. Such work would be of interest.

Based on personal observation and acquaintance with the medical literature of his day, Adams articulated concepts known now to be correct, but which could not have been established in a scientific manner at that time, because an underlying theory of heredity did not exist. Among other examples, he differentiated between "familial" and "hereditary" occurrences of some diseases. By "familial" he meant disorders occurring among siblings but not in other relatives, that is, manifesting only in the same generation. By "hereditary" he meant disorders transmitted from generation to generation. Today we can interpret his familial concept as denoting what we term autosomal Mendelian recessive disorders that require two doses of the mutant gene for manifestation, each coming from an unaffected parent. Mendelian dominants (as we term them now) are equivalent to disorders Adams identified as hereditarily transmitted from generation to generation.

Adams discussed congenital types of disease, that is, those manifesting at birth. He stated that familial disorders occurred more frequently congenitally than did disorders transmitted from generation to generation. This fits our knowledge of recessives and dominants in humans today. He also pointed out that contagion by "morbid poisoning," for example, syphilis, could also produce congenital disease that would not necessarily be familial.

Adams pointed out that it was important practically to distinguish these two types of disorder: "By confounding hereditary with familial disease we excite an unnecessary apprehension in the rising generation" (pp. vi–vi). He wrote about such matters to help advise individuals about what might happen in the future, to help overcome dread that might impose celibacy because of "the skeleton in the closet,"

1. Motulsky 1959.
2. Adams 1814.

and to overcome fears that might be magnified unnecessarily. In a sense he was the forerunner of what we term "genetic counseling" today.

He also pointed out that some familial diseases occur more frequently in what he called a "single race" of the population. Arguing from this and observations of sheep in which inbreeding had produced bony and other deformities, Adams suggested that "endemic peculiarities [in humans] may be found in certain sequestered districts"—that is, "familial" diseases might well cluster in certain areas. And he wrote that mating between near relatives would result in an increase in familial disease and would tend to deteriorate the race.

I find it particularly remarkable that Adams distinguished between "disposition" and "predisposition" (pp. 25–26). He defined a disposition as a hereditary disease occurring spontaneously at a certain age with no external cause to provoke it, while predisposition required an external cause for manifestation. A predisposition might exist for illnesses such as gout and "madness" for many generations without manifesting, until, as we would state today, some precipitating factor triggered the disorder. Medical geneticists certainly would agree with this presumption of interaction of heredity and environment. Adams even suggested that hereditary predispositions might be prevented or treated by removal of the external cause of disease expression.

Adams also wrote about what we now call intrafamilial correlation. He noted that the age at onset of hereditary disorders is similar in some families (pp. 21–22). And he hinted at what we now term a new mutation, writing that some diseases "would cease altogether, were it not that parents, free from such susceptibility, occasionally produce an offspring in whom the susceptibility originates" (p. 32).

He also called for something that is discussed nowadays: registers of familial disorders to be available for more extensive study (p. 41).

Although astute and often correct, Adams's findings were strictly empiric. We know that he was wrong about many specific disease examples. And he proposed some ideas that do not make much sense today. Yet, he set out the principles of expression of genetic disease that seem remarkably modern from a current perspective. How did he do it? He must have been an excellent clinical observer unencumbered by the medical philosophy and the medical theories of his days.

Biography

Details about Adams appear in an obituary and in the *Dictionary of National Biography*. [3] He was born in 1756 and was trained as an apothecary. Apothecaries dispensed medicines, but in the Britain of his day, apothecaries often also took on duties of primary medical care. Near the end of the eighteenth century, some apothecaries were able to obtain some medical training. Adams attended lectures

---

3. Obituary of Joseph Adams 1818; "Joseph Adams" 1886.

at St. Bartholomew's Hospital, and later at St. George Hospital, where he came under the influence of John Hunter. Adams's first book, *Morbid Poisons,* appeared in 1795.[4] In 1796 at the age of forty, he obtained an honorary M.D. in absentia from the University of Aberdeen and then moved to the island of Madeira in the Atlantic Ocean, where he practiced medicine for eight years and, with his wife, ran a convalescent home for invalids.

In 1801 he published a book on breast cancer.[5] In 1805, on returning to England, he was elected physician to the smallpox hospital, an appointment of honor. As his only formal training was that of an apothecary, this appointment required a waiver. In 1809 he was admitted as a licentiate of the Royal College of Physicians without examination, on recommendation of its president. After publication of his book on hereditary disease in 1814, mentioned earlier, he reissued in 1815 what was, in essence, the same book, under the title *A Philosophical treatise on the Hereditary Peculiarities of the Human Race: with notes illustrative of the subject, particularly in gout, scrofula, and madness.*[6] It is likely that he hoped to attract a wider public by emphasizing the general relevance of his findings by referring to the heredity of the human race rather than to the hereditary properties of disease alone. In modern terms, medical genetics became human genetics. Adams died in 1818 at the age of sixty-six. The motto "Vir justus et bonus" appeared on his tombstone.

Prematurity and Neglect

What can we learn from Adams's book regarding prematurity and neglect of new concepts? Adams was taught by the great experimental surgeon John Hunter, who stressed facts, experiments, observations, data, and directness. Under this influence, Adams ignored theory. Empirical observations in medicine may be useful and valid, but to be incorporated in the corpus of accepted scientific knowledge they must await a scientific infrastructure for explanation. Findings as remarkable as Adams's observations might exist in the old medical literature and be ignored for many years and, ultimately, be deemed correct.

If we call a mind-set a paradigm, we cannot refer to Adams as having the wrong paradigm, which might have led to biased observations. Adams had no paradigm in the Kuhnian sense. He just described, and then drew appropriate conclusions.

Was Adams's description a premature discovery? I do not think so. I would call him a precursor. His work had to await the development of theory before it could be explained. The scientific work in human genetics of the twentieth century made it possible to fit his observations into cohesive explanations.

This points up an interesting modern issue. Currently, in medical research in which there have been major advances of understanding in molecular biology, sci-

4. Adams 1796.
5. Adams 1801.
6. Adams 1815.

entists with a medical degree often work on the same fundamental research problems as their basic-scientist mentors, from whom they obtained their training. Joseph Goldstein and Michael Brown classified research by physicians into three categories: basic research, disease-oriented investigations not involving direct patient contact, and real patient-oriented clinical research.[7] Among physician-scientists today, very few work in the latter area. Pursuing such studies is more difficult than working on clean systems in the laboratory with cells, tissues, and body fluids. The example of Adams illustrates that astute clinical observations of human patients and their diseases may reveal yet unknown but significant biological and medical insights. But to be recognized, such work must be integrated or "connected," in Gunther Stent's sense, with generally accepted knowledge that the work extends.

ARCHIBALD E. GARROD (1857–1936)

Considered to be the founder of biochemical genetics and, in particular, of human biochemical genetics, Archibald Garrod had profound insights far in advance of his time about the role of genetic factors in disease and in human variation.[8] He enunciated clearly the concept of chemical individuality of every human being and stated that this uniqueness determines how the organism will interact with the environment, specifically including infections and drugs. He first described his concepts in 1902 in the British medical journal *Lancet,* in which he reported on alkaptonuria, a rare condition that he later called an inborn error of metabolism.[9] Subsequently, he studied albinism and several other inborn errors. He published a series of four lectures on inborn errors of metabolism, in 1908; the series was reissued with revisions as a book in 1909 and further expanded in 1923.[10] The 1909 book was republished by the Oxford University Press in 1963,[11] with a supplement by the British biochemical geneticist Harry Harris discussing the contemporary status of inborn errors. In 1931, near the end of his career, Garrod published a book titled *The Inborn Factors in Disease,* in which he generalized from discrete biochemical defects to chemical individuality affecting *all* diseases.[12] This book is highly modern and far ahead of its time. But it too fell flat and attracted little attention. In 1989 Charles Scriver and Barton Childs, well-known human biochemical geneticists, reissued the 1931 book with extensive commentary.[13] Another human geneticist, Alexander Bearn, in 1993 published an authoritative biography of Garrod that is well worth reading.[14]

7. Goldstein and Brown 1997.
8. Bearn 1993.
9. Garrod 1979.
10. Garrod 1908, 1909, 1923.
11. Harris 1963.
12. Garrod 1931.
13. Scriver and Childs 1989.
14. Bearn 1993.

Inborn Errors of Metabolism and Chemical Individuality

Garrod had a medical education, and worked for many years as an outpatient physician. His interest was the study of urine chemistry, and he was fully acquainted with the continental literature on chemistry. Garrod observed a baby whose diapers had been stained black by urine.[15] He showed that the blackened urine was caused by alkaptonuria and was due to a failure of oxidation of homogentisic acid. Much later, the specific enzyme defect was demonstrated directly by liver biopsy.[16] Garrod noted that the condition was congenital and often familial but seen only in one generation, with the parents having normal urine findings. Both boys and girls were affected. The condition was usually harmless and compatible with a long life. In a large proportion of patients, the parents were first cousins. Garrod sought out William Bateson, one of the early geneticists (the man who, in fact, coined the term "genetics"), to get advice about the nature of the abnormality. Bateson explained the findings as caused by a rare recessive gene in the Mendelian sense, which fitted the frequent parental consanguinity.

Garrod was careful to refer to these abnormalities as alternative courses of metabolism. In addition to alkaptonuria, he proposed other examples: cystinuria, in which the body excretes large amounts of the amino acid cystine, albinism associated with lack of skin pigmentation, and pentosuria, in which there is greater-than-normal excretion of a five-carbon sugar. He pointed out the high frequency of pentosuria among Jews, consistent with what Adams (whose writings were unknown to Garrod) had suggested almost 100 years earlier about certain familial conditions occurring more frequently in "sequestered" populations. As we know today, mutations in the genes specifying particular enzymes cause these and other inborn errors. However, Garrod did not specifically suggest enzyme alterations as an explanation.

In his 1931 book, Garrod clearly outlined the concept that our genetic determinants, which vary among individuals, result in chemical individuality that governs our responses to environmental agents.[17] Garrod suggested that inborn errors were not mere idiosyncrasies, but constituted clear examples of widespread human differences in chemical makeup. The existence of human variation led him to suspect that the phenomenon occurred widely. Indeed, he provided the conceptual background for a field that we call now pharmacogenetics. As became clear much later, there is widespread heterogeneity in drug response due to genetic variation.[18]

15. Garrod 1979.
16. La Du et al. 1958.
17. Garrod (1931).
18. Weber 1997.

Biography

Garrod was a pediatrician and a physician. His biography suggests that he was more interested in rare diseases and their mechanisms than in the common diseases in clinical practice.[19] He had an outstanding reputation in medical circles. At the age of sixty-three, he succeeded the famous William Osler as Regius Professor of Medicine at Oxford University, one of the major positions in British academic medicine. He retired at age seventy and died nine years later, in 1936.

However, while the medical establishment bestowed on Garrod an important position, invited him to give prestigious lectures, and awarded him numerous honors,[20] neither medical academics nor medical practitioners of his day appeared to understand or appreciate his concepts regarding chemical individuality. He was viewed as someone who studied rare diseases with chemical methods. The general significance of his work for what was called later the one gene, one enzyme hypothesis escaped everyone.

The One Gene, One Enzyme Hypothesis

The one gene, one enzyme concept was proposed in the early 1940s by George Beadle and Edward Tatum.[21] They, unlike Garrod, who studied human families, investigated the mold *Neurospora,* in which they could induce mutations. By removing certain nutrients from the culture medium, they could test for the loss of specific physiologic functions. With subsequent cultures, they showed some of these variants to have a genetic basis with associated loss of enzymatic activity. As a consequence they postulated that each gene specified a different enzyme, the so-called one gene, one enzyme hypothesis. Later, this concept was broadened to "one gene, one protein" and the "one gene, one polypeptide" hypothesis. This concept was a major breakthrough that recognized the role of genes in specifying protein structure.

Beadle stated that he was unaware of Garrod's conceptual formulation and had encountered his writings only after his work was done.[22] In his Nobel Prize address given in 1958, he said that Garrod had, in essence, stated it all and that he and Tatum provided more details with their experimental systems.[23] Biochemists, however, were aware of Garrod. Joshua Lederberg states that he read about Garrod at the age of thirteen in the Bodansky textbook of biochemistry in 1938.[24] The famous geneticist Sewall Wright since 1925 had devoted three of his lectures to Gar-

19. Bearn 1993.
20. On Garrod's position, lectures, and honors, see ibid.
21. Beadle and Tatum 1941.
22. See Bearn 1993 on this point
23. Beadle 1965.
24. Lederberg 1990.

rod in his course on physiologic genetics at the University of Chicago.[25] Edward Tatum, who shared the Nobel Prize with Beadle, had taught about Garrod's findings in his course on comparative biochemistry at Stanford University before the initiation of the *Neurospora* experiments.[26] Before this work, Beadle and Tatum had been unsuccessful in elucidating the problem of gene action in drosophila eye pigments. Yet, in a 1941 published lecture referring to this work, they mentioned Garrod's findings.[27] It appears that they were vaguely aware of these findings but its general significance for their work did not make an impact until later.

A revisionist historian of biologic science, Jan Sapp, suggests that contrary to current perception, the one gene, one enzyme theory was not generally accepted when first proposed in 1941.[28] Sapp feels that the prominent mention of Garrod by Beadle was a rhetorical device to strengthen the Beadle hypothesis. More generally, Sapp proposes that, when scientists appeal to history, they often do so to bolster their own place in it. He states, "The casting of Archibald Garrod as the 'father of biochemical genetics' was done by Beadle to supply his experimental evidence for the one-gene-one-enzyme theory with a plot that would parallel the rediscovery of Mendel's work."[29]

Garrod's Contributions

What were Garrod's real contributions? Did he fully enunciate or even appreciate a one gene, one enzyme relationship? Neither Garrod nor Bateson—his advisor in genetics—inferred that the normal allele of the mutant gene in the inborn errors had a "normal" function. One cannot find a full statement of the one gene, one enzyme concept in what Garrod wrote, and, as important as his insights were, one must conclude that he did not yet fully enunciate the concept.

Garrod was a physician interested in chemistry, and he followed this field closely. Despite his discoveries and consultation with the geneticist Bateson, he took no apparent interest in further developments in genetics.[30] Although he clearly saw a heredity-metabolism relationship, Garrod was not motivated to study more genetics to develop his ideas, nor were other physicians. Some of the reasons relate to disciplinary barriers. Medicine was interested in common diseases. Garrod studied very rare conditions that were hardly ever seen. No one else in medicine saw the relationship of these rare conditions to medical science and practice. Perhaps most important, genetics and medicine developed apart from each other.[31] Only recently was genetics accepted as a significant field important for both theoretical

25. Ibid.; Bearn 1993.
26. Lederberg 1990; Beadle 1965.
27. Beadle and Tatum 1941.
28. Sapp 1990.
29. Ibid.
30. Bearn 1993.
31. Motulsky 1983.

and practical advances. Medicine, therefore, was not yet receptive to a key principle that applied not just to rare inborn errors but was important for all of medicine. Most geneticists were vaguely aware of Garrod, but they ignored his work because it originated outside their field. Further, experimental manipulation in humans was not feasible, nor were there animal models available to investigate these findings.[32] J. B. S. Haldane, a geneticist who was well versed in biochemistry and who wrote a book on biochemical genetics in 1954, did not comment on Garrod until 1937.[33] While biochemists paid lip service to Garrod, his work did not enter the mainstream of biochemistry and biochemical genetics until the one gene, one enzyme hypothesis became widely accepted in the 1960s and later.

Garrod is, like Adams, a precursor. Was his work premature? In his time, appropriate methods were not available to follow up his observations. Methods for investigating enzymology and protein variation that could be applied to medicine developed only much later. The proof of molecular disease by Linus Pauling and his coworkers came only in 1949,[34] when they showed that sickle-cell anemia was caused by a mutation altering the structure of hemoglobin that could be demonstrated by electrophoresis. Many other mutations of various enzymes and proteins were found later, and the one gene, one protein concept became the cornerstone of human biochemical genetics.[35]

R. J. Williams (1893–1988)

A discussion of Archibald Garrod's role in the development of human biochemical genetics is not complete without reference to a more recent contributor to the study of human individuality. R. J. Williams was a highly productive nutritional

---

32. Harris 1963.

33. Haldane 1937, 1954.

34. Pauling et al. 1949.

35. Pauling's demonstration of a molecular disease, however, was not the first. A year before Pauling and his colleagues' 1949 publication on a molecular defect in the hemoglobin of patients with sickle-cell anemia, a German biochemist (H. Hörlein) and a medical student (G. Weber) showed that the defect in another hemoglobin disease—hereditary methemoglobinemia—was caused by a defect in globin and not in the nonprotein heme of the hemoglobin molecule (Hörlein and Weber 1948). Unlike Pauling, these investigators did not realize the fundamental and theoretical significance of their work for biochemical genetics. The article therefore did not attract general attention, and its publication in a German medical journal shortly after the war (1948) further reduced its visibility.

The intellectual scientific climate was ripe in the 1940s for testing the validity of mutational changes in a protein, as shown by a personal experience. Working in the laboratory of Karl Singer at Michael Reese Hospital in Chicago in 1948, I tried to demonstrate a defect in the hemoglobin of patients with sickle-cell anemia. Hemoglobin from control individuals and patients with sickle-cell anemia was injected into rabbits to raise antibodies that I hoped would be different in those animals injected with sickle-cell hemoglobin. However, no antibodies could be raised at all, and a comparison between normals and patients could not be carried out. Pauling and colleagues' electrophoretic demonstration of a hemoglobin abnormality in sickle-cell anemia soon afterward (1949) was clear and convinced everyone.

biochemist at the University of Texas who had a fine scientific reputation, was a member of the National Academy of Science, and was president of the American Chemical Society in 1957. In 1956, Williams published a book titled *Biochemical Individuality: The Basis for the Genetotrophic Concept.*[36] He had gathered data from many different areas, such as anatomy, enzymology, endocrinology, nutrition, and pharmacology, to document the remarkable difference between individuals, maintaining that practically every human being is a "deviate" in some sense. While he quoted Garrod as well as Beadle and Tatum, Williams wrote as a biochemist who was less knowledgeable about genetics even though he strongly stressed the role of heredity. The "genetotrophic concept" of his book's title refers to nutrition and implies that every individual has a unique genetic background with distinctive nutritional needs. Many of his publications were directed at the public, but it appears that his work had little impact on subsequent research in biochemistry, nutrition, and medicine or on the general public in the decades following the 1956 publication. Williams's career shows that valid concepts can have full resonance only if convincing research to buttress them is carried out. Correct concepts that cannot be proven with facts will not catch on. The appropriate methodology to demonstrate biochemical variation was just developing. Failure of scientists in one field (nutrition) to become acquainted with the approaches of another specialty (genetics) retards advances, since few researchers are able to work comfortably with the techniques and concepts of both fields. Williams's lack of success in influencing nutritional scientists was largely caused by the general failure to teach genetic concepts to nutritionists. His work with the traditional research approaches of biochemistry and nutrition led to no contacts with the emerging field of human biochemical genetics in the medical schools. Forty years after the publication of his book, Williams's ideas about genetics and nutrition are becoming popular again,[37] but much additional work is required to elucidate the interaction of genes and nutrition.

BIBLIOGRAPHY

Adams, J. 1796. *Observations on Morbid Poisons, Phagedena, and Cancer.* 2d ed. London: J. Callow. First published in 1795.

———. 1801. *A Treatise on the Cancerous Breast.* London: J. Callow.

———. 1814. *A Treatise on the Supposed Hereditary Properties of Disease containing Remarks on the unfounded terrors and ill-judged cautions consequent on such erroneous opinions; with notes illustrative of the subject, particularly in madness and scrofula.* London: J. Callow.

———. 1815. *A Philosophical Treatise on the Hereditary Peculiarities of the Human Race.* London: J. Callow.

Beadle, G. W. 1965. "Genes and Chemical Reactions in *Neurospora.*" In *Nobel Lectures, Including Presentation Speeches and Laureates' Biographies: Physiology or Medicine, 1942–1962,* by Nobel Foundation, pp. 587–97. Amsterdam: Elsevier, 1965.

36. Williams 1956.
37. See Motulsky 1996.

Beadle, G. W., and E. L. Tatum. 1941. "Experimental Control of Development and Differentiation: Genetic Control of Developmental Reactions." *American Naturalist* 75:107–16.

Bearn, A. G. 1993. *Archibald Garrod and the Individuality of Man.* Oxford: Clarendon Press.

Garrod, A. E. 1908. "The Croonian Lectures on Inborn Errors of Metabolism. Delivered before the Royal College of Physicians of London on June 18th, 23rd, 25th, and 30th, 1908." *Lancet* 2:1–7, 73–79, 142–48, 214–20.

———. 1909. *Inborn Errors of Metabolism.* London: Oxford University Press.

———. 1923. *Inborn Errors of Metabolism: The Croonian Lectures, Delivered before the Royal College of Physicians of London, in June, 1908.* London: Frowde, Hodder, and Soughton.

———. 1931. *The Inborn Factors in Disease: An Essay.* Oxford: Clarendon Press.

———. 1979. "The Incidence of Alkaptonuria: A Study in Chemical Individuality." In *Human Genetics: A Selection of Insights,* ed. W. J. Schull and R. Chakraborty. Stroudsburg: Dowden, Hutchinson, and Ross. First published in *Lancet* 2 (1902):653–56.

Goldstein, J. L., and M. S. Brown. 1997. "The Clinical Investigator: Bewitched, Bothered, and Bewildered—but Still Beloved." *J. Clin. Invest.* 99:2803–12.

Haldane, J. B. S. 1937. "The Biochemistry of the Individual." In *Perspectives in Biochemistry,* ed. J. Needham and D. E. Green, pp. 1–10. Cambridge: Cambridge University Press.

———. 1954. *The Biochemistry of Genetics.* New York: Macmillan Company.

Harris, H. 1963. *Garrod's Inborn Errors of Metabolism.* London: Oxford University Press.

Hörlein, H., and G. Weber. 1948. "über chronische familiäre Methämoglobinämie und eine neue Modifikation des Methämoglobins." *Dtsch. Med. Wochenschr* 72:476.

"Joseph Adams." 1886. In *Dictionary of National Biography,* ed. L. Stephen. London: Oxford University Press.

La Du, B. N., V. G. Zannoni, L. Laster, and J. E. Seegmiller. 1958. "The Nature of the Defect in Tyrosine Metabolism in Alkaptonuria." *J. Biol. Chem.* 230:251.

Lederberg, J. 1990. "Edward Lawrie Tatum, 1909–1975." *Biographical Memoirs, National Academy of Sciences, USA* 59:357–86.

Motulsky, A. G. 1959. "Joseph Adams (1756–1818): A Forgotten Founder of Medical Genetics." *AMA Arch. Intern. Med.* 104:490–96.

———. 1983. "Role of Medical Genetics in United States Academic Medicine." In *Academic Medicine: Present and Future,* ed. J. Z. Bowers and E. E. King, pp. 153–62. North Tarrytown, N.Y.: Rockefeller Archive Center.

———. 1996. "Invited Editorial: Nutritional Ecogenetics: Homocysteine-Related Arteriosclerotic Vascular Disease, Neural Tube Defects, and Folic Acid." *American Journal of Human Genetics* 58:17–20.

Obituary of Joseph Adams. 1818. Vol. 10, p. 167. London Medical Repository.

Pauling, L., H. A. Itano, S. J. Singer, and I. C. Wells. 1949. "Sickle Cell Anemia, a Molecular Disease." *Science* 110:543.

Sapp, J. 1990. *Where the Truth Lies: Franz Moewus and the Origins of Molecular Biology.* New York: Cambridge University Press.

Scriver, C. R., and B. Childs. 1989. *Garrod's Inborn Factors in Disease.* Oxford: Oxford University Press.

Weber, W. W. 1997. *Pharmacogenetics.* Oxford: Oxford University Press.

Williams, R. J. 1956. *Biochemical Individuality: The Basis for the Genetotrophic Concept.* New York: Wiley.

PART FOUR

Natural Selection and Evolution
from the Perspective of Prematurity

CHAPTER FIFTEEN

The Prematurity of Darwin's Theory of Natural Selection

Michael Ruse

It is customary and convenient when speaking of evolution to make a threefold division between the *fact* of evolution, the *path* of evolution, and the *mechanism* or *cause* (or *causes*) of evolution.[1] Defining these alternatives will lead us to the main topic of discussion: the extent to which natural selection can be considered an idea that appeared before its time. The answer is more intriguing than you might expect.

The *fact* of evolution is simply the idea that all organisms living and dead were produced by natural (that is to say, law-governed) processes from forms very different.[2] The diagram of evolution is usually thought to take the form of a tree—that is to say, it branches up and out from one or a very few forms—but for something to qualify as evolution as fact, this branching need not occur. At the beginning of the nineteenth century, the French biologist Jean-Baptiste de Lamarck proposed an evolution that saw new forms being spontaneously generated all the time from inorganic materials, and then a kind of process continuing in parallel up to the present.[3] He did in fact think that there would be some side paths leading off the main tracks, but these were very much secondary and nonessential.

The *path* of evolution (or *paths*, or, as they are known technically, the "phylogenies") is the particular way in which the evolutionary process works.[4] Strictly speaking, the discussion about whether the diagram of evolution is treelike or progresses upward in parallel falls into this category, as do discussions about issues such as the first appearance of multicellular organisms and of later occurring forms of organisms like reptiles, mammals, and primates. The much-debated question about whether birds arose from dinosaurs or came directly from other forms of reptiles also falls under this heading.

1. Ruse 1984.
2. R. J. Richards 1992.
3. Lamarck 1809; Burkhardt 1977.
4. Bowler 1976.

The *mechanism* or *cause* (or causes) of evolution is the actual process that drives organisms along their particular evolutionary paths. Included here, for instance, is the suggestion by Lamarck that a major cause of evolutionary change is the inheritance of acquired characters. Clearly, in talk about mechanisms, there will be much implicit about evolution as fact and path. If you see evolution as a branching path, then this will be reflected in your causal speculations, and, conversely, your causal thinking will influence your views on the path of evolution. Yet, for all the interconnections—because of the interconnections—it is valuable to separate the three categories, *fact, path* and *mechanisms,* even if one cannot draw a strict line between them.

THE FACT OF EVOLUTION

Claims about the fact of evolution go back to the eighteenth century. One could argue—indeed, elsewhere I have argued at great length—that evolution in this sense is very much the offspring of the ideology of progress.[5] In the eighteenth century, the time of the Enlightenment, many thinkers in Europe (primarily in Britain, France, and Germany) argued strongly that through human effort it is possible to improve intellectual understanding as well as our material, social, and industrial well-being.[6] Thanks to the triumphs of science and related areas like technology, people saw great gains being made intellectually and materially and believed that this onward and upward movement was capable of, if not indefinite extension, then at least the scaling of heights much higher than hitherto achieved. Fallout from this belief included the increasing number of figures arguing that in the organic world—the world of animal and plants—one should, and indeed did, find a matching continuous upward progression, an idea generally then used to justify claims about intellectual and social progress!

A prime example of one who argued in this way was the grandfather of Charles Darwin, the British physician Erasmus Darwin. He was a close friend of leading industrialists in the second half of the eighteenth century, and he expanded his commitments to intellectual and social progress into the world of organisms, arguing that there, too, we see upward progress, as in the poem below.

> ORGANIC LIFE beneath the shoreless waves
> Was born and nursed in Ocean's pearly caves;
> First forms minute, unseen by spheric glass,
> Move on the mud, or pierce the watery mass;
> These, as successive generations bloom,
> New powers acquire, and larger limbs assume;
> Whence countless groups of vegetation spring,
> And breathing realms of fin, and feet, and wing.

5. Ruse 1996.
6. Bury 1920.

Thus the tall Oak, the giant of the wood,
Which bears Britannia's thunders on the flood;
The Whale, unmeasured monster of the main,
The lordly Lion, monarch of the plain,
The Eagle soaring in the realms of air,
Whose eye undazzled drinks the solar glare,
Imperious Man, who rules the bestial crowd,
Of language, reason, and reflection proud,
With brow erect who scorns this earthy sod,
And styles himself the image of his God;
Arose from rudiments of form and sense,
An embryon point, or microscopic ens![7]

In turn, Darwin used the idea of progress as expressed in this poem to support his and his coterie's beliefs in the possibilities of intellectual and industrial improvement.[8]

As the nineteenth century began, similar ideas were promulgated elsewhere in Europe. Notwithstanding Lamarck's promotion of the inheritance of acquired characters (indeed, today known as Lamarckism), the chief aspect of his evolutionism was an upward progress from primitive, spontaneously produced organisms to the most sophisticated organisms seen today, namely, our own species, *Homo sapiens*.[9] In the years subsequent, like ideas were promoted elsewhere in Europe, perhaps most notoriously by the anonymous author (the Scottish publisher Robert Chambers) of the *Vestiges of the Natural History of Creation*.[10]

Yet, however widespread these ideas (that is to say, beliefs about the fact of evolution) had become by the middle of the nineteenth century, it is fair to say they had but the status of quasi or pseudo science, something akin to phrenology in those days and to scientology or transcendental meditation in our own time.[11] There were various reasons for the widespread nature of evolutionary beliefs and for their low status. In particular, evolution continued to be intimately associated with commitments to the ideology of progress. This was a philosophy that, by the middle of the nineteenth century, was almost commonplace in many circles, and yet it carried (as it always had) the taint of subversiveness or unorthodoxy. Of course, progress was seen as a challenge to the established Christian religion, which put its hopes on the providential intervention of the Creator rather than on achievements due to, and only to, human nature and effort.[12] One might fairly say it was precisely those who wanted to attack the establishment in every way who were attracted to both the philosophy of progress and its material offspring, evolution.

7. E. Darwin 1791, 1794–96, 1798, 1803; King-Hele 1963; McNeil 1987.
8. E. Darwin 1803, pp. 26–28.
9. Ruse 1979.
10. Chambers 1844.
11. Ruse 1996.
12. Sedgwick 1833.

Things changed dramatically in 1859. That year Charles Robert Darwin published his great work *On the Origin of Species*. In it, Darwin established once and for all the reasonableness of belief in evolution, transforming it from an idea dependent on the ideology of progress—and thus bearing the status of pseudo or quasi science—into an established scientific fact with much the same standing as, say, the Copernican heliocentric theory of the solar system. He did this chiefly by employing what his sometime friend and mentor William Whewell had labeled a "consilience of inductions":[13] Darwin brought beneath the hypothesis of evolution as fact many subsidiary areas of biological inquiry: behavior, paleontology, biogeography, systematics, anatomy, embryology, and more. He used the idea of evolution to explain different phenomena and the puzzling questions within these subsidiary areas, as when he explained the distributions of reptiles and birds on the Galapagos Archipelago as a product of migration from the South American mainland and subsequent divergence as the animals moved from island to island. Then, conversely, Darwin used his explanatory successes in these various areas as collective confirmation of the veracity of the fact of evolution. How could a false claim explain so much?

In a way, consilience functions much like evidential claims in a court of law.[14] One supposes that a certain suspect is the guilty culprit. Through this supposition one is able to explain various clues or pieces of evidence: bloodstains, torn fabric, broken alibis, and so forth. Then, one uses these explanatory successes collectively as overall confirmation of one's claims about the guilt of the suspect. For Darwin, the facts of biogeography, the information from paleontology, and so forth were the clues to the fact of evolution. Conversely, his ability to explain these various phenomena collectively supported the belief in evolution as a fact.

To sum up, Charles Darwin was not the first to propose the fact of evolution: claims about it go back at least a hundred years, and prominent among earlier advocates was his own grandfather, Erasmus Darwin. Nevertheless, it was Charles Darwin who made evolution as a *fact* a reasonable claim. Almost immediately, evolutionism in itself became part of the belief system of educated people.

THE PATHS OF EVOLUTION

Charles Darwin had much less to say about the second part of the evolutionary triad, the path or paths of evolution (phylogenies). Many others, however, were willing to take up this issue.[15] Indeed, from the beginning of the nineteenth century, biologists, including many who were themselves violently opposed to evolutionism as fact, had been working both on the overall structure of life's history here on earth

---

13. Whewell 1840; Ruse 1975b.
14. Ruse 1973.
15. Bowler 1976.

and on specific details. The fragmentary nature of the record was taken as a sign of real gaps rather than of evolutionary transitions that left no fossil aftermath.

One of the earliest and most important workers in this field at the beginning of the nineteenth century was among the greatest of all French biologists, the so-called father of comparative anatomy, Georges Cuvier. Although no evolutionist, Cuvier began detailed studies of the fossil record showing that, whether evolutionary or not, reptilian forms had appeared before mammalian forms. At the same time, through his work on comparative anatomy, Cuvier showed that one could relate and classify different forms on grounds of their physical similarities and differences.[16] Others took up this labor, particularly those influenced by German discoveries in embryology.[17] Hence, even during the time of Darwin, many of the basic elements of life's history were being articulated.

After the *Origin,* these elements were always interpreted within an evolutionary picture, and work proceeded at an even faster pace. This was primarily the result of fabulous fossil discoveries in the New World. Thus, for instance, Darwin's great supporter Thomas Henry Huxley used the New World fossil equine discoveries to trace out the phylogenic paths taken by the horse family, going from the earliest known form—the so-called eohippus, a doglike animal that ran on five toes—up to today's representatives, much larger animals with legs ending in a single digit.[18]

Although this is not really part of our story—at least, not yet—I should say that in the hundred years subsequent, the tracing of phylogenies has not matured in the way that the early evolutionists hoped and anticipated. It proved increasingly difficult to coordinate various pieces of evidence drawn from embryology, comparative anatomy, paleontology, biogeography, and more. Indeed, it seems fair to say that there has been a century or more of conflict, not about broad outlines but certainly about details.[19] Whether things have changed dramatically in recent years as a result of new techniques—not the least of which are those based on molecular studies—and the utilization of vastly superior databases held in computers remains to be seen. The still-unresolved debate about the evolution of birds and their possible dinosaur origins does not inspire confidence.

THE MECHANISM OF EVOLUTION

The third aspect of evolution, the mechanism or causes of evolution, is the focus of this essay. Here again, Charles Darwin comes to the fore. In the *Origin of Species,* drawing on work he had done some 20 years previously, Darwin proposed the mechanism that today is almost universally accepted as the chief causal force be-

16. Cuvier 1813, 1817; Coleman 1964.
17. Lenoir 1982; R. J. Richards 1987.
18. Huxley 1876, 1881, 1888.
19. Bowler 1996.

hind evolutionary change: natural selection, or, as it was often called (following a suggestion by Herbert Spencer), "the survival of the fittest."

In the *Origin*, Darwin introduced natural selection in a two-part argument, arguing first to the struggle for existence, and then from this (together with the variation he supposed to be widespread in the natural world), natural selection. Thus in chapter 3, he wrote:

> A struggle for existence inevitably follows from the high rate at which all organic beings tend to increase. Every being, which during its natural lifetime produces several eggs or seeds, must suffer destruction during some period of its life, and during some season or occasional year, otherwise, on the principle of geometrical increase, its numbers would quickly become so inordinately great that no country could support the product. Hence, as more individuals are produced than can possibly survive, there must in every case be a struggle for existence, either one individual with another of the same species, or with the individuals of distinct species, or with the physical conditions of life. It is the doctrine of Malthus applied with manifold force to the whole animal and vegetable kingdoms; for in this case there can be no artificial increase of food, and no prudential restraint from marriage.[20]

Following this, in chapter 4, he moved on to natural selection:

> Let it be borne in mind in what an endless number of strange peculiarities our domestic productions, and, in a lesser degree, those under nature, vary; and how strong the hereditary tendency is. Under domestication, it may be truly said that the whole organisation becomes in some degree plastic. Let it be borne in mind how infinitely complex and close-fining are the mutual relations of all organic beings to each other and to their physical conditions of life. Can it, then, be thought improbable, seeing that variations useful to man have undoubtedly occurred, that other variations useful in some way to each being in the great and complex battle of life, should sometimes occur in the course of thousands of generations? If such do occur, can we doubt (remembering that many more individuals are born than can possibly survive) that individuals having any advantage, however slight, over others, would have the best chance of surviving and of procreating their kind? On the other hand we may feel sure that any variation in the least degree injurious would be rigidly destroyed. This preservation of favourable variations and the rejection of injurious variations, I call Natural Selection.[21]

Obviously, in the century and a half since Darwin published the *Origin,* there has been much debate about natural selection. I certainly would not want to claim that our understanding today is exactly that of Charles Darwin's. In particular, much of today's thinking about natural selection is less about selection at the organismic level, which was Darwin's focus, and more about selection at the genetic level.[22] Today's evolutionists tend to think of selection as a differential reproduc-

20. C. Darwin 1859, p. 63.
21. Ibid., pp. 80–81.
22. Dawkins 1976, 1986.

tion of the ultimate units of heredity, the genes, whether these are interpreted as units or subdivided into molecular components.[23] This approach was foreign to Darwin, although in other respects we are closer to Darwin today than we were even 30 years ago. For many years, people thought in terms of selection at a group level, which was anathema to Darwin and is no favorite of today's evolutionists.[24]

However, important though the development of natural selection may be, my interest in this topic is at best secondary. What concerns me is the extent to which natural selection was an idea whose time had come. Specifically in the Darwinian context, it was an idea whose time had come with, and only with, the publication of the *Origin of Species*. This is not an entirely new question: in recent years, some of its aspects have been much discussed by historians of evolutionism.[25] Just as Darwin was not the first to think of the fact of evolution, he was also not the first to discover the idea of natural selection. Before discussing the implications, let me acknowledge that several other people came up with the idea—some did so before Darwin himself hit on it (in late autumn of 1838), and some may have hit on the idea after Darwin first thought of natural selection but before he published anything on the subject. (I am referring to the concept rather than the actual term "natural selection." As far as I know, before Darwin no one used this term, although of course the idea of and term "selection" were not new with Darwin. They were common among animal and plant breeders.)

The most significant of these anticipators was the naturalist Alfred Russel Wallace, who in early 1858 hit on the idea of selection brought on by species' struggle for survival. He sent an essay on this subject to Darwin, and as soon as it arrived (June 1858), Darwin arranged for its publication together with some writings of his own.[26] Immediately, Darwin set about writing the *Origin*, which appeared some fifteen or so months later. But let us leave Wallace for the moment and turn to others. There were at least two people (acknowledged by Darwin himself) who came to the idea of natural selection before Darwin discovered it. The first was William Wells, an American physician living in England at the beginning of the nineteenth century. In 1813 he gave a paper before the Royal Society outlining an idea that closely resembled the Darwinian notion of selection. According to Wells:

> Those who attend to the improvement of domestic animals, when they find individuals possessing, in a greater degree than common, the qualities they desire, couple a male and female of these together, then take the best of their offspring as a new stock, and in this way proceed, till they approach as near the point of view, as the nature of things will permit. But, what here is done by art, seems to be done, with equal efficacy, though more slowly, by nature, in the formation of varieties of mankind, fitted for the

23. Lewontin 1974.

24. Ruse 1980.

25. Eiseley 1958, 1959; Limoges 1971; Beddall 1972, 1973; Schwartz 1974; Sheets-Pyenson 1981; Ruse 1975b; Kohn 1980.

26. Darwin and Wallace 1858.

country which they inhabit. Of the accidental varieties of man, which would occur among the first few and scattered inhabitants of the middle regions of Africa, some one would be better fitted than the others to bear the diseases of the country. This race would consequently multiply, while the others would decrease, not only from their inability to sustain the attacks of disease, but from their incapacity of contending with their more vigorous neighbours.[27]

The paper—"An account of a white female, part of whose skin resembles that of a negro"—seems to have been ignored, despite the fact that in 1818 it was incorporated into a fairly well known volume devoted to a discussion of the nature and formation of dew. But although the discussion of dew was picked up and much admired (particularly by John Herschel in his celebrated *Preliminary Discourse on the Study of Natural Philosophy*), the discussion of natural selection seems to have been overlooked entirely until it was brought to Darwin's attention around 1860.[28] Darwin added a historical sketch to the *Origin* in which Wells was briefly, if somewhat reluctantly, noted.[29]

The second person to anticipate natural selection was a Scottish botanical writer, Patrick Matthew. In 1831 he published a work titled *On Naval Timber and Arboriculture*. Matthew wrote:

> The self-regulating adaptive disposition of organized life may, in part, be traced to the extreme fecundity of Nature, who, as before stated, has, in all the varieties of her offspring, a prolific power much beyond (in many cases a thousandfold) what is necessary to fill up the vacancies caused by senile decay. As the field of existence is limited and pre-occupied, it is only the hardier, more robust, better suited to circumstance individuals, who are able to struggle forward to maturity, these inhabiting only the situations to which they have superior adaptation and greater power of occupancy than any other kind; the weaker, less circumstance-suited, being prematurely destroyed. This principle is in constant action, it regulates the colour, the figure, the capacities and instincts; those individuals of each species, whose colour and covering are best suited to concealment or protection from enemies, or defence from vicissitude and inclemencies of climate, whose figure is best accommodated to health, strength, and support; whose self-advantage according to circumstances—in such immense waste of primary and youthful life, *those* only come forward to maturity from the strict ordeal by which Nature tests their adaptation to her standard of perfection and fitness to continue their kind by reproduction.
>
> From the unremitting operation of this law acting in concert with the tendency which the progeny have to take the more particular qualities of the parents, together with the connected sexual system in vegetables, and instinctive limitation to its own kind in animals, a considerable uniformity of figure, colour, and character, is induced, constituting species; the breed gradually acquiring the very best possible adaptation of these

27. Wells 1818; reprinted in McKinney 1971, p. 26.
28. Herschel 1831.
29. See Peckham 1959.

to its condition which it is susceptible of, and when alteration of circumstance occurs, thus changing in character to suit these as far as its nature is susceptible of change.[30]

Again, however, no great note was taken of this work, and it did not come to general attention until after the *Origin* was published, Darwin making mention of Matthew in the historical introduction that he added to the *Origin*.

I know of two other sources where natural selection is mentioned, both of which Darwin looked at but did not note. One was an essay by the naturalist Edward Blyth, writing in 1835:

> In a large herd of cattle, the strongest bull drives from him all the younger and weaker individuals of his own sex, and remains sole master of the herd; so that all the young which are produced must have had their origin from one which possessed the maximum of power and physical strength; and which, consequently, in the struggle for existence, was the best able to maintain his ground, and defend himself from every enemy. In like manner, among animals which procure their food by means of their agility, strength, or delicacy of sense, the one best organised must always obtain the greatest quantity; and must, therefore, become physically the strongest, and be thus enabled, by routing its opponents, to transmit its superior qualities to a greater number of offspring.[31]

The other was a pamphlet by the animal breeder Sir John Sebright that Darwin annotated carefully. Sebright wrote, "A severe winter, or a scarcity of food, by destroying the weak and the unhealthy, has all the good effects of the most skillful selection."[32] Darwin wrote in the margin on the previous page: "In plants man presents mixtures, varies conditions, and destroys, the unfavorable kind—could he do this last effectively and keep on the same exact conditions for many generations he would make species, which would be infertile with other species."[33]

After 1838, when Darwin discovered the idea of natural selection for himself, but before 1859, when he announced his discovery in the *Origin,* the clearest anticipation—other than that of Alfred Russel Wallace in 1858—was by Herbert Spencer, Darwin's compatriot and fellow evolutionist, who started pushing his ideas in the early 1850s. In an essay on population, Spencer clearly hit on the idea of natural selection, particularly as it applies to humans.

> And here it must be remarked, that the effect of pressure of population, in increasing the ability to multiply, is not a uniform effect, but an average one. In this case, as in many others, Nature secures each step in advance by a succession of trials, which are perpetually repeated, and cannot fail to be repeated, until success is achieved. All mankind in turn subject themselves more or less to the discipline described; they either may or may not advance under it; but, in the nature of things, only those who *do* advance under it eventually survive. For, necessarily, families and races whom this

30. Matthew 1831, reprinted in McKinney 1971, p. 38.
31. Blyth 1835, reprinted in McKinney 1971, p. 49.
32. Sebright 1809, pp. 15–16.
33. See also Ruse 1975a.

increasing difficulty of getting a living which excess of fertility entails, does not stimulate to improvements in production—that is, to greater mental activity—are on the high road to extinction; and must ultimately be supplanted by those whom the pressure does so stimulate. This truth we have recently seen exemplified in Ireland. And here, indeed, without further illustration, it will be seen that premature death, under all its forms, and from all its causes, cannot fail to work in the same direction. For as those prematurely carried off must, in the average of cases, be those in whom the power of self-preservation is the least, it unavoidably follows, that those left behind to continue the race are those in whom the power of self-preservation is the greatest—are the select of their generation. So that whether the dangers to existence be of the kind produced by excess of fertility, or of any other kind, it is clear, that by the ceaseless exercise of the faculties needed to contend with them, and by the death of all men who fail to contend with them successfully, there is ensured a constant progress towards a higher degree of skill, intelligence, and self-regulation—a better co-ordination of actions—a more complete life.[34]

There were perhaps others, including, most surprisingly, Richard Owen, who was later to become the great opponent of the Darwinians. After the *Origin* was published, poor Owen did not know whether to anathematize the whole idea of selection or to take credit for its discovery![35]

Finally we come to 1858 and Alfred Russel Wallace. Wallace, who had by this time become a professional collector traveling the Southern Hemisphere in pursuit of specimens he could bring back to England and sell, had already announced publicly his adherence to something that was at least very close to an evolutionary position. He had become an evolutionist even in the mid-1840s, under the influence of Chambers's *Vestiges,* but it was only in 1855 that he made his position public— or nearly so. In that year he published an essay (which gained some attention— Blyth was one who drew Darwin's attention to it) suggesting that new species always occur around, or replace, very similar species. One did not absolutely have to take this in an evolutionary sense, I suppose, but it would have been difficult not to. Then, in 1858 came Wallace's view of natural selection:

Let some alteration of physical conditions occur in the district—a long period of drought, a destruction of vegetation by locusts, the irruption of some new carnivorous animal seeking "pastures new"—any change in fact tending to render existence more difficult to the species in question, and tasking its utmost powers to avoid complete extermination; it is evident that, of all the individuals composing the species, those forming the most numerous and most feebly organized variety would suffer first, and, were the pressure severe, must soon become extinct. The same causes continuing in action, the parent species would next suffer, would gradually diminish in numbers, and with a recurrence of similar unfavourable conditions might also become extinct. The superior variety would then alone remain, and on a return to favourable

34. Spencer 1852a, pp. 266–67.
35. Owen 1866, pp. 1, xxxiv, quoting Owen 1850, p. 15.

circumstances would rapidly increase in numbers and occupy the place of the extinct species and variety.

The *variety* would now have replaced the *species*, of which it would be a more perfectly developed and more highly organized form. It would be in all respects better adapted to secure its safety, and to prolong its individual existence and that of the race. Such a variety *could not* return to the original form; for that form is an inferior one, and could never compete with it for existence. Granted, therefore, a "tendency" to reproduce the original type of the species, still the variety must ever remain preponderant in numbers, and under adverse physical conditions *again alone survive*. But this new, improved, and populous race might itself, in course of time, give raise to new varieties, exhibiting several diverging modifications of form, any of which, tending to increase the facilities for preserving existence, must, by the same general law, in their turn become predominant. Here, then, we have *progression and continued divergence* deduced from the general laws which regulate the existence of animals in a state of nature, and from the undisputed fact that varieties do frequently occur.[36]

There is no question that Wallace really hit on natural selection, and although there are some questions about exactly how similar it was to Darwin's vision, it was obviously close enough that now the idea had arrived—entered the scientific arena, one might say. Either Darwin had to move and publish his own ideas—which he did—or he would be left behind as one who had missed the boat. But, although this is all very interesting, it is not surprising. Darwin got the idea. Wallace got the idea. Darwin had to move.

What is both interesting and surprising are those earlier anticipations of Darwin's idea. Moving now to interpretation, the question to ask is why none of these early anticipations of natural selection really caught fire and why none of them gained the attention that Darwin's treatment did. Related to this is the question of whether Darwin was guilty of sharp practice inasmuch as he never gave credit to those whom he read—those who had already grasped the idea of selection.

THE ANTICIPATORS

Owen and Spencer I will leave on one side. I am not sure that Owen ever really did get the idea—he certainly made nothing of it. It may well be that, even before the *Origin*, Owen was an evolutionist, but if so it would have been set against his idealistic world-picture: a world picture that owed much to the German morphological movement known as *Naturphilosophie* and which had little or no place for a materialistic mechanism like natural selection.[37] And although Spencer really did get the idea, he too made nothing of it. It is true that, from the point of writing his essay, Spencer was an ardent and public evolutionist.[38] But, as he himself later ac-

---

36. Wallace 1858, reprinted in McKinney 1971, pp. 94–95.
37. Ruse 1979, 1996; E. Richards 1987.
38. Spencer 1852a,b, 1857.

knowledged, he did not pick up on this particular mechanism of selection, only realizing its full implications after Darwin had published. Even then, he never regarded natural selection as the chief mechanism of evolutionary change, preferring always to think in Lamarckian terms. In the Spencerian world it was the inheritance of acquired characteristics that really counted.[39]

What now about those who hit on selection before Darwin? Basically the answer is the same for all the anticipators and is no great mystery. Take Wells and Matthew. As with Spencer later, neither of them really made of natural selection a full-blown mechanism for evolutionary change. Moreover, even if they had, remember that at that time the very idea of evolution was looked down on as no more than a quasi or pseudo science. Neither Wells nor Matthew provided the kind of case for the fact of evolution that Darwin did. So really, without this background, it is hardly surprising that no one thought of these earlier anticipations of natural selection as significant scientific advances. It is true that Matthew was not only a believer in selection but probably also an evolutionist; but the context of the anticipation of selection was not one of advocacy for evolution. In fact, the context was that of the skills and necessities required for successful tree growing! Matthew was not really someone pushing a major evolutionary thesis.

Sebright likewise fits the pattern. I do not dismiss him as unimportant, but he offered only a fragmentary idea and certainly not one embedded in an evolutionary context—not one that represented a theory for change. This is no criticism but simply a statement about what he was, or rather what he was not, trying to do. And Blyth has to be treated the same way, even though he clearly saw selection as coming from the struggle for existence. He was no evolutionist at this time, believing rather that the struggle and consequent selection preserves the status quo rather than changes it. Let me now requote the passage given above, adding the sentences before and after:

> It is worthy of remark, however, that the original and typical form of an animal is in great measure kept up by the same identical means by which a true *breed* is produced. The original form of a species is *unquestionably* better adapted to its *natural* habits than any modification of that form; and, as the sexual passions excite to rivalry and conflict, and the stronger must always prevail over the weaker, the latter, in a state of nature, is allowed but a few opportunities of continuing its race. In a large herd of cattle, the strongest bull drives from him all the younger and weaker individuals of his own sex, and remains sole master of the herd; so that all the young which are produced must have had their origin from one which possessed the maximum of power and physical strength; and which, consequently, in the struggle for existence, was the best able to maintain his ground, and defend himself from every enemy. In like manner, among animals which procure their food by means of their agility, strength, or delicacy of sense, the one best organised must always obtain the greatest quantity; and must, therefore, become physically the strongest, and be thus enabled, by rout-

39. Spencer 1904; E. Richards 1987; Ruse 1996.

ing its opponents, to transmit its superior qualities to a greater number of offspring.
The same law, therefore, which was intended by Providence to keep up the typical
qualities of a species, can be easily converted by man into a means of raising different varieties; but it is also clear that, if man did not keep up these breeds by regulating the sexual intercourse, they would all naturally soon revert to the original type.[40]

This is not evolution through natural selection.

There is little need now to linger over the subsidiary question of whether Darwin was guilty of sharp practice in his use of Blyth and Sebright. They were simply not offering what Darwin was to offer: a mechanism for evolutionary change.
If we have learned anything from recent writings on the nature of science—one
thinks of Thomas Kuhn's *Structure of Scientific Revolutions*—it is that ideas have to be
considered in context.[41] An idea on its own or an idea in one setting is absolutely
not an idea in another setting. Now you see a duck, now you see a rabbit, to use
the famous example from Gestalt psychology. Now you see an offhand comment
perhaps proving something in one direction. Now you see a deliberate idea for a
major scientific advance, proving something in the very direction that earlier anticipations had declared blocked. It is interesting and significant that, although Blyth
became interested in evolution even to the extent of corresponding with Darwin
on the subject, not for one moment did he suggest that he had anticipated or been
unduly ignored by Darwin.

What about Wallace, the really serious and interesting case? In one respect, there
is nothing to explain. From the first, Wallace has been venerated as one of the discoverers of natural selection and has been accorded the honor that that discovery
merits. However, in another respect, explanation is demanded. For all that his partisans (who tend to be somewhat obsessive on the subject) would have us act and
believe otherwise, posterity has always given Wallace a position secondary to Darwin's. Is this fair? My sense is that it is. Since I doubt that anything that anyone says
at whatever length is going to change minds on this matter, let me say starkly that
although it is true that Wallace's essay fully grasps the idea of natural selection
and puts it in an evolutionary context, it is but an essay and it in no sense provides
an overall account of evolution. Certainly there is no attempt to explain the fact of
evolution the way that Darwin did. Wallace got credit for his discovery. It was measured credit, but looking at things from afar, it was as much credit as he warranted.

WHY DARWIN SUCCEEDED WITH NATURAL SELECTION

If others had been so unsuccessful in pushing natural selection (Wallace ex-
cepted), why was Charles Darwin so successful? There are three reasons. First,
there is the simple fact that Darwin made a much stronger case for natural selec-

40. Blyth 1835, reprinted in McKinney 1971, p. 49.
41. Kuhn 1962.

tion than anyone else. He provided the famous analogy with artificial selection (an analogy, incidentally, that Wallace denied) that showed how effective selection could be in the domestic world, and how great the change can be—and how this is precisely what one finds in the natural world. (Sebright made much of domestic selection, but the analogy was mentioned rather than discussed.) Then, when it came to specific items of explanation, Darwin made much use of natural selection, most decisively in his discussion of embryology. Why, asked Darwin, is it that—as noted by earlier biologists—organisms vastly different as adults have embryos that are nearly or completely identical? Simply because, Darwin suggested, natural selection has torn the adult forms apart but left the juvenile forms alike. Organisms within the womb experience much the same environments and stresses; and so, unless there are specific reasons for it, they are not subject to diversifying forces, even though their adult forms are. The principle is the same in the domestic world, Darwin noted, as when breeders concentrate their attentions on the adult forms, quite ignoring juvenile similarities:

> Fanciers select their horses, dogs, and pigeons, for breeding, when they are nearly grown up: they are indifferent whether the desired qualities and structures have been acquired earlier or later in life, if the full-grown animal possesses them. And the cases just given, more especially that of pigeons, seem to show that the characteristic differences which give value to each breed, and which have been accumulated by man's selection, have not generally been accumulated by man's selection, have not generally first appeared at an early period of life, and have been inherited by the offspring at a corresponding not early period.[42]

The case for selection as a mechanism was made in far, far greater detail than by any thinker previously.

The second reason for Darwin's success had to do with the context in which he introduced natural selection. He presented evolution as mechanism as a case for evolution as fact. Recall that Darwin provided an overall argument for the fact of evolution: in other words, he upgraded the status of evolution as a science far beyond what it had achieved before. Moreover and parenthetically, although there is good reason to think that Darwin was himself an ardent progressionist, he provided a picture of evolution that was in some sense independent of the ideology of progress. In fact, especially in later editions of the *Origin*, Darwin did read the ideology into his evolutionary picture; but the point is that the picture did not depend essentially on the idea of progress. Rather, evolution was justified through consilience, as discussed above.[43]

This context paid off for natural selection. On the one hand, one had the need of a mechanism, and Darwin's providing it, and arguing in the way he did, made the indirect case for evolution as fact that much stronger. On the other hand, the

---

42. C. Darwin 1859, p. 401.
43. Ruse 1996.

reasonableness of the fact of evolution led naturally to support for the proffered mechanism of evolution. It was not just a throwaway notion that could be ignored. It was prominent and given a crucial role. However you decided ultimately, you had to take natural selection seriously.

The third reason Darwin succeeded with natural selection is more sociological. It is the inverse to a major reason why his predecessors failed. By the end of the 1850s, Darwin was a well-known public figure. He had published one of the best travel books of the early Victorian period, the *Voyage of the Beagle,* and then had consolidated his reputation as a scientist not only with his work in geology but with detailed studies of barnacles.[44] He was a Royal Society prize winner and much respected as a student of the life sciences. These accomplishments prepared the way for the working of what sociologists of science refer to as the "Matthew Principle": major scientists get lots of credit for the work they have done—more so than minor scientists would for the same work—and they get this credit because they have, as it were, earned the right to be taken seriously. (The "Matthew" named here is the apostle, although in fact the pertinent words are those of Jesus: "For whomsoever hath, to him shall be given" [Matt. 13:12].)

This was paradigmatically the case with Darwin. He was a serious, respected, well-known scientist, and when he put forward the idea of evolution through natural selection, he was taken seriously for that very reason. Natural selection came with a good pedigree, and it could not be ignored. Who knew about Patrick Matthew, for instance? Why then should they take his ideas seriously? At least, why should they take seriously his ideas as science? And the same is true of others. Sebright was certainly well-known and respected in the world of breeders, but that is precisely the point. It was the world of breeders and not of science. Darwin was fortunate in that he (unlike someone like Wallace) had family connections that put him in direct contact with the breeders' world, but it was his genius and effort to make his own scientific status and to transfer to his world of science the ideas of the breeders. (It is interesting to note that critics like Owen recognized Darwin's status. Even as they criticized, they were careful to note that putting forward natural selection was no foolish thing to do, even if it did not have the full implications that Darwin suggested.)

BUT WAS NOT NATURAL SELECTION PREMATURE?

In answering the question as to why Darwin succeeded with natural selection, whereas his predecessors did not, I am of course assuming that Darwin did succeed. But surely one could argue that this question is based on a false premise. We are assuming that natural selection was in 1859 an idea whose time had come. But is this really the truth? No one would say that natural selection was a complete

---

44. C. Darwin 1851a,b, 1854a,b.

failure. As just mentioned, even critics were respectful of the idea, and indeed just about everybody—including nonevolutionists—agreed that natural selection had some role to play in life's history. Moreover, I suspect that many found it easier to accept the fact of evolution under the cover of entertaining Darwin's mechanism (if only to reject it). They could thereby preserve some facade of objectivity and an unbroken critical approach to hitherto unacceptable ideas! The spotlight was taken off the fact that they were accepting what had been rejected before.

Yet generally speaking, natural selection was something of a flop.[45] Very few people accepted natural selection completely, other than Darwin himself, and even he by the 1870s (thanks primarily to problems about the nature of heredity) was starting to invoke other mechanisms in a significant way. It is true that there were selectionists: one was the naturalist Henry Walter Bates, a sometime traveling mate of Wallace. He used selection to explain the mimicry patterns of butterflies.[46] But Bates was the exception rather than the rule. Wallace himself also did some excellent selection-based work on butterflies, but at the same time he was becoming highly dubious about natural selection as applied to our own species.[47] And others were equally skeptical about selection, although they had different reasons for reservation. Thomas Henry Huxley always downplayed the significance of selection and opted rather for some kind of evolutionary theory by jumps or, as it is better known, some kind of "saltationary theory."[48] Others, like Darwin's North American supporter Asa Gray, opted for directed variations.[49] Spencer, as I have mentioned, always preferred the inheritance of acquired characters, or Lamarckism. Then, later in the century, other mechanisms were endorsed. For instance, many North Americans, particularly those interested in the paleontological record, adopted some kind of evolution through an inner momentum: so-called orthogenesis, which apparently took organisms up to and beyond their adaptive peaks.[50]

In major respects, therefore, natural selection, even in Darwin's *Origin*, was a premature idea! There are four main reasons. First, conceptually, biology was not ready for natural selection. Somewhat paradoxically, it was less ready for natural selection when Darwin published in 1859 than it had been when he discovered the mechanism in 1838. At the earlier period, natural theology of a distinctively British kind had ruled supreme.[51] Everyone in the English-speaking biological world was looking for evidence of adaptation, characteristics that help organisms survive and reproduce. This is the chief support of function, the major premise in the so-called teleological argument for the existence of God (otherwise known as the argument from design). This form of argumentation, as represented at the beginning of the nineteenth cen-

---

45. Ruse 1979; Ellegard 1958; Bowler 1988.
46. Bates 1862, 1892.
47. Wallace 1864, 1866, 1870.
48. Huxley 1893.
49. Gray 1876.
50. Osborn 1894, 1917, 1929.
51. Ruse 1975c; Gillespie 1950.

tury by Archdeacon William Paley in his *Natural Theology*, was the background to every-one's thinking about the organic world and was the major reason most people were not evolutionists.[52] They could see no way in which blind law could explain intricate adaptation. Cuvier is the paradigm in this respect. His whole approach to the organic world, deeply antievolutionary, was functionalist through and through.[53] His major principle of the "conditions of existence" suggested that the key to understanding organic nature is function, and from this he went on to argue that evolution is not merely empirically false but conceptually impossible. (Although Cuvier was French, he had a major influence in Britain, a fact probably connected to his being a Protestant.)

Darwin spoke directly to this issue of adaptation and function when he came up with the notion of natural selection. The whole point of the mechanism is that it leads not simply to change, but to change of a particular kind. It is change that promotes adaptive complexity. (This is a point made much of by recent biologists, particularly Richard Dawkins.)[54] Yet although this was a major concern of the 1830s (as an undergraduate Darwin was exposed to Paley, and Cuvier died only at the beginning of the decade) by the 1850s, German biology had come to dominate thinking, including British thinking, about organisms. People like Owen and then Huxley (particularly the latter) were far more impressed by isomorphisms, or so-called homologies, between organisms than by their adaptive complexity.[55] If one is an anatomist or paleontologist, then generally speaking one is working with organisms that are not only dead but no longer functioning. Although it is possible to infer function, much more immediate are the similarities and differences between different sets of organisms. This is the key to successful anatomy—or at least it was considered the key to successful anatomy.

Consequently, by the end of the 1850s, the whole search for something that would explain adaptation was a lot less pressing than it had been in the 1830s, simply because the question was no longer so pressing. Hence when Darwin came up with natural selection, it was in a way a mechanism in search of an area to explain! And the failure to find this area was a major reason why, in the 1860s and indeed through the rest of the century and into this century, natural selection never caught fire. People simply had no great use for it. Combine this with the fact that the only people now enthusiastic about organic adaptation were those who tended to have religious axes to grind and who were unlikely to enthuse over a mechanism that would make God that much more remote (Asa Gray is a prime example), and natural selection was doomed.[56]

The second reason that the idea of natural selection did not catch fire after publication of the *Origin* is the internalist one of its quality and adequacy (as then per-

52. Paley 1802.
53. Coleman 1964.
54. Dawkins 1986.
55. Russell 1916.
56. Ellegard 1958.

ceived) as science. The fact is that there were epistemic problems with a selection-based evolutionism. Most significantly, the lack of an adequate theory of heredity led to unanswerable questions about the effectiveness of selection: would not its efforts be swamped by countereffects of the natural blending nature of the usual processes of descent? Would not any virtues from new variations be lost almost at once by the diluting effects of breeding through a few generations? Darwin had answers to these and related questions, but he himself acknowledged their power and, as noted, in later editions of the *Origin,* selection is supplemented with other mechanisms, notably the Lamarckian inheritance of acquired characters.[57] Add to this the other difficulties: notably, criticism by physicists who (unaware of what we now know of the heating effects of radioactive decay) were arguing that selection is far too leisurely a mechanism for the supposed time span of earth's history (a period estimated from the supposed cooling of the globe from its incandescent beginnings).[58] It is no wonder that the scientific community did not rush to embrace natural selection as the key mechanism to explain life's history.

The third reason for the problems of natural selection is one which devolves on the personalities of the people involved. I have suggested that Darwin's success—and he was very successful—depended to a great extent on his status within the scientific community. Darwin had worked long and hard as a biologist and networked from the beginning with the right people. Hence, when he published the *Origin,* he was taken seriously in a way that earlier evolutionists had not been. However, as is well-known, Darwin was a recluse, sick for most of his adult life.[59] As a result, when Darwin wanted to push natural selection, he had to rely on others. Wallace—an enthusiastic champion of barmy ideas about phrenology and spiritualism—was no great help, and would not have been even had he not been going soft on selection. Hence, Darwin had to rely on people like Thomas Henry Huxley. However, Huxley was a leading morphologist and paleontologist and, as just explained, had no great need of natural selection in his research. Naturally, Huxley himself was not prepared to make special efforts in support of the mechanism. Consequently Darwin was left to stand impotently on the sidelines, watching as evolution as fact became a great success, but evolution as mechanism—at least his own proposals—were downplayed and to a great extent discarded.

The fourth reason is perhaps the most important of all. During the 1860s and 1870s—the decades after the *Origin* was published—in particular, biology was being professionalized.[60] This was when people like Huxley in England, and others in Germany and America, were founding departments of biology in universities,

57. Vorzimmer 1970; Hull 1973.
58. Burchfield 1975.
59. Browne 1995.
60. Carron 1988.

attracting students, starting journals, carrying out studies and experiments, and so forth. In addition, they were getting support for what they were doing. In the biological world, the two areas of science that moved forward rapidly were physiology and morphology. Physiology was successful particularly because Huxley and friends sold it to the medical profession, and it became an integral part of medical training. On the one hand, doctors would now have a much more scientific background to what they were doing. On the other hand, with the explicit training, the medical profession could more easily erect barriers, keeping out untrained pretenders to their profession. Then morphology was sold by Huxley and friends to the school-teaching profession as something that would be almost a modern-day equivalent of doing the classics. Morphology entered into biology classes as the kind of hands-on activity necessary to training youth. (There is little surprise in finding that Huxley's university establishment in South Kensington was, by the 1870s, an important site of training for schoolteachers.)

The trouble was that no one could see how evolution could be fitted into this scenario. There was no money to be made out of it. If one produced evolutionists, what then would they do for a living, and how would one then support further researches in the subject? Hence, far from being incorporated into professional science, evolution remained at the level of popular science, a consequence compounded by the fact that the one place that welcomed evolution was the museum world.[61] Here, evolutionists increasingly found their niche as they prepared and conserved collections and offered displays based on evolution to visitors, particularly the young. But, even as people were doing this, there was no real call for natural selection. Selection is the kind of subject that, when studied professionally, requires experiments on fairly fast-breeding organisms. (It is significant that the early work on selection done by Bates and Wallace was composed of studies on butterflies.) Within the museum world there really was no call for any of this, since by and large, museums are collections of dead organisms. And, when it came to preparing displays, once again selection played but a minor role. Curators preparing a museum display are, if anything, most interested in portraying phylogenies. At the American Museum of Natural History, for instance, or to the British Museum of Natural History, visitors wanted scenarios showing how the horse had evolved, or discussions of the origins of the dinosaurs and their unhappy fate, or tales of human prehistory. So the idea of selection was neglected.

THE LATER HISTORY OF SELECTION

For these interrelated reasons, I suggest that natural selection in the *Origin* really was an idea before its time. But this raises a final question as to why today Dar-

61. Rainger 1991.

win is so venerated not simply as the father of the fact of evolution but also as the author of the most significant cause of evolution: natural selection. An answer is found readily. By the 1930s and 1940s, during the rise of a professional evolutionism based on natural selection—the so-called synthetic theory of evolution, or the appropriately named "neo-Darwinism"—times had changed and all the factors listed in the last section started to work in favor of recognizing natural selection rather than against it. On the one hand, as noted earlier, after some 60 or 70 years, people began to realize that all the phylogeny-tracing by means of homologies and so forth had really run out of steam. By the 1920s and 1930s people concluded that attempts to trace the paths of evolution in any definitive way were simply throwing up contradictions and unsupported hypotheses.[62] Hence, the enthusiasms of the immediate post-Darwinian period, and the conceptual ideas that had dominated it, were looking decidedly jaded. This included not only the drive to establish evolution as path (somewhat at the expense of evolution as mechanism) but also the extreme emphasis on Germanic-type ideas (such as homology) at the expense of British enthusiasms for adaptation and function.

At the same time, one had by now the discovery of the Mendelian mechanism of heredity—which opened the way in the 1920s for a number of theoretical biologists, notably R. A. Fisher and J. B. S. Haldane in England and Sewall Wright in America—to show how one could put together natural selection with genetics to provide a fundamental theoretical picture of evolutionary change.[63] There were significant differences between the British picture and the American picture, but overall, conceptually natural selection could now be given a full and significant role in the theory. No longer was it simply a mechanism in search of questions. Nor was it a mechanism surrounded by internalist scientific problems. Not only were such issues as the supposed swamping of selection by the forces of heredity dismissed, but physicists' worries about the limited length of earth history had been assuaged by new findings and theories in their own science. Natural selection was now an idea that worked.

Then again, at a more social level, in the 1930s and 1940s those individuals pushing natural selection were in a far stronger position than Darwin and the one or two others like him who had favored selection back in the 1860s. In England there was Fisher, who, although a difficult man, was astute at manipulating and making his way in the scientific community. He not only gained for himself significant university posts but was also involved in such things as the founding of a journal *(Heredity)* that could be used as a vehicle for evolutionary ideas. In addition, he networked with younger people who were prepared to try to put some empirical flesh on his theoretical spec-

---

62. Nyhart 1995; Bowler 1996.
63. Fisher 1930; Haldane 1932; Wright 1931, 1932.

ulations. Most notable in this respect was E. B. Ford at Oxford. Ford founded the whole school of ecological genetics, which was in fact a kind of neo-Batesian movement, to study the effects of natural selection in the wild, particularly as it works on such fast-breeding organisms as butterflies and snails and the like.[64] (The social and the conceptual link up here, for adaptation rather than homology is the key factor in studying such organisms.)

In America, there was not only Sewall Wright but also, most notably, the Russian-born American population geneticist Theodosius Dobzhansky.[65] Dobzhansky too was involved in networking with other biologists, in particular the ornithologist Ernst Mayr, the paleontologist G. G. Simpson, and the botanist G. L. Stebbins.[66] Together they were involved in forming a society for the study of evolution and founded a journal *(Evolution)*. Also, as Ford in England built around himself a group of students prepared to work on evolutionary ideas using natural selection, so in America Dobzhansky built around himself a group of students prepared to work on problems of selection, both in the wild and in experimental conditions. His favored organism of study was the fruit fly, and he and his students showed just how very effective natural selection could be. Once again, therefore, one had a mechanism that could answer questions being posed.[67]

Finally, there is the whole question of professionalization and support. Back in the 1860s, no one could see any way in which natural selection could lead to financially supportable science. Ford and Dobzhansky tackled this question head-on with considerable success. Both drew heavily on increasing amounts of state support, but they turned also to specific areas where they could argue selection had an economically beneficial role to play. In Ford's case, he approached the Nuffield Foundation, which was in the 1940s and 1950s looking to support projects in the life sciences, especially those with practical implications for the welfare of humankind. Ford and his students argued that their work on such problems as variation in butterflies and snails had direct implications for variations among humankind, with consequent medical benefits. Their argument convinced the foundation and they received money as a consequence.[68] Analogously, in America in the 1950s Dobzhansky and his students applied to the Atomic Energy Commission for funds, arguing that through study of fruit flies one could throw significant light on such issues as the effects of atomic fallout on the human species. Once again, selective studies were shown to be of economic and other social benefit.[69]

64. Ford 1964.
65. Dobzhansky 1937.
66. Mayr 1942; Simpson 1944; Stebbins 1950.
67. Cain 1993, 1994.
68. Ruse 1996.
69. Ibid.

CONCLUSION

Before publication of the *Origin,* natural selection was an idea before its time; and in significant respects it was an idea before its time *after* publication of the *Origin!* It was not until this century, until the 1930s and 1940s, that natural selection became an idea whose time had come, an idea whose story from then on has been one of almost unbroken success, for no one today doubts that natural selection is a major player in the evolutionary scenario. In fact, most active evolutionists agree that it is by far the most significant player.

But my moral is not the present glory of selection. Rather it is that the success of an idea in science depends on many different factors. Some of these are straightforward internalist factors—that is, the success of an idea (and no doubt the failure of its rivals) in the epistemic arena: how well does it function to solve the problems that one is now facing? As we have seen in the case of selection, success at this level depends not only on the science within which a new idea is introduced but also on the constraints and implications of the state of other impinging sciences. Until the physicists had made their moves on the age of the earth, natural selection simply was condemned to remain under a cloud.

Yet there is more to success than this. The history of the idea of natural selection suggests strongly that, whether or not one judges the idea to be true or the most significant of all evolutionary mechanisms, its success was by no means guaranteed simply by its own nature or even by its own merits. There were many other factors surrounding natural selection, both in its birth and in its subsequent development. Some of these were conceptual, like the extent to which adaptation and function were considered crucial questions for an evolutionist to address and answer. At a point like this, the scientific and the nonscientific (like religion, for example) blend. Some of these factors were more social and cultural, like the personalities of the players involved and the status of the science at the particular moment a new idea is introduced. All these things had their part to play, not only in the success that Darwin had with natural selection but also in the success that Darwin did not have with natural selection.

I would not presume to generalize from this one case to all other instances of scientific discovery—even to only those that posterity judges premature—but I would be surprised if the tale of natural selection were unique.

BIBLIOGRAPHY

Bates, H. W. 1862. "Contributions to an Insect Fauna of the Amazon Valley." *Transactions of the Linnaean Society of London* 23:495–566.
———. 1892. *The Naturalist on the River Amazons.* 1863. Reprint, London: John Murray.
Beddall, B. O. 1972. "Wallace, Darwin, and Edward Blyth: Further Notes on the Development of Evolution Theory." *Journal of the History of Biology* 5:153–58.
———. 1973. " 'Notes for Mr. Darwin': Letters to Charles Darwin from Edward Blyth at Calcutta: A Study in the Process of Discovery." *Journal of the History of Biology* 6:69–95.

Blyth, E. 1835. "An attempt to classify the 'varieties' of animals, with observations on the marked seasonal and other changes which naturally take place in various British species, and which do not constitute varieties." *Magazine of Natural History, and Journal of Zoology, Botany, Mineralogy, Geology, and Meteorology* 8:40–53.

Bowler, P. 1976. *Fossils and Progress.* New York: Science History Publications.

———. 1984. *Evolution: The History of an Idea.* Berkeley and Los Angeles: University of California Press.

———. 1988. *The Non-Darwinian Revolution: Reinterpreting a Historical Myth.* Baltimore: Johns Hopkins University Press.

———. 1996. *Life's Splendid History.* Chicago: University of Chicago Press.

Browne, E. J. 1995. *Charles Darwin: Voyaging. Vol. 1 of a Biography.* New York: Knopf.

Burchfield, J. 1975. Lord Kelvin and the Age of the Earth. New York: Science History Publications.

Burkhardt, R. W. 1977. *The Spirit of the System: Lamarck and Evolutionary Biology.* Cambridge: Harvard University Press.

Bury, J. B. 1920. *The Idea of Progress: An Inquiry into Its Origin and Growth.* Limited ed. 1924. London: Macmillan.

Cain, J. A. 1993. "Common Problems and Cooperative Solutions: Organizational Activity in Evolutionary Studies, 1936–1947." *ISIS* 84:1–25.

———. 1994. "Ernst Mayr as Community Architect: Launching the Society for the Study of Evolution and the Journal *Evolution.*" *Biology and Philosophy* 9:387–428.

Carron, A. 1988. " 'Biology' in the Life Sciences: A Historiographical Contribution." *History of Science* 26:223–68.

Chambers, R. 1844. *Vestiges of the Natural History of Creation.* London: Churchill.

Coleman, W. 1964. *Georges Cuvier, Zoologist: A Study in the History of Evolution Theory.* Cambridge: Harvard University Press.

Cuvier, G. 1813. *Essay on the Theory of the Earth.* Trans. R. Kerr. Edinburgh: W. Blackwood.

———. 1817. *Le règne animal distribué d'après son organisation, pour servir de base à l'histoire naturelle des animaux et d'introduction à l'anatomie comparée.* Paris: Déterville.

Darwin, C. 1851a. *A Monograph of the Fossil Lepadidae; or Pedunculated Cirripedes of Great Britain.* London: Palaeontographical Society.

———. 1851b. *A Monograph of the Subclass Cirripedia, with Figures of All the Species. The Lepadidae; or Pedunculated Cirripedes.* London: Ray Society.

———. 1854a. *A Monograph of the Fossil Balaniae and Verrucidue of Great Britain.* London: Palaeontographical Society.

———. 1854b. *A Monograph of the Sub-Class Cirripedia, with Figures of All the Species. The Balanidge (or Sessile Cirripedes; the Verrucidae, etc., etc., etc.* London: Ray Society.

———. 1859. *On the Origin of Species.* London: John Murray.

———. 1871. *The Descent of Man.* 2 vols. London: John Murray.

Darwin, C., and A. R. Wallace. [1858] 1958. *Evolution by Natural Selection.* Foreword by G. de Beer. Cambridge: Syndics of the Cambridge University Press.

Darwin, E. 1791. "The Economy of Vegetation." Pt. 1 of 2. In *The Botanic Garden.* London: J. Johnson.

———. 1794–1796. *Zoonomia; or, the Laws of Organic Life.* London: J. Johnson.

———. 1798. "The Loves of the Plants." Pt. 2 of 2. In *The Botanic Garden.* London: J. Johnson.

———. 1803. *The Temple of Nature.* London: J. Johnson.

Dawkins, R. 1976. *The Selfish Gene*. Oxford: Oxford University Press.

———. 1986. *The Blind Watchmaker*. New York: Norton.

Dobzhansky, T. 1937. *Genetics and the Origin of Species*. New York: Columbia University Press.

Eiseley, L. 1958. *Darwin's Century: Evolution and the Men Who Discovered It*. New York: Doubleday.

———. 1959. "Charles Darwin, Edward Blyth, and the Theory of Natural Selection." *Proceedings of the American Philosophical Society* 103:94–158.

Ellegard, A. 1958. *Darwin and the General Reader*. Goteborg: Goteborgs Universitets Arrskrift.

Fisher, R. A. 1930. *The Genetical Theory of Natural Selection*. Oxford: Clarendon Press.

Ford, E. B. 1964. *Ecological Genetics*. London: Methuen.

Gillespie, C. 1950. *Genesis and Geology*. Cambridge: Harvard University Press.

Gray, A. 1876. *Darwiniana*. New York: D. Appleton. Reprint edited by A. H. Dupree. Cambridge: Harvard University Press, 1963.

Haldane, J. B. S. 1932a. *The Causes of Evolution*. New York: Cornell University Press.

Herschel, J. F. W. 1831. *Preliminary Discourse on the Study of Natural Philosophy*. London: Longman, Rees, Orme, Brown, and Green.

Hull, D. 1973. *Darwin and His Critics*. Cambridge: Harvard University Press.

Huxley, T. H. 1876. "Lectures on Evolution." In *Science and the Hebrew Tradition*, pp. 46–138. London: Macmillan.

———. 1881. "The Rise and Progress of Paleontology." In *Science and the Hebrew Tradition*, pp. 24–45. London: Macmillan.

———. 1888. *American Addresses, with a Lecture on the Study of Biology*. New York: D. Appleton and Co..

———. 1893. *Darwiniana*. London: Macmillan.

King-Hele, D. 1963. *Erasmus Darwin: Grandfather of Charles Darwin*. New York: Scribners.

Kohn, D. 1980. "Theories to Work By: Rejected Theories, Reproduction, and Darwin's Path to Natural Selection." *Studies in the History of Biology* 4:67–170.

Kuhn, T. S. 1962. *The Structure of Scientific Revolutions*. Chicago: University of Chicago Press.

Lamarck, J. B. 1809. *Zoological Philosophy*. Translated by H. Elliot. New York: Hafner, 1963.

Lenoir, T. 1982. *The Strategy of Life: Teleology and Mechanics in Nineteenth Century German Biology*. Dordrecht: Reidel.

Lewontin, R. C. 1974. *The Genetic Basis of Evolutionary Change*. Columbia Biology Series no. 25. New York: Columbia University Press.

Limoges. C. 1971. *La Sélection Naturelle*. Paris: Presses Universitaires de France.

Matthew, P. 1831. *On Naval Timber and Arboriculture; with Critical Notes on Authors who have Recently Treated the Subject of Planting*. London: Longman, Rees, Orme, Brown, and Greene.

Mayr, E. 1942. *Systematics and the Origin of Species*. New York: Columbia University Press.

McKinney, H. L. 1972. *Wallace and Natural Selection*. New Haven: Yale University Press.

McKinney, H. L., ed. 1971. *Lamarck to Darwin: Contributions to Evolutionary Biology, 1809–1859*. Lawrence: Coronado Press.

McNeil, M. 1987. *Under the Banner of Science: Erasmus Darwin and His Age*. Manchester: Manchester University Press.

Nyhart, L. K. 1986. "Morphology and the German University, 1860–1900." Ph.D. diss., University of Pennsylvania.

———. 1995. *Biology Takes Form: Animal Morphology and the German Universities*. Chicago: University of Chicago Press.

Osborn, H. F. 1894. "The Hereditary Mechanism and the Search for the Unknown Factors of Evolution." In *Defining Biology: Lectures from the 1890s,* ed. J. Maienschein, pp. 83–104. Cambridge: Harvard University Press.

———. 1917. *The Origin and Evolution of Life on the Theory of Action Reaction and Interaction of Energy.* New York: Charles Scribner's Sons.

———. 1929. *The Titanotheres of Ancient Wyoming, Dakota, and Nebraska.* U.S. Geological Survey Monograph 55. Washington, D.C.: U.S. Geological Survey.

Ospovat, D. 1995. *The Development of Darwin's Theory: Natural History, Natural Theology, and Natural Selection, 1838–1859.* 1981. Reprint, Cambridge: Cambridge University Press.

Owen, R. 1850. "On the genus *Dinornis.* part 4: containing the restoration of the feet of that genus and of *Palapteryx,* with a description of the sternum in *Palapteryx* and *Aptornis.*" *Transactions of the Zoological Society* 4:1–20.

———. 1866. *Comparative Anatomy and Physiology of Vertebrates.* London: Longmans and Green.

Paley, W. 1802. *Natural Theology.* London: Rivington.

Peckham, M., ed. 1959. *The Origin of Species by Charles Darwin: A Variorum Text.* Philadelphia: University of Pennsylvania.

Rainger, R. 1991. *An Agenda for Antiquity: Henry Fairfield Osborn and Vertebrate Paleontology at the American Museum of Natural History, 1890–1935.* Tuscaloosa: University of Alabama Press.

Richards, E. 1987. "A Question of Property Rights: Richard Owen's Evolutionism Reassessed." *British Journal for the History of Science* 20:129–71.

Richards, R. J. 1987. *Darwin and the Emergence of Evolutionary Theories of Mind and Behavior.* Chicago: University of Chicago Press.

———. 1992. *The Meaning of Evolution: The Morphological Construction and Ideological Reconstruction of Darwin's Theory.* Chicago: University of Chicago Press.

Rupke, N. A. 1994. *Richard Gwen: Victorian Naturalist.* New Haven: Yale University Press.

Ruse, M. 1973. *The Philosophy of Biology.* London: Hutchinson.

———. 1975a. "Charles Darwin and Artificial Selection." *Journal of the History of Ideas* 36:339–50.

———. 1975b. "Darwin's Debt to Philosophy: An Examination of the Influence of the Philosophical Ideas of John F. W. Herschel and William Whewell on the Development of Charles Darwin's Theory of Evolution." *Studies in the History and Philosophy of Science* 6:159–81.

———. 1975c. "The Relationship between Science and Religion in Britain, 1830–1870." *Church History* 44:505–22.

———. 1979. *The Darwinian Revolution: Science Red in Tooth and Claw.* Chicago: University of Chicago Press.

———. 1980. "Charles Darwin and Group Selection." *Annals of Science* 37:615–30.

———. 1984. "Is There a Limit to Our Knowledge of Evolution?" *BioScience* 34:100–104.

———. 1993. "Evolution and Progress." *Trends in Ecology and Evolution* 8:55–59.

———. 1996. *Monad to Man: The Concept of Progress in Evolutionary Biology.* Cambridge: Harvard University Press.

Russell, E. S. 1916. *Form and Function: A Contribution to the History of Animal Morphology.* London: John Murray.

Schwartz, J. S. 1974. "Charles Darwin's Debt to Malthus and Edward Blyth." *Journal of the History of Biology* 7:301–18.

Sebright, J. 1809. *The Art of Improving the Breeds of Domestic Animals, in a Letter Addressed to the Right Hon. Sir Joseph Banks, KB.* London: Privately published.

Sedgwick, A. 1833. *A Discourse on the Studies of the University.* London: Parker.

Sheets-Pyenson, S. 1981. "Darwin's Data: His Reading of Natural History Journals, 1837–1842." *Journal of the History of Biology* 14:231–48.

Simpson, G. G. 1944. *Tempo and Mode in Evolution.* New York: Columbia University Press.

Spencer, H. 1852a. "A Theory of Population, Deduced from the General Law of Animal Fertility." *Westminster Review* 1:468–501.

———. 1852b. "The Development Hypothesis." In *Essays: Scientific, Political, and Speculative,* pp. 377–83. London: Williams and Norgate.

———. 1857. "Progress: Its Law and Cause." *Westminster Review* 67:244–67.

———. 1904. *Autobiography.* London: Williams and Norgate.

Stebbins, G. L. 1950. *Variation and Evolution in Plants.* New York: Columbia University Press.

Vorzimmer, P. J. 1970. *Charles Darwin: The Years of Controversy.* Philadelphia: Temple University Press.

Wallace, A. R. 1855. "On the Law Which Has Regulated the Introduction of New Species." *Annals and Magazine of Natural History* 16:184–96.

———. 1858. "On the Tendency of Varieties to Depart Indefinitely from the Original Type." *Journal of the Proceedings of the Linnaean Society, Zoology* 3:53–62.

———. 1864. "The Origin of Human Races and the Antiquity of Man Deduced from the Theory of Natural Selection." *Journal of the Anthropological Society of London* 2: clvii–clxxxvii.

———. 1866. "On the Phenomena of Variation and Geographical Distribution as Illustrated by the Papillionidae of the Malayan Region." *Transactions of the Linnean Society of London* 25:1–27.

———. 1870. "The Limits of Natural Selection as Applied to Man." In *Contributions to the Theory of Natural Selection,* p. 332–71. London: Macmillan.

Wells, W. C. 1818. "An account of a female of the white race of mankind, part of whose skin resembles that of a negro; with some observations on the causes of the differences in colour and form between the white and negro races of men." In *Two Essays: One upon Single Vision with Two Eyes; the Other on Dew,* pp. 425–39. London: Archibald Constable and Co.

Whewell, W. 1840. *The Philosophy of the Inductive Sciences.* 2 vols. London: Parker.

Wright, S. 1931. "Evolution in Mendelian populations." *Genetics* 16:97–159.

———. 1932. "The Roles of Mutation, Inbreeding, Crossbreeding, and Selection in Evolution." *Proceedings of the Sixth International Congress of Genetics* 1:356–66.

CHAPTER SIXTEEN

Prematurity, Evolutionary Biology, and the Historical Sciences

Michael T. Ghiselin

A number of Charles Darwin's scientific discoveries are major components of modern theory. Some were immediately accepted by the scientific community, whereas others had to wait over a hundred years for acceptance. Still others may yet have their day. Darwin's case is noteworthy partly as a result of the remarkably detailed historical record that allows us to document his accomplishments. As a consequence of that record, we have an excellent opportunity for testing the utility of Gunther Stent's seminal insights about prematurity in scientific discovery.

Stent terms a discovery as "premature if its implications cannot be connected by a series of simple logical steps to contemporary canonical knowledge."[1] I think that the Darwinian corpus fits this criterion admirably, with suitable qualifications. However, when I tried years ago to apply Thomas Kuhn's paradigm theory to Darwin, I managed to do so only by denying some of the most basic features of Kuhn's theory, and by making science evolutionary rather than revolutionary, individualistic rather than social—in short, by emending it virtually beyond recognition.[2] I find myself doing somewhat the same with Stent's ideas.

CANONICAL KNOWLEDGE

Not everybody agrees on the definition of canonical knowledge or under what circumstances a discovery's connection to it is strong enough to qualify for inclusion. Also, if a minority of scientists accept a discovery, or even pay serious attention to it, then the discovery is not altogether premature in the Stentian sense. When we look at the scientific community as a whole, of course, we find a wide diversity of

---

1. Stent 1972b, p. 84; see also Stent 1972a, p. 435, reprinted in part as chapter 2 in this volume.
2. Ghiselin 1971.

abilities and inclinations that would allow a scientist to make such a connection, as well as a scientific community that is by no means homogeneous. Likewise, if a connection is in fact made, then obviously it can be made—but negative evidence does not establish that it cannot be made. Suppose, for example, one could connect a specific discovery to the canon, but only by an act of creative reasoning possible only for a truly extraordinary intellect. From Stent's statement it is not clear what kind of possibility he has in mind: one that is logically possible and one that is humanly possible are by no means the same thing. Stent, of course, tried to make a categorical distinction between that which is premature and that which is not. We can accept Stent's distinction as an idealized model, but it seems to me that, just as babies are not absolutely and without qualification premature, relative prematurity can teach us something. Likewise the connection might admit of degrees with respect to the strength of the connection. Stent's model suffers from a lack of realism, partly because of its typological and nonquantitative formulation.

DARWIN

Darwin's theory about coral reefs made him a famous young man as soon as it was announced.[3] It was anything but premature, and it is easy to see why. Darwin explained the morphology of coral formations such as atolls and barrier reefs in terms of skeletonized animals growing and depositing calcium carbonate on sinking islands and continents. It fitted quite nicely with his own discoveries about the uplift of South America and, more important, with the geological theories of his role model, Charles Lyell. There was a nice mechanism for such crustal movements, one that everybody could understand, even if not everyone accepted it. Yet, although Darwin's coral reef theory was always the majority view among geologists, there were some holdouts, especially among zoologists, until two things happened. First, it became possible to observe the submarine landscape by sonic techniques and to drill through thousands of feet of limestone. Second, the theory of plate tectonics emerged only a few years later, and Darwin's basic ideas were readily incorporated into the new global geology. I conclude that Darwin's coral reef theory not only fitted what was canonical knowledge from the outset but also fitted the canonical knowledge that evolved later.

A more problematic case is Darwin's theory of evolution by natural (and artificial and sexual) selection. The basic historical pattern that requires explanation is roughly as follows. *The Origin of Species* was published in 1859. Within about ten years, evolution was generally accepted as a fact by the scientific community, and it served as a basis for numerous research programs. On the other hand, natural selection, which is widely considered Darwin's most important contribution to knowledge, had no such immediate success. Only a minority of scientists went along

3. Darwin 1837, 1842, 1874.

with Darwin's view that natural selection was, as he put it, the main, though not the exclusive, cause of evolutionary change. It was much more generally treated as a minor causal influence. Perhaps the most important consideration is that only a handful of scientists, including Darwin himself, based major research projects on it. General acceptance of natural selection did not occur until the emergence of the synthetic theory of evolution during the period from just before World War 2 until around 1950. Perhaps it is worth mentioning parenthetically that the exceptions just alluded to included some truly outstanding and highly influential biologists, especially Fritz Müller and August Weismann, as well as Darwin's co-discoverer Alfred Russel Wallace and Wallace's friend and traveling companion Henry Walter Bates.

Indeed, Darwin's theory of sexual selection, which together with natural selection forms part of a more general theory, was almost entirely neglected and misunderstood until about 1969. I pointed out at that time that sexual selection provided compelling evidence for Darwin's more general selection theory, of which it is a corollary.[4] Because so many people continue to treat sexual selection as if it were a more particular case of natural selection, I wonder if this insight was itself perhaps premature. However, the time had obviously come for a breakthrough based on treating the evolution of reproductive strategies from the point of view of pure reproductive competition, and at least my own effort along such lines definitely was not premature.[5] In general, what was going on at the time was a reconsideration of the point that Darwin understood very well indeed but had to explain to Wallace, that selection works by reproductive competition between individuals, including individual families (which, as explained below, is not paradoxical). Making a connection in such cases requires an appreciation of the metaphysical issues, and whether such understanding is part of canonical knowledge is a most intriguing question.

So we must ask whether evolution was accepted more or less from the outset because it was not premature, and whether natural selection was premature in 1859 but ceased to be so by the middle of the twentieth century. (And *fortiori,* we must ask whether sexual selection was premature for another 20 or 30 years, and whether the discovery of its philosophical importance remains premature.) Moreover, we must ask whether Darwin's version of evolutionary biology is still premature. It seems to me that we can answer all these questions more or less in the affirmative, but not without all sorts of qualifications, which I hope prove illuminating.

DARWIN'S PRECURSORS

One very important task is to dispose of some mythology, beginning with Darwin's real and imagined precursors. Tradition has it that there were a lot of evolu-

4. Ghiselin 1969a.
5. Ghiselin 1969b.

tionists before Darwin. A closer examination shows that many of these were really concerned with something else.[6] Good examples are Goethe, Lorenz Oken, and many other so-called *Naturphilosophen,* who were widely considered evolutionists even by such luminaries as Darwin and Ernst Haeckel. The remainder were generally concerned with the possibility of evolution rather than with creating a science of evolutionary biology.[7] I have yet to find any scientist before Darwin whose empirical research program was based on evolutionary principles and, therefore, comparable to the work he did on barnacles between 1846 and 1854. Such quasi evolutionists as Jean-Baptiste Lamarck and the elder Geoffroy Saint-Hilaire of course had something like evolution in mind, and that affected how they did their work. Moreover, I would not be surprised were a concerted effort to turn up a few marginal or perhaps trivial instances. One partial exception of course was Wallace, who was looking for geographical clues to evolution, and found them.

As to natural selection, there were likewise a host of putative anticipations, few of which really anticipated natural selection. Some of these concerned a kind of selective breeding in a state of nature that kept species constant. Others provided means of adaptation to local circumstances.[8] Even when there was speculation about possible long-term and directional effects of something like natural selection, it was never accompanied by evidence that made natural selection credible as a mechanism for evolution or that showed how it could be made the basis for a viable research program. Darwin was able to accomplish all that in 1844 when he wrote a preliminary but extended "essay" on his theory. In 1859 he presented the world with natural selection as a plausible mechanism of change. He showed how it could explain much of what was considered canonical knowledge, such as embryology mirroring taxonomy. He adduced facts that could not be otherwise explained, such as the behavior of neuter castes in social insects. He showed how his theory gave quite unexpected predictions, such as revealing the prevalence of sex. And he explained how to do research in various kinds of what we now call evolutionary biology. The joint paper by Darwin and Wallace (1858) that was read to the Linnaean Society was not perceived as a real breakthrough, precisely because it failed to make those connections.

For some people, including not only Darwin himself but also that small minority of biologists who were his real supporters, making those connections was enough. There was now plenty of linkage to canonical knowledge. They, at any rate, *did not* perceive it as premature. One might then want to argue that, with the passage of time, more and more of biology became canonical, so that ever more linkages were established and selection became increasingly attractive to a growing proportion of biologists.

---

6. Ghiselin 1969a; see also Ruse, chapter 15 in this volume.
7. For example, Chambers 1844.
8. Shryock 1944 provides a good discussion on William Wells.

DARWIN AND HEREDITY

The usual mythology dished up as undergraduate pablum is that Darwin had no theory of heredity, and then one was found, so that finally genetics came to the rescue. Many are the objections to that argument and to the less naive variants of it presented by historians such as Peter Bowler.[9] The early pioneers of genetics were hostile to natural selection and to evolution in general. More important, it was the refutation of misconceptions about genetics, not anything lacking in Darwin's theory, that led to the recrudescence of natural selection. Lamarckism increasingly failed to mesh with what was becoming canonical knowledge, and there were no serious competitors to Darwinism. Even so, the case of Richard Goldschmidt shows how determined was the resistance among geneticists.

The best we can say with respect to the role of genetics is that Darwinism appropriated genetics, much as it has appropriated molecular biology of late. Chromosomes are great materials for doing phylogenetics. One interesting alternative to having geneticists be the heroes is that a new kind of systematics emerged, complete with a better species concept, and that that was what made the synthesis work. This is, roughly speaking, Ernst Mayr's position.[10] Although basically true, this alternative was in a sense premature because it was metaphysically incorrect. It could not be connected by a series of logical steps to canonical assumptions about the nature of ultimate reality. People therefore had a hard time making the connections to alternative assumptions, ones that, although widely accepted, were generally rejected.

DARWINISM AND CHANGE

Let me abuse the Stentian model a bit and argue that there was a fundamental metaphysical cleavage between Darwin's biology and that of his predecessors. Given the traditional outlook, it was only too easy to relate evolution to what was then canonical knowledge, thereby providing at least an excuse for not undertaking the sort of reconstruction at a fundamental level that was necessary if Darwin's theory was to be understood, much less accepted. Briefly, pre-Darwinian science had learned to deal with some kinds of change, but Darwin had in mind a kind of change that was different and, one might say, more profound. So when Darwin's contemporaries tried to cope with evolution, they modeled their conception of it on the sort of change that was familiar to them. They found connections, but all too often these were misleading ones.

Darwin was able to invoke the kind of change that occurs in domesticated plants and animals as an example of evolution by a selective mechanism. This made natural selection seem more or less plausible, an acceptable scientific hypothesis. A major consequence of having a plausible mechanism was that it legitimized evo-

9. Bowler 1988.
10. Mayr 1980.

lutionary theory in general, so that other mechanisms were also open to discussion. Such mechanisms might go completely unspecified, or be only vaguely suggested. Also, the option of accepting selection was open, but only as a minor causal influence. Perhaps natural selection could cull out the unfit or produce adaptation to local circumstances, but the real explanation had to lie elsewhere. Biologists had generally believed that species, like inorganic objects, occurred as so-called natural kinds that were sometimes variable, but only within circumscribed limits. For a kind itself to have the capacity for indefinite change was unthinkable—indeed, a contradiction in terms. Under such circumstances natural selection could account for the origin of varieties but not species.

Another kind of change familiar to everybody was learning. If evolution is like learning, we get Lamarckism. Some versions of Lamarckism are based on inherited memory. The versions that stress the inheritance of acquired somatic changes may be interpreted as variants on that theme. Darwin, as is generally known, accepted the inheritance of acquired characters, and even the efforts of his great follower Weismann failed to dampen the enthusiasm for various versions of so-called Lamarckism.

Then we have the sort of change that occurs in the development of an embryo. The science of embryology advanced a great deal during the first half of the nineteenth century, and much descriptive work was available. A general correlation was recognized between the taxonomic hierarchy, the succession of fossils, and the stages of embryonic development. The developing organisms seemed to follow well-ordered pathways toward a particular terminal state. Therefore, one might want to treat organic evolution as a kind of embryology on a geological scale, with a change from potentiality to actuality. Under such a conception there would be no origin of something ontologically new, for the "kinds" of organisms would have existed in a prenascent condition from the beginning, if indeed they were not downright eternal. The result was orthogenesis, or evolution along predetermined lines, of which there are several versions.

ORTHOGENESIS

One version of orthogenesis held that God created a common ancestor that contained within itself the potentiality to differentiate into all the diverse forms of living beings, much as a developing embryo gives rise to an increasing diversity of cells, tissues, and organs. Another version had God foreseeing all such developments from the beginning, but miraculously superintending the execution of his masterpiece. Yet another version of orthogenesis would have God ordain laws of nature, such that secondary causes would do the work. This is a version of design on the installment plan, as John Dewey so aptly termed it.[11] The efforts of Tory

---

11. Dewey 1910.

historians such as Nicolaas Rupke to convert Richard Owen, for example, into an evolutionist strike me as more than just a little bit forced.[12]

One might indeed try to do without the Creator, but attributing development and "evolution" to such laws still meant a kind of orthogenesis and, again, the coming into being of nothing ontologically new. The fundamental physical model for the development of an organism was the formation of a crystal. Both kinds of change would be the result of laws of nature. Accordingly, plant and animal species would be like mineral species, and the analogy could be extrapolated to higher taxonomic levels as well. Instances of different kinds of organisms would then form whenever and wherever the necessary and sufficient physical conditions were met. Of course, it might take any number of generations for this to happen, but it was perfectly reasonable to assume that planets in far-distant parts of the universe would be populated with vertebrates, mammals, and even human beings. This, of course, is precisely what we see on television, where the extraterrestrials not only speak English, but the English of the present day, not of Shakespeare's.

POSITIVISM

Many pre-Darwinian biologists were looking for a major breakthrough in biological theory. But they expected the crucial discovery to be laws of nature rather than historical contingencies, and that suggests a kind of prematurity in the sense that the world was not fully prepared for what happened. Many had expected embryology to supply the laws they had been seeking. In reaction to Darwinism, the opposition fell back on the same basic idea: we explain things in terms of law, not chance (the pejorative equivalent of history).[13] Embryology remains to this very day the last refuge of antievolutionism. Of course, present-day antievolutionists do not deny that evolution exists. Rather, they insist that what really counts are the so-called laws of form, which are generally assumed to be fundamentally physicochemical.

Now, such an attitude toward what is an appropriate way of understanding the world—that is, in terms of laws of nature rather than in terms of history—can easily be connected to canonical knowledge. Indeed, it is a basic part of the metaphysics of positivism, including such versions as are still being taught by professors of science, and of its philosophy. The effect of such positivism on the research programs of biologists has lately been noticed by V. B. Smocovitis, though it is noteworthy how little she perceives its effect on her own research program in the history of science.[14] Metaphysics tells people that history is not important, so, as does Smocovitis, they ignore systematics. Or, it tells them that it is second-rate science, and, as does Michael Ruse, they project that value judgment into their historical

12. Rupke 1994.
13. A fine example is Hertwig 1922.
14. Smocovitis 1996.

narratives.[15] Whereas August Comte wanted a version of Catholicism minus Christianity, Ruse advocates a version of Darwinism minus evolution.[16]

SPECIES AS INDIVIDUALS IN AN ONTOLOGICAL SENSE

The alternative metaphysics that has emerged over the last few decades is of course the view that species and other spatiotemporally restricted entities such as clades and languages are not natural kinds at all, but individuals in a broad ontological sense.[17] As in the cases of George Washington and our solar system, there are no laws of nature for such entities. Credit for the individuality thesis is generally given to David Hull and me.[18] When I first proposed it, it was premature, at least insofar as Hull rejected it out of hand, though only for a while. Once it had ceased to be premature, plenty of so-called precursors were exhumed from the literature.

A serious look at such anticipations of the individuality thesis is of some interest from the point of view of prematurity; the "discoverers" seem not to have considered it an important idea. Insofar as the philosophers go, we find the possibility mentioned by Woodger, Gregg, and even the very early Hull.[19] But they treated it as no more than an intellectual curiosity and screened it out. From a Stentian point of view, we might say that the discovery was premature because it could not be connected with the metaphysics accepted as canonical in the philosophy of science of the day. Insofar as logic went, there were no apparent difficulties, but an idea has to be more than just a logical possibility to attract much attention. The discovery ceased to be premature when, and only when, a serious effort was made to construct a philosophy that would do justice to evolutionary biology. But that meant rejecting the metaphysics of positivism, something that was supposed not even to exist.

EVOLUTIONARY BIOLOGY AND THE HISTORICAL SCIENCES

Evolutionary biology, as it rests on its new metaphysical foundation, can now serve as a model for other historical sciences, for we now can appreciate that the goal of such sciences is the creation of an explanatory narrative that synthesizes the historically contingent with the nomologically necessary. The laws of nature in biology refer to classes of individuals, including classes of species, but the species and other taxa themselves are of a purely historical character. So the goal of biology is to find out what, say, the population genetical laws might be and conjoin them with historical material to produce an explanatory history of the lineages that interest us—to produce, in other words, a phylogeny.

15. Ruse 1996.
16. Ghiselin 1997a.
17. For a latest statement, see ibid.
18. Hull 1975; Ghiselin 1966, 1969a, 1974.
19. Woodger 1952; Gregg 1954; Hull 1969.

It may now be clear why I am sympathetic to Stent's idea yet am not much bothered by a need to undertake all sorts of intellectual gymnastics to make it work. Neither in biology nor in history are the connections between the particular events and the laws of nature a straightforward and simple matter. Practitioners tend to think in terms of principles rather than the laws of nature that provide their more fundamental rationale. And they repeatedly find themselves overwhelmed by the particulars. Stent is a good scientist. He wants to find laws of epistemology: spatiotemporally unrestricted truths, having reference to classes of discoveries, that are necessarily true of everything to which they apply. He then turns, again like a good scientist, to epistemography for historical events that might tend to confirm or disconfirm his bold hypothesis. And the question that comes next is whether such a law, or something like it, might serve as one basis for an ultimate synthetic science, or in other words, an epistemogeny.

In this light, we can clearly see how a discovery that was perhaps a bit premature fifteen years ago can now be connected with canonical knowledge. But then again, the body of knowledge in question is generally accepted mainly by theoreticians of systematic biology and by philosophers who are seriously interested in such matters. Indeed, what converted Hull to the individuality thesis was his reading of a book by J. J. C. Smart, in which it was asserted that there are no laws of nature in biology.[20] Smart asked what laws there are for *Homo sapiens*. Hull then realized not only that there are no laws for *Homo sapiens* but also that the reason there are no such laws is that *Homo sapiens* is an individual, and that there are no laws for any individual in any science.[21]

Smart, like a lot of people, wanted to treat science as if it were purely a body of laws of nature. Accordingly, a historical science would be a contradiction in terms. But if we accept such a definition, then an astronomer is not doing science when discussing the earth, our solar system, the Milky Way, the universe, or the big bang, for all of these are individuals and the laws of nature refer to none of them. These laws can refer to big bangs and lesser bangs, but not to any particular bang. Likewise, plate tectonics is a theory that refers to Pangea and the Pacific Plate, and therefore it is not science, but history.

The notion that history is something other than science is a straightforward consequence of bad metaphysics and intellectual bigotry. But it is a widespread attitude among physical scientists, and might even be considered canonical knowledge in what is called the philosophy of science. I am by no means sure of the extent to which this attitude has affected Stent's thinking about prematurity and other aspects of scientific discovery. The kind of biology that he has practiced is not the sort that would predispose one to think like a geologist or a comparative anatomist when dealing with intellectual history. He got out of molecular biology while it was

20. Smart 1963.
21. Hull 1975.

still very much like chemistry, and before the emergence of molecular and developmental phylogenetics.

When the history of science is practiced as a natural science, one of the goals is to find laws of nature that refer to classes of discoveries. But without the individuals and the historical narratives, we wander off into vacuous abstraction. Science is about something, and that something is always concrete particular things. I suspect from some of his remarks that Stent has considered the uniqueness of individual events to mean that the historical sciences do not generalize. But particular events have multiple consequences, the publication of *On the Origin of Species* being a ready example. Furthermore, in passing from the part to the whole—for example, from a species to its phylum—we do shift from the more particular to the more general.

The goal of science is not just to connect particulars to recurrent patterns but also to place them within the context of the larger wholes. What is or is not premature is contingent on historical circumstances, including those peculiar scientists and disciplines. Stent's model may have been just a little bit premature, given what knowledge was viewed as canonical at the time when it was proposed.

BIBLIOGRAPHY

Bowler, P. J. 1988. *The Non-Darwinian Revolution: Reinterpreting a Historical Myth*. Baltimore: Johns Hopkins University Press.

Chambers, R. 1844. *Vestiges of the Natural History of Creation*. London: John Churchill.

Darwin, C. 1837. "On Certain Areas of Elevation and Subsidence in the Pacific and Indian Oceans, as Deduced from the Study of Coral Formations." *Proceedings of the Geological Society of London* 2:552–54.

———. 1842. *The Structure and Distribution of Coral Reefs, Being the First Part of the Geology of the Voyage of the Beagle*. London: Smith and Elder.

———. 1874. *The Structure and Distribution of Coral Reefs*. 2d ed. London: Smith, Elder.

Darwin, C., and A. R. Wallace. 1858. "On the Tendency of Species to Form Varieties; and on the Perpetuation of Varieties and Species by Natural Means of Selection." *Journal of the Proceedings of the Linnaean Society (Zoology)* 3:45–62.

Dewey, J. 1910. *The Influence of Darwin on Philosophy, and Other Essays in Contemporary Thought*. New York: Henry Holt and Company.

Ghiselin, M. T. 1966. "On Psychologism in the Logic of Taxonomic Controversies." *Systematic Zoology* 15:207–15.

———. 1969a. *The Triumph of the Darwinian Method*. Berkeley and Los Angeles: University of California Press.

———. 1969b. "The Evolution of Hermaphroditism among Animals." *Quarterly Review of Biology* 44:189–208.

———. 1971. "The Individual in the Darwinian Revolution." *New Literary History* 3:113–34.

———. 1974. "A Radical Solution to the Species Problem." *Systematic Zoology* 23:536–44.

———. 1997a. "*Monad to Man: The Concept of Progress in Evolutionary Biology*, by Michael Ruse." *Quarterly Review of Biology* 72:452.

————. 1997b. *Metaphysics and the Origin of Species.* Albany: State University of New York Press.

Gregg, J. R. 1954. *The Language of Taxonomy: An Application of Symbolic Logic to the Study of Classificatory Systems.* New York: Columbia University Press.

Hertwig, O. 1922. *Das Werden der Organismen: Zur Widerlegung von Darwins Zufallstheorie durch das Gesetz in der Entwicklung.* 3d ed. Jena: Verlag von Gustav Fischer.

Hull, D. L. 1969. "What the Philosophy of Biology Is Not." *Journal of the History of Biology* 2:241–68.

————. 1975. "Central Subjects and Historical Narratives." *History and Theory* 14:253–74.

Mayr, E. 1980. "Prologue: Some Thoughts on the History of the Evolutionary Synthesis." In *The Evolutionary Synthesis: Perspectives on the Unification of Biology,* ed. E. Mayr and W. B. Provine, pp. 1–48. Cambridge: Harvard University Press.

Rupke, N. A. 1994. *Richard Owen: Victorian Naturalist.* New Haven: Yale University Press.

Ruse, M. 1996. *Monad to Man: the Concept of Progress in Evolutionary Biology.* Cambridge: Harvard University Press.

Shryock, R. H. 1966. "The Strange Case of Wells' Theory of Natural Selection (1813): Some Comments on the Dissemination of Scientific Ideas." In *Medicine in America: Historical Essays,* pp. 259–72. Baltimore: Johns Hopkins University Press.

Smart, J. J. C. 1963. *Philosophy and Scientific Realism.* London: Routledge and Kegan Paul.

Smocovitis, V. B. 1996. *Unifying Biology: The Evolutionary Synthesis and Evolutionary Biology.* Princeton: Princeton University Press.

Stent, G. S. 1972a. "Prematurity and Uniqueness in Scientific Discovery." *Advances in the Biosciences* 8:433–49.

————. 1972b. "Prematurity and Uniqueness in Scientific Discovery." *Scientific American* 227 (December): 84–93.

Woodger, J. H. 1952. "From Biology to Mathematics." *British Journal for the Philosophy of Science* 3:1–21.

PART FIVE

Perspectives from the Vantage Point
of the Social Sciences

CHAPTER SEVENTEEN

The Prematurity of "Prematurity" in Political Science

George Von der Muhll

Shortly after the Second World War, the "Behavioral Revolution" swept through the academic study of politics.[1] From that point onward, professional students of the subject have searched for a single organizing paradigm that would provide their field with the shared concepts and established propositional canon they see in the natural sciences. None has yet emerged. Instead, several proposed theoretical perspectives have competed for attention within the various disciplinary subfields of political "science." Their proliferation has so far served mainly to emphasize a conspicuous deficiency of logical integration within the discipline.

In such a setting, it is impossible to say that any one paradigm has displaced another. Nor can one say that an important finding is ignored because it appears anomalous within a currently accepted framework. Instead, various proposed models of inquiry—frequently drawn by analogy from the much more logically integrated field of economics or from one or another of the natural sciences—compete for attention, enjoy a brief half-life of attention as their theoretical architecture

---

1. See, among the numerous proclamations from this period, Eulau 1963 for a concise statement of its widely shared premises. The term "Behavioral Revolution" became commonly adopted in the 1950s as a means of drawing attention to several coterminous and loosely intercorrelated developments—some innovative, some a marked acceleration in previous trends—that had come to characterize the systematic study of politics after the World War. The most distinctive features of this "revolution" included (1) the call for replacing a disciplinary focus on historically unique governmental configurations by an identification of those properties of human behavior in politics that, through their simplicity, universality, and frequent recurrence, better lent themselves to analytic modeling and statistical generalization; (2) a corresponding shift of attention from salient actors within manifest political structures (most preeminently, the state) to the more anonymous fields of social forces conditioning outcomes in overtly "political" arenas; and (3) explicit acknowledgement of the need to turn to the other, more generic "behavioral" social sciences, especially sociology and psychology, for propositions and evidence concerning the key characteristics and determinants of such behavior.

is assessed, then yield to other approaches that seem more promising even before their research implications have been seriously explored. The fate of any given approach appears more a function of its resonance with extradisciplinary theoretical developments and with shifts in societal concerns than with success or failure in attempting to elaborate logical connections from the approach to a mainstream of established knowledge. The contemporary importance of a proposed paradigm's adherents within the discipline offers little protection against such exogenous determinants of acceptability. In all these respects political "science" offers instructive contrasts with the other cases discussed in this volume.

I exposit three such models of inquiry to help substantiate these points. I propose to consider, in turn, one relatively successful paradigm—*rational-choice* (or *public-choice*) theories of politics—that after a delay of nearly two decades established a solid niche within the field; *general systems* theory, which made its way into the study of politics from biology by way of an influential school of sociologists but then became discredited as a supposedly nonrevolutionary equilibrium model; and *cybernetic* theory, imported directly from Norbert Wiener's modeling of informational circuitry at Massachusetts Institute of Technology (MIT) but too exotic to serve as a readily comprehensible image to political scientists at a time when computers were still largely confined to laboratories.

Of the three approaches, rational-choice theory most closely approximates the cases discussed by Gunther Stent in its outcome. Outsiders might find self-evident its basic premise that politics can be fruitfully and rigorously interpreted as competition for the power to make collectively binding decisions on behalf of comprehensive societal units and that survival in the political arena requires subordination of other concerns to that imperative. In some sense, such a perspective is traceable to the Renaissance political philosophers Niccolò Machiavelli and Thomas Hobbes. From the outset, however, it has been overlaid by a disposition of most political analysts to view this approach in terms of the apparently cynical tone of its prudential injunctions and a corresponding stress on the contents of such visions. Theories predicated on the rational character of political actors were also easy to ignore in the period between the twentieth century's two world wars, when both the growing popularity of Freudian theory and Fascist ideology and practice suggested the predominant irrationality of mass behavior.

The rational-actor approach attained a well-defined autonomous theoretical focus only with its link—significantly, by the Austrian economist Joseph Schumpeter in his *Capitalism, Socialism, and Democracy* (1954)—to the more specific contention that democratic politics can be perceived as directly analogous to economic theories of entrepreneurial choice in free-market conditions.[2] Schumpeter grounded his thesis in the contention that, whatever the particular ends sought, the disciplining *condition* for entrepreneurial survival in both arenas is solvency—consumer support in

---

2. Schumpeter 1954, pp. 269 ff.

the free market; votes and voter commitment in the democratic "marketplace." But Schumpeter's thesis concerning the transferability of economic theorizing to politics was buried in 3 among nearly 30 chapters of a treatise seeking to explain a purported exhaustion of entrepreneurial vigor and technological innovation. Partly for that reason, it was another full decade and more before another economist, Anthony Downs, picked up Schumpeter's suggestive metaphor. But Downs went far beyond Schumpeter in transforming a metaphor into a systematically exposited treatise with careful technical definitions, propositions that flowed logically from the explicitly defined starting point of the inquiry, and an orderly application of these propositions to various strategies of politics unfolding as responses to *problems* generated by that starting point.[3]

For still another decade, Downs's *An Economic Theory of Democracy* (1957) was treated by most political scientists as more entertainingly provocative than compellingly paradigmatic. Those who took it seriously swiftly reduced its scope to a narrow focus on some paradoxes of majority rule in committee voting procedures.[4] Ironically, Downs's approach was also threatened by the premature mathematicization of rational-actor coalition theory that left a large majority of political scientists convinced that years of disciplinarily uncongenial investment in the acquisition of advanced levels of calculus would be required to track a debate concerning a small subset of peripheral issues.[5] In the mid-1960s, however, yet another economist, Mancur Olson, demonstrated the continued staying power and wide-ranging fruitfulness of a Schumpeterian-Downsian approach in *The Logic of Collective Action* (1965), while the soon-to-be Nobel Prize–winning economist James Buchanan and his political-scientist colleague Gordon Tullock began producing a succession of volumes and essays that large numbers of political scientists concluded they could no longer afford to ignore.[6] Rational-choice theory (sometimes designated by economists as public-choice theory because of its implications for nonmarket governmental decision-making) has by no means become adopted as the organizing paradigm for research in political science. Its proponents, however, have attained a critical mass of theoretical elaborations and applications sufficient to induce whole departments in the field—for example, the political science department at UCLA—to restructure their entire graduate curriculum in its light.

In a period when Downs's *Economic Theory of Democracy* could still be viewed as an isolated sport, general systems theory seemed on the verge of becoming the unifying paradigm political scientists sought for their field.[7] The basic premise of this ap-

3. Downs 1957.

4. The groundwork for these narrowly focused discussions of certain paradoxes in rational choice was laid in two works that appeared before Downs's own treatise: Black 1948, pp. 245–61; and Arrow 1951.

5. Though clearly foreshadowed in the immediate postwar work of the mathematician John von Neumann on formalized theories of "games," the trend toward mathematicization of Downs's political analysis was given impetus by Riker 1962.

6. Olson 1965; Buchanan and Tullock 1962; Tullock 1970.

7. For a distillation of these trends in a single volume, see Bertalanffy 1968.

proach—that social structures should be viewed as patterned adaptations to specified environments—was again familiar in more general form to many thinkers both in and outside the field. With the acceptance of Darwinian doctrines, biologists—and later, many influential sociologists—had become accustomed to taking survival value as the starting point for the explanation of structural patterns. Cross-boundary transactions between inner structural units and encompassing outer residual settings, and homeostatic adjustments of these transactional flows within a critical range defining conditions for patterned survival, correspondingly became the accepted mode for organizing inquiry in these fields. Certain political scientists—most notably, David Easton of the University of Chicago (soon to be president of the American Political Science Association)—saw a dazzling potential for systematic extension of these biological concepts to the study of political patterns, while such *structural-functional* theorists as the sociologist Talcott Parsons of Harvard became similarly and simultaneously intrigued by the possibility of projecting their elaborately articulated theoretical approach onto the "underdeveloped" study of politics.[8]

Once again, however, this simple—even simplistic—perspective had to compete with a more widespread disposition to analyze political structures in terms of the *purposes* of those who established and acted within their framework of rules.[9] General systems theorists had also to contend with the charge that their highly abstract formulations at most contributed insights accessible through less technical prose.[10] Proclaimed affiliation with general systems theory tended to be confined to prolegomena of research projects thereafter governed by conventionally formulated questions drawn from a miscellaneous stock of historical proverbs concerning the motivations and skills of political actors. Reactions to systems theorists included devastatingly influential purported "translations" of their theoretical approach to commonsense prose; and systems theorists themselves had increasingly to admit difficulties in finding unambiguous boundaries to their units paralleling those of biological cells.[11] Inability to move convincingly from a framework of analysis to detailed explication of internal coding mechanisms and responses plagued social systems theorists at the very point at which biologists were conspicuously beginning to unravel the secrets of DNA and other genetic material to transform the research orientation of their field.

But the fatal blight on the initial promise of systems theory in the study of politics came less from such difficulties, which sharpened reformulations and elaborations were

8. Parsons's incorporation of political "systems" into his general systemic paradigm is perhaps easiest to appreciate and assess in Mitchell 1967.

9. The flourishing of "policy" studies in political science in the last several decades is indicative of the greater congeniality of this orientation.

10. This contempt for the elaborate abstractions of general systems theory was given its most memorable, if frequently misleading, form in the Marxist sociologist C. Wright Mills's purported "translation" of Parsonian theorizing. See Mills 1959, chap. 2.

11. The conceptual and empirical difficulties of establishing unambiguous system boundaries is well discussed by David Easton (1965a)—himself a strong advocate of applying systems theory to the study of politics.

beginning to cure. Rather, it sprang from a growing suspicion that systems theory, with its emphasis on homeostatic reequilibration, concealed an unacknowledged conservative political agenda. Younger political scientists—fueled by revelations concerning Vietnam and American corporate interests in Latin America—claimed that systems theory was a politically obscurantist doctrine systematically masking internal conflict and the possibility for revolutionary outbreak. To such critics, the characteristically high level of abstraction of systems theory also seemed designed to preclude embarrassing identification of beneficiaries from the processes maintaining such equilibria. The nearly complete displacement of general systems theories in the 1970s by conceptually more conventional, less theoretically ambitious investigations of the determinants of public policy is largely attributable to these concerns.

Cybernetic theory, the last case I examine here, is also least among the three in terms of its general influence on the study of politics. In the late 1950s another future Noble Prize–winning economist, Herbert Simon, created a certain stir among political scientists with a succession of lucid, readable essays—"The Architecture of Complexity" was in some respect the most comprehensive of these—in which he reported on the early results of research he had undertaken with several associates at Carnegie Institution of Technology on computer modeling of human cognitive processes.[12] Although on these occasions Simon made no direct effort to apply his conclusions to the study of politics, he had once been a major figure in triggering the Behavioral Revolution in political science through his highly influential work on the cognitive and motivational determinants of administrative decision-making. His more recent essays, then, inevitably raised the question of whether they likewise foreshadowed a new theoretical direction within the field.

These questions seemed to receive a definitive answer with the publication of Karl W. Deutsch's *The Nerves of Government.*[13] Deutsch, on the verge of election as president of the American Political Science Association and author of a major textbook in the discipline, was already very well known within the subfield of international politics for his highly original work on measuring communications flows across national borders and assessing their political implications. *The Nerves of Government* was an attempt to generalize from this research. More important, it was a proposal to reconceptualize the organizing terms of his discipline—political action, power, systems, comparative influence—in a unified manner explicitly deriving from the feedback-response model of his MIT colleague Norbert Weiner.[14]

After reviewing systematically the deficiencies of all previous attempts to produce an integrated theory of politics, Deutsch proposed for consideration by his fellow practitioners a strikingly simple cybernetic paradigm of government as a "steering" mechanism based on environmental monitoring, informational storage

---

12. Simon 1962, pp. 467–92.
13. Deutsch 1963.
14. See Weiner's classic early collaborative article with Rosenblueth and Bigelow (1943, pp. 18–24); and, for a much fuller statement, Weiner 1961.

of abstracted patterns, comparison of these patterns with those prescribed as ideal rules, and subsequent correction of public policy activity in light of this matching process. More comfortable than most political scientists with mathematics and the research demands of the natural sciences, Deutsch took great care to formulate operational definitions explicitly including—in principle—quantitative dimensions for measurement and comparison. He was able to spell out with far greater lucidity an earlier contention by systems theorist Talcott Parsons that the core concept of politics—political power—was mistakenly restricted by virtually all political scientists to zero-sum impositions of will by individual actors on one another, whereas its most important manifestations were often positive-sum synergies within the governmental process in relation to collective goals. In a concluding tour de force of unparalleled audacity, Deutsch set out to demonstrate that even terms scarcely heard in politics since the Middle Ages—"grace," "redemption," "natural law"— could be given usefully measurable meaning within his paradigm.[15]

Deutsch's proposed paradigm was apparently *too* audacious. Despite its often stunning originality, despite its broad sweep and insistent operationality, despite Deutsch's repeated identification of readily researchable problems flowing directly from its central premises, the aptly titled *Nerves of Government* is perhaps the clearest instance in the history of political science of a major treatise that has generated no successors. In notable contrast to general systems theory, it not only failed to guide subsequent fieldwork in politics, but it did not even receive acknowledgement as a theoretical directive in the prefaces to such works. Critics, seemingly struck silent by its manner of rendering the familiar unrecognizable, neither endorsed nor disputed its premises. After producing a textbook retaining only passing appropriation of a few of its metaphors, Deutsch himself went on to more conventional themes. "Feedback," "monitoring," "informational networks," "self-correcting behavior," and the like have become staples of mid-1990s popular and academic discourse in politics. Their casually unsystematic usage, however, suggests seepage from the recent computer "revolution" rather than any architectonic vision implying conceptual indebtedness to Deutsch.

The three paradigmatic theoretical approaches to the systematic study of politics discussed above experienced very different fates. These differing outcomes, in turn, were at least partly rooted in differing developments within their ancillary disciplinary sources. Economists, whose rational-actor predictive model has served them well since the late eighteenth century, have shown a sustained interest in recent decades in applying their organizing concepts and propositions to the neighboring study of politics, even if only a limited number of political scientists have joined them in that endeavor. Biologists, entranced by the explosively productive results of highly technical laboratory work on genetic material, have shown no great recent interest in the macroanalytic speculations of Ludwig von Bertalanffy. The conceptual linkage between biology and social studies, always somewhat remote, has thinned noticeably in

15. Deutsch 1963, pp. 229–42.

recent decades, apart from semipopular adaptations of E. O. Wilson's work on competitive genetic selection, while both sociologists and political scientists have turned away from the search for unifying architectonic theorizing to an engagement with the policy issues of the day. As for cybernetics, its mathematical and mechanistic origins in engineering and the explosive evolution of its technology have always assured its distance from the theoretical concerns and capacities of most political scientists, while its transformative incorporation into practical life has become so pervasive as to leave little room or role for a distinctively autonomous metaphorical extension into political theory. The search for a unifying paradigm of politics therefore now seems likely to proceed, if at all, in other directions. And any attempts to label anomalous findings in the field as "premature" will, correspondingly, remain premature.

BIBLIOGRAPHY

Arrow, K. 1951. *Social Choice and Individual Values.* New York: John Wiley and Sons.

Bertalanffy, L. von. 1968. *General System Theory.* New York: Braziller.

Black, D. 1948. "The Decisions of a Committee Using a Special Majority." *Econometric* 16 (July): 245–61.

Buchanan, J. M., and G. Tullock. 1962. *The Calculus of Consent.* Ann Arbor: University of Michigan Press.

Deutsch, K. W. 1963. *The Nerves of Government.* New York: Free Press of Glencoe.

Downs, A. 1957. *An Economic Theory of Democracy.* New York: Harper and Brothers.

Easton, D. 1965a. *A Framework for Political Analysis.* Englewood Cliffs, N.J.: Prentice-Hall.

———. 1965b. *A Systems Analysis of Political Life.* New York: John Wiley and Sons.

Eulau, H. 1963. *The Behavioral Persuasion in Politics.* New York: Random House.

Mills, C. W. 1959. *The Sociological Imagination.* New York: Oxford University Press.

Mitchell, W. C. 1967. *Sociological Analysis and Politics: The Theories of Talcott Parsons.* Englewood Cliffs, N.J.: Prentice-Hall.

Moore, B., Jr. 1955. "The New Scholasticism and the Study of Politics." *World Politics* 8:1–19.

Olson, M. 1965. *The Logic of Collective Action.* Cambridge: Harvard University Press.

Parsons, T. 1937. *The Structure of Social Action.* New York: McGraw-Hill.

———. 1951. *The Social System.* New York: Free Press of Glencoe.

———. 1963. "On the Concept of Political Power." *Proceedings of the American Philosophical Society* 107. Reprinted as chap. 14 of T. Parsons, *Politics and Social Structure* (New York: Free Press, 1969).

———. 1969. *Politics and Social Structure.* New York: Free Press.

Riker, W. 1962. *The Theory of Coalitions.* New Haven: Yale University Press.

Schumpeter, J. 1954. *Capitalism, Socialism, and Democracy.* 4th ed. London: George Allen and Unwin.

Simon, H. 1962. "The Architecture of Complexity." *Proceedings of the American Philosophical Society* 106:467–92.

Tullock, G. 1970. *Private Wants, Public Means.* New York: Basic Books.

Weiner, N. 1961. *Cybernetics.* 2d ed. New York: John Wiley.

Weiner, N., A. Rosenblueth, and J. Bigelow. 1943. "Behavior, Purpose, and Teleology." *Philosophy of Science* 10:18–24.

CHAPTER EIGHTEEN

The Impact and Fate of Gunther Stent's Prematurity Thesis

Lawrence H. Stern

Gunther Stent's concept of premature discovery in science has become a part of the lexicon in science studies. But to what extent, and how, have scholars investigating the processes of scientific development used Stent's concept? Has it been fruitful? Has the concept had a tangible impact on the work of scholars investigating processes of scientific development? To address these questions I used the *Science Citation Index (SCI)* and the *Social Science Citation Index (SSCI)* and identified 76 papers that cited either of Stent's two prematurity articles between the years 1973 and 1997.

The analysis proceeds in two parts. First, I present quantitative measures—the number of times Stent's papers on prematurity were cited and how that number varied over time—and compare them with average citation rates of (1) all scientific papers covered in the *SCI* and *SSCI* databases, and (2) papers located in the more specialized and relevant field of social studies of science. Second, I provide a fine-grained qualitative analysis of the citing works.

CITATION ANALYSIS: A CAUTIONARY NOTE

Although the *SCI* and *SSCI* databases were originally devised as tools to improve the identification and retrieval of scientific information, scholars and science administrators have used them for quite different purposes. The *SCI* database has been put to good use, for example, to map and examine the cognitive structure of research areas and has been instrumental in analyses of the stratification, reward, and evaluation systems in science.[1] Citation analysis also plays a prominent role in

1. Co-citation analysis is an excellent means of identifying active research fronts in science. A series of yearly maps can trace dynamically the shifting foci of research. See Garfield 1998a for a brief review. Citation analysis has also been used to good effect to measure the degree of consensus in fields

"science indicators" studies and is often used to assess the performance of countries, institutions, fields, departments, and individuals.

There are of course limitations to the indexes, and their potential misuse has been widely addressed in the literature.[2] Among the problems of interpretation in citation analysis are those of overcitation and undercitation. Overcitation may occur if a work is cited inappropriately, or in a perfunctory manner simply to provide a token of support, or to invoke an example of fallacious or fraudulent work. While overcitation of a work may be detected relatively easily by examining the original reference, detecting undercitation is more difficult, for many reasons. Previous work that has had a significant influence on an author and is highly pertinent may not appear in the references. Authors may choose deliberately not to cite a pertinent work in order to enhance the perception of the originality of their own contribution. Perhaps more widespread, however, is a pattern that Robert Merton calls "obliteration by incorporation."[3] After becoming part of canonical knowledge, the source that contains the original formulations of ideas, methods, concepts, or findings may become obliterated and hence uncited.

Another potential difficulty is that an influential scientific idea may be cited, but not registered in the citation indexes. This depends in part, of course, on whether the citing source is included in the *SCI* and *SSCI* databases. While journal coverage has always been a cause of concern, over the past 35 years coverage has grown from 600 to over 8,000 journals. It is certainly possible that, especially early on, some important citations escaped detection.[4]

A more serious shortcoming, however, results from the fact that the coverage of source items includes only journals and some symposium or "series" volumes. Citations found in most books and monographs are not registered in the database. The various sciences and humanities disciplines make differing use of journal articles, monographs, and books to report work at the research front. Citations in journals alone may adequately reflect the influence of specific ideas in such disciplines as physics and molecular biology. But in disciplines such as history, philosophy, and sociology, citation counts most certainly underrepresent to a significant extent the use of an idea.

---

and the codification and immediacy of literature. See Cole et al. 1978. On analyses of the stratification, reward, and evaluation systems in science, see, for example, Cole and Cole 1973.

2. Although a number of studies (see Garfield 1998a) have validated the use of citation counts as a reasonable measure of the quality of scientific work, this remains a contentious issue. Most widely disputed—and thus perhaps best known—is the use of mere citation counts in assessing the performance of individual scientists, especially in the context of promotion or grant decisions. Even Eugene Garfield, the "father" of the *SCI/SSCI*, warned against the possible "promiscuous and careless use of quantitative citation data for . . . evaluation, including personnel and fellowship selection" (1963).

3. Merton 1968, p. 28.

4. Erroneous omission, presumably through oversight, may also occur. Four of the articles identified that cite Stent, though published in journals covered by the indexes, were not included on the *SCI/SSCI*-generated list.

Given these circumstances, one would expect Stent's prematurity concept to be undercited. An appreciable number of publications that appear in the field that Stent's contribution most readily "fits" into—social studies of science—are monographs and volumes of edited articles. Moreover, Stent's prematurity concept may be an example of the "obliteration by incorporation" phenomenon. Although Eugene Garfield has cited Stent's work on prematurity on seven separate occasions, if the term "premature discovery" is keyed into the search engine tied to Garfield's *Essays of an Information Scientist,* volumes 1–15, the search yields only 25 documents.[5] Still, with all their limitations, citation counts provide a useful, if crude, approximation of the impact of research on subsequent scientific development. To refine the analysis, however, one must supplement it with an examination of both the contents of the citation and the context in which it appears.

QUANTITATIVE ANALYSIS

Stent's article" Prematurity and Uniqueness in Scientific Discovery" was cited 76 times from 1973 to 1997.[6] Of these, 8 cited Stent for his discussion of uniqueness in scientific discovery and will not be considered further in this analysis.[7]

Stent's analysis of prematurity in science has generated interest consistently throughout the 25 years covered in this analysis. Although variability exists among fields, a large proportion of papers contained in the *SCI* database were never cited at all. Of those that were, the average number of times a paper was cited in a given year ranged between 1.8 and 2.1. Eugene Garfield recently analyzed the distribution of citations of the 32,728,729 papers that received at least one citation between the years 1945 and 1988.[8] He reports that well over half (55.8 percent) of these papers were cited only once in subsequent years, while nearly 80 percent (79.9 percent) were cited no more than 5 times. Only 1.5 percent of all cited papers were cited more than 50 times over their lifetime. During a typical 5-year period, an average paper cited in the *SCI* is mentioned roughly 3.5 times. Stent's article, in comparison, was cited between 12 and 16 times during each 5-year interval.

5. The search engine is linked to Eugene Garfield's Web page (http://www.garfield.library.upenn.edu). The term "delayed recognition"—which, for Garfield, is a kindred concept to prematurity—led to an even larger number of documents.

6. Seventy-two of these citations were registered in the *SCI* and *SSCI*. Of these, 27 appeared only in the *SCI,* 232 appeared only in the *SSCI,* and 22 were listed in both indexes. I came upon the remaining 4 citations serendipitously.

7. It is, however, interesting to note that 5 of these 8 citations appeared in the 5 years following the publication of Stent's article, with only 1 appearing after 1981. In contrast, Stent's prematurity concept was cited 43 times after 1981.

8. Garfield 1998a. See also the various annual *Journal Citation Reports,* published by the *Institute for Scientific Information* in conjunction with the *SCI.*

The citation rates of highly cited articles ordinarily peak at 5 years after publication and then diminish steadily thereafter.[9] But Garfield identifies three other citation patterns: (1) "rockets," which start out with a bang and keep rising rapidly; (2) "shooting stars," which achieve spectacular levels early but then fade quickly; and (3) "perennials," which achieve medium- to high-citation rates that remain fairly stable over 20 or more years. Stent's prematurity article appears to fit the last category.

A more suitable frame of reference, however, consists of other publications in the history, philosophy, and sociology of science. By 1972, the year Stent published his prematurity thesis, the field was marked by heated (and quite healthy) debates about the conduct of scientists and scientific change. How innovative claims—premature claims fit into this category—should be, or actually are, dealt with occupied an important place in the competing theoretical models that appear in the monographs listed in table 18.1.

Each monograph has been cited on average more often each year than has Stent's prematurity thesis. It is perhaps unfair to compare Stent's brief article with the more fully elaborated arguments set out by the other authors listed. If one compares the citation rate of Stent's prematurity article to those of other articles published in journals devoted to the history, philosophy, and sociology of science, it fares very well indeed. The average "impact factors" of the dozen most-cited journals in science studies over the years 1980 to 1988 are listed in table 18.2.[10] Impact factors of journals are determined by dividing the number of times a journal is cited by the number of source items it has published. The result is a measure of the average number of times an article published in that journal is cited in that year. Although considerable variability exists, most articles receive less than one citation in a given year. The highest citation rates accrue to articles published in *Scientometrics* (.933) and *Social Studies of Science* (1.289), journals attended to more by sociologists. Set against the average article published in science studies, then, Stent's prematurity paper, cited an average of 3 times each year over a 25-year period, did very well.

The disciplinary backgrounds of authors citing Stent's prematurity articles are presented in table 18.3. Forty-eight authors account for the 68 citing documents.[11] As can be seen, the diversity of backgrounds is quite large, as 24 different university departments are represented. This broad range is likely due to the appearance of one version of Stent's article in *Scientific American*, a semipopular journal that boasts a large interdisciplinary readership. The largest cluster represented is in life sciences, with 18, or 37.5 percent, of the citing authors affiliated with such departments. If those from departments of medicine are included in this category, this raises the percentage to 45.8 percent.

9. Garfield 1990b.

10. See the listings and rankings of journals by their impact factors in the annual *Journal Citation Reports*, published by the *Institute for Scientific Information*.

11. Tobias, an anthropologist, cites Stent in 8 separate publications; Garfield, founder of the *SCI*, cites Stent on 7 occasions; Zuckerman, a sociologist of science, does so 3 times; and 6 others do so twice.

TABLE 18.1 Citations of Selected Monographs in the History, Philosophy, and Sociology of Science, 1972–1999

Publication	Years	Total Number of Citations	Mean Number of Citations per Year
Thomas Kuhn, *The Structure of Scientific Revolutions*	1972–1999	5,266	188.07
Karl Popper, *The Logic of Scientific Discovery*	1972–1999	1,519	54.25
Michael Polanyi, *Personal Knowledge*	1972–1999	853	30.46
Ernest Nagel, *Structure of Science*	1972–1999	792	28.29
Larry Laudan, *Progress and Its Problems*	1977–1999	579	26.30
Robert K. Merton, *The Sociology of Science*	1973–1999	606	22.44
David Bloor, *Knowledge and Social Imagery*	1976–1999	414	17.25
Norwood Russell Hanson, *Patterns of Discovery*	1972–1999	464	16.57
Ludwik Fleck, *Genesis and Development of a Scientific Fact*	1979–1999	238	11.33
Stephen Toulmin, *Human Understanding*	1972–1999	312	11.14
Karl Popper, *Conjectures and Refutations*	1972–1999	238	8.50
Barry Barnes, *Interests and the Growth of Knowledge*	1977–1999	186	8.09
Gerald Holton, *Thematic Origins of Scientific Thought*	1973–1999	217	8.04
Paul Feyerabend, *Against Method*	1975–1999	186	7.44
Imre Lakatos, *Proofs and Refutations*	1976–1999	161	6.71

TABLE 18.2 Average Impact Factors of Most-Cited Science Studies Journals, 1980–1988

Journal	Average
Annals of Science	0.375
Archive for History of Exact Sciences	0.361
British Journal for the History of Science	0.548
British Journal for the Philosophy of Science	0.395
Bulletin of the History of Medicine	0.388
Impact of Science on Society	0.106
ISIS	0.715
Journal of the History of Medicine and Allied Sciences	0.316
Medical History	0.398
Philosophy of Science	0.671
Scientometrics	0.933
Social Studies of Science	1.289

TABLE 18.3 Disciplinary Affiliation of Authors
Citing Stent's Prematurity Analysis

Discipline	Number	Percentage
Life sciences (biochemistry, biology, genetics, molecular biology, neuroscience, zoology)	18	37.5
Science studies (history of science, information science, science studies, sociology)	12	25.0
Medical science	4	8.3
Psychology	2	4.2
Anthropology	1	2.1
Chemistry	1	2.1
Education	1	2.1
English	1	2.1
Futurology	1	2.1
Humanities	1	2.1
Journalism	1	2.1
Law	1	2.1
Library science	1	2.1
Organizations	1	2.1
Political science	1	2.1
Technological development	1	2.1

The next-largest disciplinary cluster—composed of historians, philosophers, sociologists, and others who specialize in science studies—accounts for 25 percent of those citing Stent. The remaining 29.2 percent of citing authors are spread evenly among the remaining 13 disciplines (with the exception of psychology, which has 2 citing authors). No citation was found in a publication in the physical sciences.

That only one-quarter of the scholars citing Stent are located in science studies is quite surprising. Nevertheless, the general concept of prematurity, if not its precise substantive meaning, appears to have diffused across a number of disciplinary boundaries.[12]

QUALITATIVE ANALYSIS

How, then, has Stent's analysis of prematurity in science been taken up and used by fellow scientists? One of the patterns long recognized in the development of

12. That this work was widely diffused across disciplinary boundaries *and within* disciplines is further indicated by the fact that the 68 citing documents appeared in 57 separate journals. The "Subject Category Listings" and "Number of Citing Journals" (the latter in parentheses) are: behavioral sciences (14), medical sciences (11), science studies (10), multidisciplinary (10), life sciences (6), communications (1), computer applications and cybernetics (1), education—English (1), law (1), management (1), and planning and development (1).

knowledge is the partial incorporation of selected aspects of scientific claims. That is to say, as knowledge claims from the research front are selectively absorbed and incorporated into the core, they often are not taken all of a piece. Some aspects of the innovation are considered more important and are accentuated, while other aspects are stripped away. Through use, these claims are often transformed as different shades of meaning are attached to the innovation. Before examining how prematurity has been cited, then, we must consider the main thrust, or "central message," of Stent's thesis.[13]

Stent argues that prematurity, as he defines it, is a useful historical concept that contributes to an understanding of the development of scientific knowledge. His overall argument contains five interrelated claims and includes cognitive, behavioral, structuralist, temporal, and prescriptive components.

1. A premature discovery is a generic type of cognitive claim: it cannot be "connected" to canonical knowledge. Moreover, this cognitive attribute—its disconnectedness—is the reason or explanation why it is not appreciated and/or accepted by the relevant practitioners in the field at the time it is presented. So-called sociological factors, though perhaps part of the relevant context, are not to be considered determinative.

2. That the discovery is *not* appreciated and/or accepted—that, as a rule, scientists choose not to pursue it—is the behavioral component of Stent's argument.

3. Structuralism provides the basis for understanding why a discovery cannot be appreciated: until it can be connected logically to contemporary canonical knowledge, the discovery is quite literally perceived to be meaningless.

4. The prematurity thesis is not restricted to past cases; it can be applied to present-day cases as well. Instances of "here-and-now prematurity" may be identified. Temporally, it is open-ended. Moreover, Stent is agnostic concerning the eventual outcome of any particular case. It may or may not be accepted at a later date.

5. The prematurity of an alleged discovery is a rational reason for its neglect or rejection. Moreover, it is wholly appropriate that the scientific community ignore or reject such claims until they may be connected to canonical knowledge. This is the prescriptive component of Stent's argument.[14]

13. The expression "central message" is from Patinkin 1983.

14. In Stent's earlier writings (1968; 1969, p. 31), he refers to as the Eddington Rule: "It is also a good rule not to put overmuch confidence in the observational results that are put forward until they have been confirmed by theory." In the prematurity article, Stent makes no mention of Eddington and now cites Polanyi in support of this claim. Moreover, it is of some interest to note that by invoking the structuralist argument, Stent transforms the reasons for the nonacceptance of a premature discovery from a methodological mandate to a cognitive constraint that renders it meaningless.

TABLE 18.4 Ways in Which Stent's "Prematurity and Uniqueness in Scientific Discovery" Has Been Utilized by Citing Authors, 1973–1997

Type of Use	Number of Articles	Percentage
1. Used as an ex post facto interpretive device		
a. Casually applied without discussion	15	22.1
b. Applied to nonacceptance of citing author's work	8	11.8
c. Extended analysis of case	7	10.3
2. Challenges applicability to case of Avery et al.	6	8.8
3. Extends or modifies the concept of prematurity or used as part of author's own approach	4	5.9
4. Premature discovery and citation analysis		
a. Potential cases of premature discovery identified by citation analysis	3	4.4
b. Implications of the concept of prematurity for the use of citation analysis	4	5.9
5. Existence of prematurity cited as general context in discussion of related issues; no explicit role in analysis	10	14.7
6. Commentary on prematurity; science news report	2	2.9
7. Idiosyncratic or irrelevant usage	1	1.5
8. Citation for concept of uniqueness	8	11.8

[a]Eight of the 76 citing articles are excluded from this analysis. Seven were unavailable, and one is a self-citation by Stent.

Various ways in which authors use the term "prematurity" in citing Stent are listed in table 18.4.[15] Here, I discuss categories 1 through 4.[16]

The expressed purpose of Stent's prematurity thesis is to explain why certain scientific discoveries are not appreciated at the time they are initially introduced to the scientific community. As seen in table 18.4, prematurity is used as an ex post facto interpretative device in 44 percent of the articles citing it (categories 1a, b, and c combined).[17] Of these, however, fully half (15 of the 30 articles) apply the con-

15. Although a phenomenology of citation practices has not yet been developed, a number of schemes (e.g., Moravcsik and Murugesan 1975; Chubin and Moitra 1975) and qualitative analyses (e.g., Cole 1975) have appeared in the literature. The typology presented here is derived from an examination of the citing articles. Since I was the only coder, it is possible that an independent classification by another coder would exhibit differences.

16. The concept of prematurity plays no significant role in the analyses contained in the articles listed in category 5. For example, Paul and Charney (1995) simply state that the "prematurity" of a knowledge claim will affect the rhetorical strategy adopted by authors as they argue their point. Portis (1986), in a discussion of authority in science, argues that prematurity is one possible negative consequence of this state of affairs.

17. The list of citing articles and authors keyed to this table may be obtained from the author.

cept casually and with virtually no discussion. Typically, an author treats the existence of the general phenomenon of prematurity as given—as meaning, in essence, simply no more than "ahead of its time"—and, in what nearly amounts to a throw-away line, affixes the term, without any analysis, to a particular case. For example, discussing their discovery of direct evidence for the existence of fluid membranes, S. L. Tamm and S. Tamm write, "After 'discovering' this unusual phenomenon, we found that Kirby had noticed it in devescovinids more than 20 years ago. It is an interesting example of prematureness in scientific discovery that Kirby's view of the 'fluidity and lability of the surface layer' did not attract attention at the time."[18]

Or consider R. Sekuler, who, after briefly commenting on controversial reports in research on vision, writes, "We may be approaching the point where it will no longer be premature (Stent, 1972) to explore bimodal effects in spatial vision."[19] In neither of these cases, nor in the 14 others listed in table 18.4, category 1a, does the author provide any supporting evidence that the concept of prematurity as enunciated by Stent is applicable to the case in hand.

For years, scientists have lamented either the nonacceptance or what they believe to be the slow recognition of their work. Not surprisingly, a number have invoked Stent's prematurity thesis as an explanation of their plight. However, much the same can be said of the eight documents listed in category 1b as has been said of the papers in the preceding category. In all but two, the authors simply affix the tag of prematurity to their case with little or no discussion. The lone exception is Phillip V. Tobias, an eminent paleoanthropologist who, in two papers, presents an extended discussion of evidence that supports his assertion that the joint discovery by Louis Leakey, Tobias, and J. R. Napier of a new species of human being, *Homo habilis*, is a clear example of Stent's prematurity thesis. Tobias also presents evidence that another of his claims, that *Homo habilis* was the first speaking primate, may qualify as an example of what Stent has called here-and-now prematurity.[20]

Seven papers, 10.3 percent of all citing articles, present more extended discussions of the reception of scientific innovations and raise the issue of prematurity in their analysis. Five do so in a positive way. Of these, three focus on specific historical cases: C. T. Sawin discusses Arnold Adolph Berthold's experiments on the transplantation of testes, J. H. Comroe examines Kurt von Neergaard's discovery of factors that affect the collapse of lungs in newborns, and P. V. Tobias provides an analysis of Raymond Dart's discovery of the new hominid species *Australopithecus africanus*.[21] Each suggests that the concept of prematurity offers a likely explanation of the initial neglect of his or her case at hand.

Two authors examine relatively recent cases. O. E. Landman uses prematurity to account for the nonacceptance of experimental evidence in support of the in-

18. Tamm and Tamm 1974.
19. Sekuler 1974.
20. Tobias 1992, 1996a; Leakey et al. 1964.
21. Sawin 1996; Comroe 1977; Tobias 1985.

heritance of acquired characteristics, while A. C. Wardlaw argues that the development of acellular pertussis vaccines is best understood in this light.[22]

It is relevant to note that these five papers are neither authored by nor addressed to specialists in the history, philosophy, or sociology of science. Three appear in medical journals *(Endocrinologist, American Review of Respiratory Disease,* and *Vaccine)*, the fourth in a semipopular science magazine *(Bioscience)*, and the fifth in a disciplinary text devoted to review articles *(Yearbook of Physical Anthropology)*. Although more than casual attention is paid to the circumstances surrounding the reception of these claims, detailed and systematic analysis that would unambiguously establish these as bona fide instances of premature discoveries is absent.

The two remaining papers in this category, however, were written by historians of science and published in a journal devoted to this field *(Bulletin of the History of Medicine)*. Ilana Löwy examines the reception of James Bumgardner Murphy's early discovery of the role of lymphocytes in immune reactions; W. H. Schneider considers the reception of Karl Landsteiner's discovery of blood groups.[23] Their analyses lead them to altogether different conclusions. Although each raises the issue of prematurity in the introductions of their papers, both Löwy and Schneider argue that it does not apply in the specific case under scrutiny. But more to the point, each expresses serious reservations about the heuristic potential of the prematurity thesis in historical analysis.[24]

Reservations are also expressed in the six papers listed in category 2 of table 18.4. Here, although the legitimacy of the prematurity thesis is, for the most part, not questioned, its applicability to the case Stent used as his prime example—O. T. Avery, C. M. MacLeod, and M. McCarty's demonstration of the transformation of pneumococci by DNA—is clearly challenged.[25] Bentley Glass, for example, writes, "The criterion [of Stent's prematurity] may well be valid even though Stent's . . . example [Avery and his colleagues] of long neglect may not be well cho-

---

22. Landman 1993; Wardlaw 1992.

23. Löwy 1989; Schneider 1983.

24. Schneider (1983, p. 562), for example, objects that the concept of prematurity "carries with it implications of the old 'whig' interpretation of the history of science and medicine." He argues, further, that the "concept of prematurity in medical or scientific discovery carries with it another notion that confuses an understanding of the process. It implies an inevitability or predictability—even if interrupted—like fruit ripening on the vine. Progress is seen as normal and logical, delays as extraordinary and unexpected. Such a deterministic model of historical explanation is not in keeping with the usual facts of historical reality." Löwy's reservations center on the difficulties involved in identifying actual cases of prematurity. This is not to say, however, that either Löwy or Schneider depict Stent's argument in a wholly accurate fashion. As shall be seen, each in his or her own way uses the concept of prematurity somewhat differently than Stent intended.

25. Cohen and Portugal 1975 is the lone exception that *does* question it. While the authors refer to prematurity as a "provocative concept," their analysis of the Avery case has them conclude that the discovery might just as well have been considered "belated" as premature. Thus, they conclude, "since a discovery can hardly be 'premature' and 'belated' simultaneously, it is preferable to discard such labels as being unnecessarily deterministic" (p. 207).

sen. Also, the criterion may be valid, yet not account entirely for the long neglect that occurs."[26]

For Glass, as well as for Joshua Lederberg and R. D. Hotchkiss, the "touchstone," or prime indicator, of a case of prematurity is the behavioral response of scientists working in the same or closely related fields. As Lederberg notes, and Hotchkiss affirms, Avery, MacLeod, and McCarty's demonstration was well-known and elicited active discussion and inquiry on the part of many scientists.[27] Their paper, Lederberg reports, "enjoyed almost 300 citations between 1945 and 1954, not to mention many more earned by McCarty's elaborations." Far from being ignored, the work by Avery and his colleagues was challenged on several fronts just as it challenged other scientists to explore its implications.[28]

Combining categories 1 and 2 from table 18.4, then, reveals that 36 articles, roughly 53 percent of those articles citing Stent, either apply (in varying degrees) or question the applicability of prematurity to 22 separate cases.[29] This seems to indicate that fellow scientists found Stent's argument both evocative and provocative. But in 15 of these papers (category 1a) prematurity is used in passing as a simple and perhaps convenient descriptive tag with no substantiating arguments. Extended analyses appear in only 13 articles, and, of these, only 5 contain some measure of support. The two scholars trained in the history of science that conducted extended analyses were critical of the prematurity concept.

26. Glass, 1974, p. 105. In an earlier work, Glass discusses the long neglect of discoveries in genetics and states that it "is not all unusual" (1965, p. 227).) As examples, he presents the cases of Mendel, Fredrich Miescher's discovery of the chemical basis of heredity, and Sir Archibald Garrod's discovery of inborn errors of metabolism. Here, he argues that Avery's work was neither neglected nor delayed for an unreasonable length of time. Scientists simply "suspended judgment" as additional studies were carried out. He concludes, "One may say that the criterion of prematurity, as defined by Stent, without question applies well to the classic cases of neglect of the work of Mendel, Miescher, and Garrod. It leads, on the contrary, to the rejection of the proposition that the work of Avery, MacLeod, and McCarty should be added to their number" (1974, p. 110).

27. Lederberg 1994; Hotchkiss 1979, 1995. After outlining both the varied responses to works by Avery et al. and the various investigations sparked by the discovery, Hotchkiss (1979, p. 339) concludes, "It was a time of maturation of a new field. But I see this maturation as the growth of an infant science and not the delicate nurturing of a 'premature' one as Stent has suggested and others have tried to explain. As a midwife and nurse in the upbringing of the infant, I want to report that it was a normal healthy one. The fact that the infant by the age of twenty or so had composed brilliant concertos, sonatas, and symphonies need not raise in anyone's mind the question, why didn't he compose a rondo or cadenza before the age of ten?" Instead of a case of premature discovery, Hotchkiss argues that this case is in keeping with a broad pattern characteristic of the history of genetics: one of "discovery, consolidation, and reorientation."

28. Lederberg's diary entry upon reading the paper by Avery et al. and his unpublished letter to *Scientific American* in response to Stent's prematurity article appear on the Web site of the National Library of Medicine (see http://profiles.nlm.nih.gov). For two additional criticisms of Stent's use of prematurity with regard to Avery by "insiders," see Dubos 1976 and McCarty 1985.

29. Eleven cases occurred in the history of medicine and physiology, 3 each in the neurosciences and paleoanthropology, 2 in genetics, and 1 each in psychology, chemistry, and population genetics.

Of the 48 scholars citing Stent, only 4 (located in categories 3 and 4a in table 18.4) have incorporated one or more aspects of the prematurity thesis into their ongoing work. In each instance the analyst has recast the phenomenon to fit more readily into his or her developing perspective. Using as his case the discovery by Avery and his colleagues' that DNA is the carrier of genetic material, H. V. Wyatt, a professor of biology, extends Stent's thesis by introducing a technical dimension into the operational definition of prematurity. In his view, the reason Avery's work was not fully appreciated when first introduced was not because it was cognitively disconnected. Rather, the chief difficulty lay in the fact that "the technical means were not yet available to extend the work into other systems and confirm the universal nature of the phenomena."[30] Broadening the concept of prematurity to include this technical dimension, Wyatt argues more generally that, regardless of its cognitive fit, a "discovery can be premature if it is not capable of being extended experimentally because of technical reasons."[31] Wyatt has not pursued this line of inquiry further; nor, it appears, have others attempted to follow his lead.

Tobias is another who picks up and elaborates Stent's formulation of prematurity. In addition to his earlier work noted above, in November 1994 at the American Philosophical Society, Tobias turned his attention to the more general issue of the acceptance and rejection of scientific discoveries.[32] Considering prematurity as a special case of "delayed acceptance," Tobias stresses that other factors, in addition to and in concert with the proposed discovery's lack of fit with canonical knowledge, may account for delays. He discusses, in turn, linguistic, political, personality, and theological concerns.

Prematurity, however, remains Tobias's main focus, and he suggests that, given enough cases, field-specific indexes of prematurity—the average lapse of time between the first publication and the general acceptance of the discovery—can be constructed. These indexes, then, may be compared, and the relationship between a field's index and its rate of cognitive development may be assessed. As a preliminary step, Tobias identifies ten cases drawn from five different fields and assigns them index values.[33]

---

30. Wyatt 1975, p. 152.

31. Wyatt 1975, p. 149. McCarty (1985, p. 231) is no more impressed with Wyatt's analysis than he is with Stent's.

32. Tobias 1996b.

33. These ten candidate cases are (1) Benjamin Franklin's remark in 1751 that "there is no bound to the prolific nature of plants or animals but what is made by their crowding and interfering with each other's means of subsistence," which finally "bore fruit 107 years later when Darwin and Wallace presented the theory of natural selection," (2) Frere's hypothesis on the great antiquity of humankind published in 1800, (3) Mendel's announcement of the laws of inheritance in 1865, (4) Snider's discussion of continental drift in 1858, (5) Chagas's discovery of *Trypanosoma cruzi* in 1909, (6) Dart's discovery of *Australopithecus africanus* in 1924, (7) Fleming's discovery of penicillin in 1929, (8) Avery and his colleagues' discovery in 1944 that DNA is the basic hereditary substance, (9) Watson and Crick's announcement of the double helix in 1953, and (10) Tobias's joint discovery, with Leakey and Napier, of *Homo habilis* in 1964. It must be noted that

Tobias clearly accepts both the cognitive and behavioral components of Stent's operational definition of prematurity, and he is the only analyst in this study who makes reference to here-and-now prematurity. But Tobias makes no mention of the structuralist component of Stent's argument, and, moreover, he signals his departure from Stent's formulation when he recasts prematurity as a type of delayed acceptance. By doing so, Tobias takes the position that a discovery must gain eventual acceptance to qualify. By abandoning Stent's agnostic stance, then, Tobias actually throws the whole notion of here-and-now prematurity into question.

The concept of prematurity also figures prominently in the research of sociologists working in the so-called Columbia tradition in the sociology of science. Indeed, a vigorous program of research devoted to what Robert Merton identifies as "the problem of identifying the conditions and processes making for continuity and discontinuity in science" was established well before the appearance of Stent's prematurity papers. Merton's analysis of priority disputes in science and the resistance to such studies, Bernard Barber's analysis of "resistance by scientists to scientific discovery," and Stephen Cole's empirical investigation of the phenomenon of "delayed recognition" in physics all provide ample evidence of this previous interest.[34] Harriet Zuckerman's more recent work on postmature discoveries in science—"scientific contributions that presumably could have been made some time before they actually were, if only their narrowly *specific* cognitive ingredients had been sufficient for the outcome"—reflects a continuing interest in these issues.[35] It is thus no surprise that Stent's concept of prematurity struck a resonant chord with this group.[36]

The concept of prematurity was also much discussed by those in this tradition after the publication of Stent's thesis. As Zuckerman notes,

---

Tobias casts his net rather wide in his search for candidate cases. Although he states that "it is very likely that all of [the] cases listed here are premature discoveries," others will surely challenge this assertion.

34. On Merton's analysis of priority disputes, see Merton 1957, 1973; on resistance to the study of such disputes, see Merton 1963, 1973; on Barber's analysis of scientists' resistance to scientific discovery, see Barber 1961; on Cole's empirical investigation of delayed recognition, see Cole 1970.

35. Zuckerman 1978, p. 80. In a later work in which Zuckerman collaborated with Lederberg, they write, "For a discovery to qualify as postmature, for it to evoke surprise from the pertinent scientific community that it was not made earlier, it must have three attributes. In retrospect, it must be judged to have been technically achievable at an earlier time with methods then available. It must be judged to have been understandable, capable of being expressed in terms comprehensive to working scientists at the time, and its implications must have been capable of having been appreciated" (Zuckerman and Lederberg 1986, p. 629).

36. In fact, Stent acknowledges his debt to both Merton and Zuckerman "for helping me to focus my ideas more sharply." Each was in attendance when Stent presented his prematurity and uniqueness argument in May 1970 at a conference sponsored by the American Academy of Arts and Sciences, titled "History of Biochemistry and Molecular Biology." Zuckerman called Stent's attention to Polanyi's work and its relevance to Stent's prematurity concept, while Merton's comments centered on the notion of uniqueness. Stent's acknowledgment may be found at the end of his 1972 paper, published in *Advances in the Biosciences,* and then again in "About the Author," which accompanies the version published in *Scientific American.*

The concept [of postmaturity] was designed to round out the family of kindred concepts of "premature" and "mature" discoveries as developed in the course of a year's work (1973–74) in the "historical sociology of scientific knowledge" by Yehuda Elkana, Joshua Lederberg, Robert Merton, Arnold Thackray, and Harriet Zuckerman at the Center for Advanced Study in the Behavioral Sciences. Premature contributions are those that, once made, are not immediately followed up and developed by the pertinent community of scientists; scientific significance is attributed to them only later (sometimes after independent rediscoveries). These are in retrospect, sometimes described as having been "ahead of their time." Mature discoveries are those appearing in their apt time, being recognized and taken up at once; a special subset of such contributions appear in the form of multiple, independent, and more or less simultaneous discoveries. On cognitive sources of prematurity, see Stent (1972); on social sources, see Barber 1961); on maturity, see Merton (1973, chapters 14–17).[37]

It seems clear, then, that the basic concept of prematurity was found to be of sufficient interest to be incorporated into the ongoing research program of the group. Yet rather than fully embrace Stent's total formulation of the prematurity thesis, members chose to transform the concept of prematurity, to broaden it in some ways and strip it down in others. For example, in stating that scientific significance is attributed to such claims at a later date—sometimes after independent rediscoveries—the members discard Stent's agnostic stance. So, too, the temporal dimension is discarded when Zuckerman and Lederberg assert that cases of prematurity, like postmaturity, can be recognized only retrospectively.[38]

Perhaps most important, prematurity is now said to have "social sources." This point is made more explicit in Zuckerman's later work with Lederberg. Here, they list, in addition to Stent's operational definition, five other reasons why a discovery may be premature: "Discoveries can be premature because they are conceptually misconnected with 'canonical knowledge,' are made by an obscure discoverer, published in an obscure place, or are incompatible with prevailing religious and political doctrine. Barriers between disciplines imposed by specialization of inquiry also contribute to neglect or resistance."[39]

Garfield, president of the Institute for Scientific Information and creator of the citation indexes, has noted the heuristic potential of Stent's formulation of prematurity and, further, has sought to operationalize the concept. In Garfield's view, premature discoveries, which he considers a type of "delayed recognition," constitute strategic research materials through which to examine those factors affecting the reception of innovative claims.[40] With the vast database of the *Science Citation Index* at his disposal, Garfield develops a systematic, quantitative algorithm to

37. Zuckerman 1978, p. 88.
38. Zuckerman and Lederberg 1986, p. 629.
39. Ibid.
40. Conceptually, Garfield considers Barber's research on resistance, with its emphasis on social factors, and Stent's ideas on prematurity, with their sole emphasis on cognitive fit, to be special subsets of the general phenomenon of "delayed recognition," as set out by Cole 1970.

identify possible examples of prematurity by tracing the citation histories of key papers associated with these discoveries.[41]

As a first approximation, those heavily cited papers *(Citation Classics)* that had low citation frequencies for the first five or more years after publication were used to identify cases of delayed recognition. Low citation frequency was initially defined as being near the average of one cite per year for a typical paper.[42] In this first attempt, Garfield identifies five papers and is clearly disappointed with the results.[43] No systematic analysis is included, and, judging from the hints offered by the authors of the papers under consideration, it is unlikely that any of these cases qualify as examples of prematurity in Stent's sense.

Garfield's second effort, published the following year, was somewhat more successful. Starting with a list of the 100 most frequently cited papers in the 1945–1988 *Science Citation Index,* Garfield set new criteria for inclusion. Here, to qualify as a candidate case of delayed recognition a paper had to have 10 or fewer citations per year at age 10 and at least a tenfold increase in citations at age 20.[44] Garfield discusses 4 of the 20 papers identified in this manner and in two of the cases provides comments by the authors that note perceived resistance to their work when it was initially introduced. A third case, in Garfield's view, provides a good example of the delayed application of a research method.[45]

Although Stent's notion of prematurity—together with Barber's work on resistance and Cole's on delayed recognition—motivated Garfield's efforts, it was not his intent to provide the systematic analysis required to determine whether or not these cases qualify as instances of prematurity in Stent's sense. Though clearly suggestive, the cases Garfield identifies have not, to my knowledge, been followed up in any systematic way.

DISCUSSION

Since the data are drawn solely from those published papers that cited either of Stent's two prematurity articles, our analysis is necessarily restricted—perhaps severely. Careful examination of the many monographs, edited volumes, and con-

41. Cole was the first to use citation records as a means of identifying examples of what he called "delayed recognition" (ibid.). Limited by the data available to him at the time, Cole considered as delayed those papers that received at least 10 citations in 1966 and 3 or fewer citations in 1961.

42. Garfield 1989.

43. Garfield concludes, "The phenomenon of delayed recognition in the classic sense appears to be relatively unusual. But clearly such papers do exist. Undoubtedly there are dozens of other examples that may or may not be identified by citation analysis. However, where the expert systems may fail, the human brain may succeed. So if you know of a scientific contribution that belongs in the category of delayed recognition, please send me the details. I hope to review such new examples and comment upon them in a future essay" (ibid., p. 159).

44. Garfield 1990a.

45. Garfield also discusses 2 additional cases suggested by readers.

ference proceedings that have been published in the social studies of science during this period would undoubtedly yield additional evidence. Moreover, not all instances of cognitive influence leave an archival trace. Despite the limited data, however, it is clear that prematurity has attracted the attention of many scholars. Stent's paper is near the top 1 percent of all cited papers.

To the extent that citations reflect scientists' interest in a particular contribution, the 68 citations spread over the 25-year period suggests that the idea was initially found to be stimulating and continued to be so for longer than those first proposed in the average scientific publication. However, Stent's contribution was cited by only a handful of scholars devoted to the systematic investigation of the development of scientific knowledge. And it is this audience, I think, that Stent most wished to attract.

Moreover, citing authors often altered—in subtle and not-so-subtle ways—the meaning of Stent's prematurity. They selectively shaped and molded Stent's analysis to fit their own concerns. In some ways, Stent's argument has been severely compressed. Stripped away are the structuralist, temporal, and prescriptive components of his overall argument. In other ways, his argument it has been broadened. For some, the fate of an alleged premature discovery, which is retrospectively identified as being ahead of its time, may be explained by *either* cognitive factors—its "fit" with canonical knowledge—*or* social factors including, but not limited to, the professional standing of the author of the claim, publication outlet (including the language in which it is written), political concerns, and religious doctrine.

For the most part, however, Stent's notion has been interpreted by scholars simply to designate episodes in the history of science in which a scientific claim, neglected or rejected by the scientific community when first introduced, subsequently gained acceptance. His paper has become a symbolic marker of the general phenomenon of some proposal being, for any reason, "ahead of its time," the usual understanding of "premature." But this, in itself, is no small thing.[46] Scientific concepts are designed to fix one's perception on selected aspects of reality and to cause one to think about them in certain ways. Stent's prematurity thesis does precisely that as it forces us to consider processes that are basic to the workings of science.

Stent, of course, was neither the first nor the only one to do so. As mentioned above, a number of competing models of scientific change were introduced during this time, and heated debates were rampant. Despite obvious differences, there was nevertheless agreement that the introduction of scientific claims that depart

46. Henry Small (1978) has noted that cited documents often come to be taken as "symbols" of the scientific ideas expressed in the text. At the same time, he continues, this reflects the citing authors' selection and interpretation of these ideas. Although in most cases, the intended message is the one received, Small argues that the possibility of the social transformation of meaning must be recognized. The "uniformity of usage," which Small defines as "the percentage of citing contexts which share a particular view (the most prevalent) of the cited item," is an empirical matter and likely to vary under specifiable conditions.

in significant ways from prevailing cognitive frameworks is ubiquitous to science and poses formidable problems. At least one dozen kindred but not identical concepts have been used to describe this generic type of cognitive claim, including anomalous, extraordinary, pathological, unorthodox, unconventional, heterodox, controversial, deviant, revolutionary, monster, and pseudoscientific. But none, I think, has struck as resonant a chord as prematurity. The concept appears to have moved effortlessly into mainstream scientific discourse.

That scholars with such different disciplinary backgrounds have picked up the term "prematurity" stems, I suggest, from the basic familiarity of the word. It does not sound like jargon, and its everyday connotations are so well-known that specialists and nonspecialists alike, when hearing it for the first time, recognize it instantly and immediately grasp some sense of its meaning.[47] Can the same be said about "anomaly"?

Through the years, then, Stent's concept of premature discovery in science has been transformed in the literature from his original formulation. Some of the components of his thesis have been accentuated—perhaps even exaggerated—while other components have faded from view. This winnowing process, through which scientific claims are selectively absorbed and then partially incorporated into ongoing work, exemplifies one pattern in the development of scientific knowledge.

BIBLIOGRAPHY

Avery, O. T., C. M. MacLeod, and M. McCarty. 1944. "Studies on the Chemical Nature of the Substance Inducing Transformation of Pneumococcal Types." *Journal of Experimental Medicine* 79:137.

Barber, B. 1961. "Resistance by Scientists to Scientific Discovery." *Science* 134:596–602.

Carroll, L. 1871. *Through the Looking-Glass, and What Alice Found There*. Special ed. New York: Random House, 1946.

Chubin, D. E., and S. D. Moitra. 1975. "Content Analysis of References: Adjunct or Alternative to Citation Counting." *Social Studies of Science* 5:423–40.

Cohen, J. S., and F. H. Portugal. 1975. "Comment on Historical Analysis in Biochemistry." *Perspectives in Biology and Medicine* 18:204–7.

Cole, J. R., and S. Cole. 1973. *Social Stratification in Science*. Chicago: University of Chicago Press.

Cole, S. 1970. "Professional Standing and the Reception of Scientific Discoveries." *American Journal of Sociology* 76:286–306.

———. 1975. "The Growth of Scientific Knowledge." In *The Idea of Social Structure: Papers in Honor of Robert K. Merton,* ed. L. Coser, pp. 175–220. New York: Harcourt Brace Jovanovich.

47. The fact that prematurity or premature discovery appears to have moved rather easily into scientific discourse suggests this might be an example of Merton's "obliteration by incorporation" phenomenon (1968, pp. 28, 35, 38), discussed earlier. If so, the limitations of relying solely on published citations to Stent's work to assess its impact are, of course, considerably enlarged.

Cole, S., J. R. Cole, and L. Dietrich. 1978. "Measuring the Cognitive State of Scientific Disciplines." In *Toward a Metric of Science: The Advent of Science Indicators*, ed. Y. Elkana, J. Lederberg, R. K. Merton, A. Thackray, and H. Zuckerman, pp. 209–52. New York: John Wily and Sons.

Comroe, J. H. 1977. "Premature Science and Immature Lungs. 1. Some Premature Discoveries." *American Review of Respiratory Disease* 116:127–35.

Dubos, R. J. 1976. *The Professor, the Institute, and DNA: Oswald T. Avery, His Life and Achievements.* New York: Rockefeller University Press.

Fleck, L. 1979. *Genesis and Development of a Scientific Fact.* 1935. Reprint, Chicago: University of Chicago Press.

Garfield, E. 1963. "Citation Indexes in Sociological and Historical Research." *American Documentation* 14:289–91.

———. 1979. *Citation Indexing—Its Theory and Application in Science, Technology, and Humanities.* Philadelphia: Institute for Scientific Information Press.

———. 1985. "Uses and Misuses of Citation Frequency." In *Essays of an Information Scientist*, pp. 403–9. Vol. 8. Philadelphia: Institute for Scientific Information Press.

———. 1987. "A Different Sort of Great-Books List: The 50 Twentieth-Century Works Most Cited in the Arts and Humanities Citation Index, 1976–1983." In *Essays of an Information Scientist*, pp. 101–5. Vol. 10. Philadelphia: Institute for Scientific Information Press.

———. 1989. "Delayed Recognition in Scientific Discovery–Citation Frequency Analysis Aids the Search for Case Histories." In *Essays of an Information Scientist*, pp. 154–61. Vol. 12. Philadelphia: Institute for Scientific Information Press.

———. 1990a. "More Delayed Recognition 2. From Inhibin to Scanning Electron-Microscopy." In *Essays of an Information Scientist*, pp. 68–74. Vol. 13. Philadelphia: Institute for Scientific Information Press.

———. 1990b. "The Most-Cited Papers of All time, SCI 1945–1988. Part 1B. Superstars New to the SCI Top 100." In *Essays of an Information Scientist*, pp. 57–76. Vol. 13. Philadelphia: Institute for Scientific Information Press.

———. 1998a. "From Citation Indexes to Informetrics: Is the Tail Now Wagging the Dog?" *Libri* 48:67–80.

———. 1998b. "Random Thoughts on Citationology: Its Theory and Practice." *Scientometrics* 43:69–76.

Glass, B. 1965. "A Century of Biochemical Genetics." *Proceedings of the American Philosophical Society* 109:227–36.

———. 1974. "The Long Neglect of Genetic Discoveries and the Criterion of Prematurity." *Journal of the History of Biology* 7:101–10.

Hotchkiss, R. D. 1979. "Identification of Nucleic Acids as Genetic Determinants." *Annals of the New York Academy of Sciences* 325:321–42.

———. 1995. "DNA in the Decade before the Double Helix." *Annals of the New York Academy of Sciences* 758:55-73.

Kuhn, T. S. 1970. *The Structure of Scientific Revolutions.* 2d ed. Chicago: University of Chicago Press.

Landman, O. E. 1993. "Inheritance of Acquired Characteristics Revisited." *Bioscience* 43:696–705.

Leakey, L. S. B., P. V. Tobias, and J. R. Napier. 1964. "A New Species of the Genus *Homo* from Olduvai Gorge." *Nature* 202:7–9.

Lederberg, J. 1994. "The Transformation of Genetics by DNA: An Anniversary Celebration of Avery, MacLeod, and McCarty (1944)." *Genetics* 136:423–26.

Lowry, O. H., N. J. Rosebrough, A. L. Farr, and R. J. Randall. 1951. "Protein Measurement with the Folin Phenol Reagent." *Journal of Biological Chemistry* 193:265–75.

Löwy, I. 1989. "Biomedical-Research and the Constraints of Medical-Practice—James Bumgardner Murphy and the Early Discovery of the Role of Lymphocytes in Immune-Reactions." *Bulletin of the History of Medicine* 63:356–91.

McCarty, M. 1985. *The Transforming Principle: Discovering That Genes Are Made of DNA.* New York: W. W. Norton and Company.

Merton, R. K. 1957. "Priorities in Scientific Discovery." *American Sociological Review* 22:635–59. Reprinted in R. K. Merton, *The Sociology of Science: Theoretical and Empirical Investigations* (Chicago: University of Chicago Press, 1973), chap. 14.

———. 1963. "Resistance to the Systematic Study of Multiple Discoveries in Science." *European Journal of Sociology* 4:237–49. Reprinted in R. K. Merton, *The Sociology of Science: Theoretical and Empirical Investigations* (Chicago: University of Chicago Press, 1973), chap. 17.

———. 1968. "On the History and Systematics of Sociological Theory." In *Social Theory and Social Structure,* pp. 1–38. Enlarged ed. New York: Free Press.

———. 1973. *The Sociology of Science: Theoretical and Empirical Investigations.* Chicago: University of Chicago Press.

———. 1979. Forward to *Citation Indexing—Its Theory and Application in Science, Technology, and Humanities,* by E. Garfield. Philadelphia: Institute for Scientific Information Press.

———. 1981. "Foreword: Remarks on Theoretical Pluralism." In *Continuities in Structural Inquiry,* ed. P. M. Blau and R. K. Merton. London: Sage Publications. Reprinted in *Robert K. Merton: On Social Structure and Science,* ed. P. Sztomka (Chicago: University of Chicago Press, 1996).

Moravcsik, M. J., and P. Murugesan. 1975. "Some Results on the Function and Quality of Citations." *Social Studies of Science* 5:86–92.

Patinkin, D. 1983. "Multiple Discoveries and the Central Message." *American Journal of Sociology* 89:306–23.

Paul, D., and D. Charney. 1995. "Introducing Chaos (Theory) into Science and Engineering—Effects of Rhetorical Strategies on Scientific Readers." *Written Communication* 12:396–438.

Portis, E. B. 1986. "Theoretical Authority in Social-Science." *Social Science Journal* 23:397–410.

Sawin, C. T. 1996. "Arnold Adolph Berthold and the Transplantation Testes." *Endocrinologist* 6:164–68.

Schneider, W. H. 1983. "Chance and Social Setting in the Application of the Discovery of Blood Groups." *Bulletin of the History of Medicine* 57:545–62.

Sekuler, R. 1974. "Spatial Vision." *Annual Review of Psychology* 25:195–232.

Small, H. 1978. "Cited Documents as Concept Symbols." *Social Studies of Science* 8:327–40.

Stent, G. 1968. "Letter to the Editor: DNA Discovery in Perspective." *Science* 160:1397–98.

———. 1969. *The Coming of the Golden Age: A View of the End of Progress.* New York: Natural History Press.

———. 1972a. "Prematurity and Uniqueness in Scientific Discovery." *Advances in the Biosciences* 8:433–49.

———. 1972b. "Prematurity and Uniqueness in Scientific Discovery." *Scientific American* 227 (December): 84–93.

Tamm, S. L., and S. Tamm. 1974. "Direct Evidence for Fluid Membranes." *Proceedings of the National Academy of Sciences* 71:4589–93.

Tobias, P. V. 1985. "History of Physical Anthropology in Southern Africa." *Yearbook of Physical Anthropology* 28:1–52.

———. 1992. "Piltdown—An Appraisal of the Case against Arthur Keith." *Current Anthropology* 33:243–94.

———. 1996a. "The Dating of Linguistic Beginnings." *Behavioral and Brain Sciences* 19:789.

———. 1996b. "Premature Discoveries in Science, with Especial Reference to *Australopithecus* and Homo habilis." *Proceedings of the American Philosophical Society* 140, no. 1: 49–64.

Wardlaw, A. C. 1992. "Multiple Discontinuity as a Remarkable Feature of the Development of Acellular Pertussis Vaccines." *Vaccine* 10:643–51.

Weber, M. 1946. "Science as a Vocation." In *From Max Weber: Essays in Sociology,* ed. C. W. Mills; trans. H. H. Gerth, pp. 129–56. New York: Oxford University Press.

Wyatt, H. V. 1975. "Knowledge and Prematurity: The Journey from Transformation to DNA." *Perspectives in Biology and Medicine* 18:149–56.

Zuckerman, H. 1978. "Theory-Choice and Problem-Choice in Science." *Sociological Inquiry* 48:65–95.

Zuckerman, H., and J. Lederberg. 1986. "Postmature Scientific Discovery?" *Nature* 324:629–31.

CHAPTER NINETEEN

Premature Discovery Is Failure of Intersection among Social Worlds

Elihu M. Gerson

The notion of premature discovery exerts a kind of dramatic fascination, encouraging thoughts of insightful and creative scientists struggling to articulate their ideas and convince indifferent communities of their new truths. But there are many difficulties with the notion, melodrama aside, and it is time to sort them out and identify what is useful about the notion. I focus here primarily on the organizational and institutional issues that arise in thinking about prematurity. My approach stems from the Pragmatist philosophy of John Dewey and George Herbert Mead and from the Chicago school of sociology that it influenced.[1]

Let me begin with some difficulties of terminology. The term "premature" seems to imply that there is a developmental sequence to the course of research, that this sequence can go wrong somehow, and that discoveries can thus appear "out of turn." Frederic Holmes put it very nicely when he referred to the possibility of a "normal rate of scientific activity."[2] A closely related difficulty is the idea that discovery occurs in some straightforward fashion, so that we might reasonably speak of scientists as being on track or on schedule. This view of things seems to have metaphysical commitments in it that few of us want. To speak, then, of premature discovery is to introduce a kind of post hoc view, the kind of hindsight that historians have condemned as "presentism" or the "Whig theory of history."[3] This condemnation is not without its difficulties in turn, but it is clear that our interpretation of discoveries cannot rest on anachronism. I will say more about this problem in the last section.

Gunther Stent's approach avoids the worst of these difficulties but raises fresh problems in its turn. A discovery is premature, in Stent's view, if it cannot be con-

---

1. Cf., for example, Dewey 1916, 1938; Mead 1934.
2. Holmes, chapter 12 in this volume.
3. Butterfield 1931.

nected to canonical knowledge by a series of simple steps.[4] Stent is at pains to emphasize that the lack of connection is not simply a matter of personal failings on the part of scientists but rather is "structural." This is a major improvement over the notion that discoveries have clocks and schedules (or worse yet, fates and destinies) attached to them. Let us capture this improvement by abandoning the term "premature," and speak instead of "unconnected" discovery. This leaves us with three additional difficulties: What constitutes discovery? What is canonical knowledge? What does it mean to be connected by simple steps?

WHAT CONSTITUTES DISCOVERY?

Stent's argument assumes that discoveries can be isolated conceptually from the flow of questions and results that make up the research enterprise. We need to clarify what is meant by "discovery." Discoveries have often been seen as particular events, as emerging rather suddenly from the work of a single scientist or team. There has also been a tendency to think of discoveries as consisting of ideas or insights—that is, as mental events. At the limit, these two tendencies result in a cartoon vision of the canonical discovery: the apple falls and hits Newton on the head; Newton thereupon considers this data, deduces a generalization from the facts, and thus we have the three laws of motion. This view of discovery has been especially strong in much of twentieth-century philosophy of science, where the conceptions of Hans Reichenbach and Karl Popper have been especially influential.[5]

In recent years, this view has become increasingly untenable.[6] Discovery is not a psychological process; it takes place not in the minds of individual scientists but as part of an interactive back-and-forth among scientists. We need to distinguish sharply between the individual-level experience of insight or realization, on the one hand, and the development of a new understanding available for use by many communities, on the other. Nor do scientists typically make large leaps in their research—they tend to go from one problem to another related problem, working outward from the edges of their settled knowledge. Hence, we usually have a complex chain or network of related discoveries that amount (in effect) to a larger discovery. Discoveries are thus quite complex, both in timing and in character.

Discoveries have careers. A career begins when a scientist notices some relationship or occurrence and begins wondering about it, and ends some time later when the results of inquiry are turned over to engineers, physicians, and other users for exploitation. Between those two points, there are many phases and possibili-

---

4. Stent 1972a,b; also see chapter 2 in this volume.
5. Reichenbach 1938; Popper 1957.
6. See, for example, Brannigan 1981; Nickles 1980a,b, 1985, 1997; Sapp 1990b.

ties. G. Buchdahl, for example, distinguishes three stages: formulation, probation, and consolidation.[7] John Dewey's discussion of inquiry suggests a similar list:[8]

Puzzle or problem: Research begins with a puzzle, question, or problem—an unsettled question.

Trial: One or more (typically, multiple) alternative partial solutions are proposed and tried. Often, this involves many steps and repetitions. It also involves subdiscoveries, false starts, clarifications, revisions of the question, and so on. Innovations in procedure (e.g., refinements of instrument design) are often made, along with development of new concepts and/or specification of new theories. This process may take a long time and involve the work of many scientists, technicians, and other specialists. Those who start the project may not be among those who finish it.

Tentative success: Eventually, researchers come up with a tentative solution that seems to work, and they publish it.

Probation, revision, refinement: The tentative solution gets more settled as it is replicated, reproduced, and tested, both within the research group that first produced it and by other groups that try to reproduce, refine, and extend the solution.

Sometimes the proposed solution does not work, and falls into obscurity; sometimes there are debates about validity, the interpretation of results, and so on.

Often refinements of technique and expression transform the solution in significant ways. For example, Maxwell's equations as taught in physics textbooks do not appear in Maxwell's Treatise of 1873; they are the result of substantial efforts by Maxwell's students and followers.[9]

Acceptance: Eventually, the solution (typically greatly modified since its first proposal) becomes broadly accepted in the scientific community.

Researchers come to assume the validity of the solution as established fact, and move on to other problems.

Needless to say, the course of discovery does not necessarily run in the nice linear fashion this scheme describes. There are blind alleys, false starts, reworkings and reconstructions of the problem, and partial results. Often enough, the discovery is that scientists were asking the wrong question in the first place.

Moreover, there are often multiple research groups working along approximately parallel lines but at different rates and with different emphases and resources. As a result, different groups make partial discoveries that overlap. Sometimes, the connections among these partial discoveries are not understood until after the discoveries have been extensively reworked and refined.

7. Buchdahl 1991.
8. Dewey 1938.
9. Maxwell 1873; Hunt 1991; Morrison 2000.

Where in all this is "the discovery"? Clearly, the implicit definition contained in the notion of premature discovery is short-term, small-scale, and psychologistic. But histories and sociological studies repeatedly show that we do not have a discovery until the scientific community accepts it as such and stops debating about it.[10] Until then the proposed solution is in an intermediate state. It is rather like pregnancy: something wonderful has happened; something wonderful is going to happen; but meanwhile there is a lot of low back pain and nausea. Discovery, like pregnancy, is a matter of social organization, not psychological insight.

WHAT IS CANONICAL KNOWLEDGE?

Stent's definition is that a discovery is premature if it cannot be connected to canonical knowledge. What does "canonical" mean? Stent's vision is worth quoting at length:

> Knowledge about the world enters the mind not as raw data but in already highly abstracted form, namely as structures. In the preconscious process of converting the primary data of our experience step by step into structures, information is necessarily lost, because the creation of structures, or the recognition of patterns, is nothing else than the selective destruction of information. Thus since the mind doesn't gain access to the full set of data about the world, it can neither mirror nor construct reality. Instead for the mind reality is a set of structural transforms of primary data taken from the world. This transformation process is hierarchical, in that "stronger" structures are formed from "weaker" structures through selective destruction of information. Any set of primary data becomes meaningful only after a series of such operations has so transformed it that it has become congruent with a stronger structure preexisting in the mind.[11] ...
>
> Canonical knowledge is simply the set of preexisting "strong" structures with which primary scientific data are made congruent in the mental-abstraction process. Hence data that cannot be transformed into a structure congruent with canonical knowledge are a dead end; in the last analysis they remain meaningless.[12]

This view, with its focus on mind and on abstract structure, provides a conception of canonical knowledge that is both highly idealistic and highly reductionist. Whether or not it is successful at either task is beside the point here, for neither way of thinking about knowledge is in accord with experience. Once again, this is to say that the discovery process is a matter of social organization over time, and of reliable results that can be incorporated in new lines of effort. The issue is not one of changing minds but of changing or inventing conventional practices.

What, then, does "canonical" mean? Usually, to say that something is canonical means that it is in finished, reduced, or authoritative form. Such knowledge

---

10. For example, Brannigan 1981; Latour and Woolgar 1979.
11. Stent 1972b, p. 92.
12. Ibid., p. 93.

receives the kind of presentation characteristic of undergraduate textbooks. This implies that knowledge cannot be canonical if it is not settled. Canonical knowledge is established—which is to say that it is knowledge lagging behind the frontier or cutting edge of scientific work. But to state that some bit of knowledge is established in this sense, is simply to say that the scientific community accepts and uses that knowledge unproblematically. This, once again, is a matter of social organization, not abstract mind, and not the mental state of particular individuals.

The real problem here is how relatively tentative knowledge can come to influence or be part of settled knowledge. To understand this, we must look at the ways in which connections among different producers and stewards of knowledge take place.

WHAT DOES IT MEAN FOR A DISCOVERY TO BE CONNECTED BY SIMPLE STEPS?

We are concerned here with results that are (ultimately) accepted as a valid discovery, not those that are tentative or permanently rejected for some reason. This last point is important enough to bear repeating. To be "in contention" as potentially connectable, results must already be quite far along in their developmental career.

Ordinarily, scientists assimilate new results in their fields in a straightforward manner, even when these results are quite striking. Yet sometimes this process seems to fail and results are ignored or rejected, even though they later turn out to be "good" discoveries. By what process are new results connected to canonical knowledge, and, in particular, how does this process fail?

Stent portrays the connections of interest as conceptual, saying, for example, "Until it is possible to connect ESP with canonical knowledge of, say, electromagnetic radiation and neurophysiology no demonstration of its occurrence could be appreciated."[13] Thus, Stent argues that some results are so far from the accepted "structure" of thought that they simply cannot be assimilated. There are many difficulties with this view. For example, it makes any kind of major conceptual or theoretical change in science very difficult to explain. I discuss some of these difficulties in the final section. For now, I suggest that failures of connection be understood first as matters of social organization—that in order to understand how failures of connection occur, we should focus on the actual work of connecting.

There are several ways of looking at this. For example, we might focus on individual scientists or research teams as they go about their work in their laboratories and offices. Alternatively, we might concentrate instead on lines of work (i.e., specialties and subspecialties) and their relationships. I adopt this latter approach here.

Let me begin with some abstractions. A "social world" consists of all the activities that make up a line of work or a way of living. Worlds thus include industries,

13. Ibid., p. 88.

hobbies, residential communities, scientific disciplines, professions, and ethnic groups. Worlds are a kind of social organization, like bureaucracies, voluntary associations, markets, and small groups. Social worlds (or just "worlds") are related activities organized around a common subject-matter. Activities, of course, have actors—organizations and people—to conduct them. But every actor participates in multiple worlds, just as every world has multiple actors. Hence, worlds cannot be delimited by particular actors' participation in them: activities, not actors, define the boundaries of a world.[14]

Worlds contain more specialized activities as parts, or subworlds. Biology, for example, is a scientific specialty that contains the subworlds genetics, ecology, botany, paleontology, and so on. Two worlds intersect when a particular activity becomes part of both of them at once. Paleontological research, to take another example, is at the intersection of biology and geology. Certain places—such as university campuses—serve as points of intersection for many worlds.

A scientific field, then, can be seen as many relatively large worlds (major disciplines or areas such as cell biology or biochemistry), each made up of many smaller segments (for example, molecular genetics), each of which is in turn made up of still smaller segments (such as the study of heat-shock proteins). Segments form and re-form constantly as research leads to more intense focus on some problem areas and the abandonment of others. Other worlds cut across these specialties, creating additional subworlds. National boundaries and cultures, for example, create national subworlds within disciplines. Intersection among specialties and between specialties and other worlds plays an important role in weaving together an extraordinarily complex pattern of connections based on common problems, common techniques, common theoretical approaches, or common audiences outside the system of specialties.

The pattern of segmentation and intersection in research worlds means that each line of research is not connected to every other. Rather, there is relatively dense interaction among some subworlds, but relatively little interaction among most. Scientists in a specialty may know all the details of work going on across the country in that specialty but be unaware of work in a different specialty being carried out in the same building. Similarly, scientists learn about progress in a specialty whose work is complementary to their own but may never hear of efforts in areas with which they have no ties.

This pattern suggests an obvious hypothesis: *Discoveries will remain unconnected to canonical knowledge in a given field if they arise outside the field, and if the intersections that carry knowledge of the discovery into the field do not exist or do not function.* In short, ideas do not travel by themselves, but instead travel via interaction among scientists. The interaction may be face-to-face, via print, or lately, via the Internet. Without some such interaction, however, a discovery cannot be assimilated, because it is just such

14. This discussion draws on the ideas of Anselm Strauss (1978, 1982, 1984) and Howard Becker (1982). I have discussed scientific social worlds in more detail in Gerson 1983 and 1998.

interactions that instantiate the "simple steps" that link discovery to canonical knowledge.

"Simple steps" means something more than abstract logical or conceptual connection. For example, in their papers announcing the discovery of the structure of DNA, James Watson and Francis Crick noted that the mirror-image organization of the molecule's strands could easily be linked to the problems of heredity.[15] This kind of insight relies on knowledge of DNA's chemical structure *and* on familiarity with problems of research on inheritance. One must know about segregation and sorting of genes among progeny, but one must also know about amino acids, centrifuging, crystal structure, and denaturing. That is, the "simple steps" are embodied in the work practices and arrangements of the scientists who make them.

Without suitable intersections, then, a discovery from outside a field cannot be connected to the canonical knowledge of a field, because it is literally strange— that is, it comes from somewhere else, and the practices of the specialty are not (yet) equipped to deal with it adequately.

Several important points must be made in connection with this view:

1. Conceptually, we have substituted concrete interaction among scientists for the abstract logical connection of "simple steps." Steps may be logically simple in retrospect, but they are often not simple in terms of material accomplishment.
2. Discovery is always premature in relation to some particular subworld. It does not make sense to talk about prematurity in general.
3. This hypothesis has a clear implication: discoveries that arise within a given subworld are never premature *within* that subworld. Discoveries ultimately judged as premature always originate *outside* the given subworld.

All of the examples cited in Stent's article meet this test. In particular, the discovery by O. T. Avery, C. M. MacLeod, and M. McCarty's that DNA is the genetic material illustrates the point nicely.[16] The work of Avery and his colleagues was part of the bacteriology world, focusing on problems of virulence, especially in pneumonia. Their work was not part of the world of genetics. In 1944, the powerful connections between bacteriology and genetics that later proved so effective had only begun to form. Norton Zinder tells us that the Avery group was not well connected to the phage group led by Max Delbrück.[17] From the perspective of the phage group's ongoing and rapidly elaborating research program, the work of Avery and his colleagues may have been premature. From the perspective of bacteriologists, it was not.

The lesson here is that we have to chart the relevant audiences for a discovery carefully. Discoverers may be well connected to some audiences to whom the dis-

---

15. Watson and Crick 1953.
16. Avery, MacLeod, and McCarty 1944. For this story, see, for example, Dubos 1976; McCarty 1985.
17. Zinder, chapter 5 in this volume.

covery is uninteresting or unproblematic, and poorly connected to other audiences to whom the discovery is important.

The real issue, then, is understanding the conditions under which effective intersections take place. We should look for and analyze circumstances that block or retard the formation of fruitful intersections. Many of these circumstances cannot be identified from a post hoc analysis of the reconstructed discovery, because they involve institutional arrangements and historical circumstances that are largely independent of the intellectual content of discovery.

Mary Jo Nye gives us one telling example when she mentions that Michael Polanyi gave up teaching his own theory and taught Irving Langmuir's instead, because it was Langmuir's theory on which his students would be tested.[18] Thus the academic training system intervened in the relationship between Polanyi's efforts and Langmuir's: the university's requirements restricted Polanyi as a side effect of its own concerns. This is an example of an intersection (between academia and chemistry) blocking or retarding a line of work. Yet there is nothing in the reconstructed intellectual content of Polanyi's work, or Langmuir's, that would explain Polanyi's action. Similarly, it would be interesting to reconstruct the effect of mobilization in World War 2 on the work of Avery and his team—for surely the very practical problems of working on pneumonia under wartime pressures meant that the Avery team had less opportunity to follow up the genetic implications of their research.

WHAT ABOUT DISCOVERIES "BEFORE THEIR TIME"?

A premature discovery, in Stent's view, requires three conditions. First, a discovery candidate (an argument, theoretical model, concept, specimen, etc.) is presented to a community of scientists. Second, the discovery candidate cannot be connected to accepted knowledge in that community. That is, the interpretive apparatus needed to bring the new information into line with accepted knowledge is unavailable or nonfunctional. Third, at some later date, a new interpretive apparatus is developed, and the discovery candidate fits with the new interpretive context.

Clearly, the notion of premature discovery makes sense only after the new interpretive apparatus is available. The notion of "here-and-now prematurity" suggested by Stent does not stand up.[19] In Stent's view, a discovery is here-and-now premature if there is no way of carrying out the inference steps that will connect it to canonical knowledge. But we cannot decide if a discovery is premature, wrongly posed, or simply irrelevant without the after-the-fact knowledge of whether or not a suitable interpretive apparatus has been developed and applied successfully.

Prematurity is thus a frustrating concept: one cannot tell if it applies to something until long after the discovery. Indeed, any assessment of irrelevance can, in

18. Nye, chapter 11 in this volume.
19. Stent 1972a, p. 438; 1972b, p. 87.

principle, be overthrown by the development of new ways of thinking—which is to say, by the emergence of a new technical specialty. So, what is the value of the notion of premature discovery?

As David Hull notes, we all have seen instances when a new discovery is recognized as re-creating the content (or something very like the content) of an older discovery that has fallen into disuse.[20] We say these older studies are "before their time" or "premature," and we have the sense that there is something to be explained. But in rejecting the inadequate concept of prematurity, do we not lose sight of the "before its time" character of some studies?

Not at all. We can see the phenomenon of "before its time" as composed of two separate steps. The first takes place when a new discovery does not get tied to the conventional knowledge of its day and remains unconnected in the literature. The second step occurs when new events lead to the "rediscovery" of the unconnected results in a changed context that enables or even facilitates its connection to the conventional knowledge of the rediscovering context.

Scientists often find it useful and convenient to reconstruct the histories of their disciplines,[21] and recovering lost precursors and premature discoveries is a good way to do this. That is, the notion of premature discovery is particularly useful for scientists in organizing their work and their relationships to one another. The notion (and its cousin, simultaneous discovery) aids scientists in establishing their relationships to one another, distribute credit, maintain relationships with other lines of work, illustrate good and bad behavior for students, explain the purpose and value of their work to others, and justify contested lines of action as authentically scientific. Jan Sapp, for example, suggests that many of these purposes have been served by scientists' historical reconstructions of Mendel and his work.[22] These are all perfectly reasonable and very ordinary activities, ones that appear in every line of work, intellectual or otherwise—including those that have as their object the criticism of science.

The notion of premature discovery is one of those concepts—let us call them preceptive concepts, or simply precepts—that help to organize and conduct the public administration or cameralistics of science, just as the term "frictionless plane" helps to organize the substantive work of mechanics, and "rational consumer" helps to organize the substantive work of economics.

Everyone knows these ideas are false if taken literally; that is not the point. Rather, they offer a starting point for conceptualizing, designing, and negotiating practical solutions to problems that arise in the course of the work. Stent, for example, begins his *Scientific American* article by discussing "Oswald Avery's identification of DNA as the active principle in bacterial transformation and hence as genetic material."[23] The

---

20. Hull, chapter 22 in this volume.
21. Strauss 1984; Winsor 2001.
22. Sapp 1990a.
23. Stent 1972b, p. 84.

immediate stimulus for this discussion, for Stent's article as a whole, and for the concept of premature discovery was Stent's desire to respond to criticism that he had omitted a citation to the work on bacterial transformation in a previous historical account.[24] So the concept of premature discovery is a device to manage a debate about the proper conduct of research. Just as literally false models may lead to truer theories, so literally false histories, philosophies, or sociologies may lead to more effective and/or more peaceful research.[25]

Since I can easily imagine that last sentence being quoted out of context in support of all sorts of misconception, let me expand. Just as the designer of an automobile tire may well make use of frictionless planes in the early part of the work, so scientists may well use notions like premature discovery in organizing and conducting their relationships among themselves and with others. But neither the tire design nor the conduct of research ends with precepts: they are only aids to the work, not definitive of it.

Scientists are not unscholarly or irresponsible when they create and use preceptive concepts, any more than engineers are irresponsible when they use notions like "frictionless plane." That is simply part of the way the work gets done. Nor is this process unique to science: concepts such as "Whig interpretation of history," "master narrative," and "Hobbesian state of nature" operate the same way in the humanities and social sciences.

Similarly, historians, philosophers, sociologists, and other students of the research process are not trying to second-guess scientists when they look at research; they have different tasks in hand. Just as it is pointless to hold them to scientists' standards of research performance (because that is not what they are doing), so it is equally pointless to hold scientists to the performance standards of historians, philosophers, and sociologists.[26]

Scientists and the scholars who observe them (let us not forget, these groups overlap) do need to work out a modus vivendi. This will not be easy, nor will it ever be entirely comfortable. Like clinical psychologists who flinch at the lawyers' notion of "prudent layperson," many science-studies scholars are going to flinch at notions like "premature discovery." Conversely, scientists will object to portrayals that do not coincide with their views of what happened. Some of these disagreements will be resolved by a combination of all the usual things that scholars do: patient argument, character assassination, amassing of facts, intimidation, careful review of logic, mischaracterization of opponents' views, scrupulous analysis, and ridicule. Some disagreements will simply become obsolete and fade away. And some will not be resolved at all.

24. Stent 1968.

25. On false models leading to truer theories, see Wimsatt 1987.

26. This line of thought was inspired by Winsor's discussion of historians' concerns with scientists' use of history (Winsor 2001).

BIBLIOGRAPHY

Avery, O., C. M. MacLeod, and M. McCarty. 1944. "Studies on the Chemical Nature of the Substance Inducing Transformation in the Pneumococcus." *Journal of Experimental Medicine* 79:137–58.

Becker, H. S. 1982. *Art Worlds.* Berkeley and Los Angeles: University of California Press.

Brannigan, A. 1981. *The Social Basis of Scientific Discoveries.* New York: Cambridge University Press.

Buchdahl, G. 1991. "Deductivist versus Inductivist Approaches in the Philosophy of Science as Illustrated by Some Controversies between Whewell and Mill." In *William Whewell, a Composite Portrait,* ed. M. Fisch and S. Schaffer, pp. 311–44. New York: Oxford University Press.

Butterfield, H. 1931. *The Whig Interpretation of History.* London: Bell.

Dewey, J. 1916. *Essays in Experimental Logic.* Chicago: University of Chicago Press.

———. 1938. *Logic: The Theory of Inquiry.* New York: Henry Holt.

Dubos, R. J. 1976. *The Professor, the Institute, and DNA: Oswald T Avery, His Life and Scientific Achievements.* New York: Rockefeller University Press.

Gerson, E. M. 1983. "Scientific Work and Social Worlds." *Knowledge* 4:357–77.

———. 1998. "The American System of Research: Evolutionary Biology, 1890–1950." Ph.D. diss., University of Chicago.

Hunt, B. J. 1991. *The Maxwellians.* Ithaca, N.Y.: Cornell University Press.

Latour, B., and S. Woolgar. 1979. *Laboratory Life.* Beverly Hills: Sage Publications.

Maxwell, J. C. 1873. *Treatise on Electricity and Magnetism.* Oxford: Clarendon.

McCarty, M. 1985. *The Transforming Principle: Discovering That Genes Are Made of DNA.* New York: Norton.

Mead, G. H. 1934. *Mind, Self, and Society.* Chicago: University of Chicago Press.

Morrison, M. 2000. *Unifying Scientific Theories: Physical Concepts and Mathematical Structures.* New York: Cambridge University Press.

Nickles, T. 1985. "Beyond Divorce: Current Status of the Discovery Debate." *Philosophy of Science* 52:177–206.

———. 1997. "A Multi-pass Conception of Scientific Inquiry." *Danish Yearbook of Philosophy* 32:11–44.

Nickles, T., ed. 1980a. *Scientific Discovery, Logic, and Rationality.* Boston: D. Reidel.

———. 1980b. *Scientific Discovery: Case Studies.* Boston: D. Reidel.

Popper, K. R. 1959. *The Logic of Scientific Discovery.* New York: Basic Books.

Reichenbach, H. 1938. *Experience and Prediction.* Chicago: University of Chicago Press.

Rogers, E. M. 1994. *A History of Communication Study: A Biographical Approach.* New York: Free Press.

Sapp, J. 1990a. "The Nine Lives of Gregor Mendel." In *Experimental Inquiries,* ed. H. E. Le Grand, pp. 137–66. Dordrecht: Kluwer.

———. 1990b. *Where the Truth Lies: Franz Moewus and the Origins of Molecular Biology.* New York: Cambridge University Press.

Stent, G. S. 1968. "That Was the Molecular Biology That Was." *Science* 160:390–95.

———. 1972a. "Prematurity and Uniqueness in Scientific Discovery." *Advances in the Biosciences* 8:433–49.

———. 1972b. "Prematurity and Uniqueness in Scientific Discovery." *Scientific American* 227 (December): 84–93.

Strauss, A. L. 1978. "A Social Worlds Perspective." *Studies in Symbolic Interaction* 1:119–28.

———. 1982. "Social Worlds and Legitimation Processes." *Studies in Symbolic Interaction* 4:171–90.

———. 1984. "Social Worlds and Their Segmentation Processes." *Studies in Symbolic Interaction* 5:123–79.

Watson, J. D., and F. H. C. Crick. 1953. "Genetical Implications of the Structure of Deoxyribonucleic Acid." *Nature* 171:964–67.

Wimsatt, W. C. 1987. "False Models as Means to Truer Theories." In *Neutral Models in Biology*, ed. M. H. Nitecki, pp. 23–55. Chicago: University of Chicago Press.

Winsor, M. P. 2001. "The Practitioner of Science: Everyone Her Own Historian." *Journal of the History of Biology* 34:229–45.

PART SIX

Philosophical Perspectives

CHAPTER TWENTY

Fleck, Kuhn, and Stent

Loose Reflections on the Notion of Prematurity

Ilana Löwy

I present here some reflections on the possibility of linking prematurity with the ideas developed by Thomas Kuhn and Ludwik Fleck (especially the latter) on the structure of scientific communities and the organization of scientific work.

STENT

According to Gunther Stent, "A discovery is premature if its implication cannot be connected by a series of simple logical steps to contemporary canonical [or generally accepted] knowledge."[1] This definition includes several terms that need to be clarified.

Discovery: one may assume that Stent uses the term to describe a collective process of recognition of the importance of a theory, a set of observations, or an experimental study by a given community of specialists, not to indicate the unique experience of the individual scientist at the bench or the process of clarification of his or her ideas through writing or talking.

Canonical knowledge: this may be defined, I propose, as the knowledge and practices accepted by a given scientific community.

Simple, logical steps: one may suppose that these are steps that correspond to the target community's ways of asking relevant questions, of producing experimental results, and of examining new evidence.

Premature: this may be seen as the delayed integration of knowledge or evidence into the accepted knowledge and practices of a given scientific community.

---

1. See Stent, chapter 2 in this volume.

The concept of premature discovery is related, but not identical, to the notion of the nonrecognized precursor. Once an important discovery is made, there is often a flurry of searching for its precursors.[2] Such a search, the French philosopher and historian of science Georges Canguilhem explains, is a dangerous practice for a historian. If certain scientists—the "precursors"—could be extracted from their historical background and relocated at will in another one, this would mean that science has no historical dimension. Therefore, "before joining together two segments of a road one would be well advised to make sure that one is really dealing with the same road."[3] Canguilhem recommends carefully examining the road traveled by the putative precursor and its followers to determine whether it is indeed the same. If the road is not the same—if the work of the precursor is too distant from the later discovery, she or he formulated and answered the problems in a very different way, or the precursor and the followers inhabited different cognitive universes—the term "precursor" is meaningless. If the road is the same (or to be more precise, sufficiently similar to argue that a close relationship exists), one may distinguish between two cases. The putative precursor may in fact have made an important, direct, and quasi-immediate contribution to the later discovery (for example, some scientists explain that O. T. Avery's results directly stimulated Erwin Chargaff's studies on nucleic acids, and possibly even James Watson and Francis Crick's studies). In this case the precursor is but one of the researchers involved in the collective process of scientific discovery, and his or her work does not qualify as premature.

The other case is of a true precursor: a person who made an observation or developed a theory that was not accepted by the relevant scientific community but later was integrated into this community's disciplinary matrix (either through belated recognition of the original discovery or through independent rediscovery). Only the latter is a premature discovery. This is, however, a rather restrictive definition. True similarity seldom occurs if discoveries are separated by a long span of time (a historian would be reluctant to accept the existence of precursors of modern genetics in the seventeenth century), and even if events are close in time, the conceptual frameworks of the premature and the mature discoveries may be quite different. I have argued elsewhere that when Avery and his colleagues proposed that DNA induced hereditary changes in pneumococci, they believed DNA to be a substance able to modify the function of enzymes (a "specific mutagen"). Only later was Avery's discovery perceived in light of the understanding of DNA's func-

---

2. The search for precursors is not limited to the sciences. In his essay "Kafka and His Precursors" (1964), Jorge Luis Borges explains that every important work of art creates its own trail of precursors: works that otherwise would not be perceived as having something in common become associated because of their affinities to later artistic creations. A posteriori, the writings of Kafka displayed hidden relationships between previously unrelated works of art that could be characterized as "kafkaesque." Borges does not try, however, to explain history but the way cultural associations are created, maintained, and transformed.

3. Canguilhem 1974, p. 21.

tion as a molecule coding for the structure of enzymes.[4] The definition of the nature of the discovery (i.e., the meaning of the sentence "DNA carries genetic information") was a part of the process of discovery itself. This case may illustrate the complexities of the classification of an event as a premature discovery.

The replication of a discovery in a temporally different context is probably rare, and many among the candidates for classification as premature would probably fail to withstand a critical examination. Does that mean, however, that the notion of premature discovery should be abandoned? I do not think so: in some cases there is enough circumstantial evidence or sufficient "family resemblance" between an earlier and a later event to justify a claim that a discovery made in a given time and place was belatedly integrated into the corpus of mainstream science, and to legitimate the assumption that the investigation of this delay can teach us something useful about the functioning of science.

Ernest Hook proposes that one reason why Stent's concept of prematurity did not receive wider attention was that it was conflated with Kuhn's discussion of paradigm. A premature discovery was viewed simply as one that did not fit into the existing paradigm.[5] I will present Kuhn's and then Ludwig Fleck's perception of science, and argue that Stent's view of prematurity may be better adapted to the conceptual framework proposed by Fleck, within which it may gain a more precise meaning.

KUHN

There are multiple ways of understanding Kuhn's term "paradigm," and Kuhn himself admitted in the preface to the second edition of *The Structure of Scientific Revolutions* that his use of this term was occasionally obscure.[6] Nevertheless, in later writing Kuhn seems to have a preference for two related meanings of the term: *exemplar,* meaning a model of professional conduct; and *disciplinary matrix,* meaning the totality of the accepted ideas and practices of a given discipline or specialty. Such uses of the term "paradigm" stress the importance of patterns of learning and of fixed disciplinary frameworks in the functioning of science. This is an important point. I will show later that Fleck discussed the importance of the training of young scientists and of the acquisition of a specific way of seeing the external world through the lenses of disciplinary practices. Fleck did not stress, however, as Kuhn did, the role of fixed exemplars in teaching and in the socialization of scientists. Typical samples of mathematical problems, homogenized experimental models, and oft-repeated "discovery accounts" are important devices in training beginners in the proper ways of doing scientific investigations.[7] One of Kuhn's most

4. Löwy 1990b.
5. See Hook, chapter 1 in this volume.
6. Kuhn 1970.
7. As Simon Schaffer has put it, "Discovery is a retrospective label attributed to candidate events by communities, a technique of marking technical practices that are prized by the community" (1986).

important innovations is to have displayed the conservatism of the scientific enterprise. Science is usually defined as a permanent search for new knowledge. Historians, philosophers, and sociologists of science therefore focus as a rule on innovation in science, on change, and on "great discoveries." The opposition between the (rare) "scientific revolutions" and the (usual) functioning of "normal science" in Kuhn's study brings to the fore the importance of the conservative element in science. The great majority of scientists, Kuhn explains, are not busy contesting accepted knowledge or falsifying major claims but instead repeat—with relatively small variants—the work of their predecessors. Moreover, scientists are organized in distinct and incommensurable communities, each shaped by a different disciplinary matrix, and they work exclusively within the framework of this matrix. Only occasionally does a great upheaval take place: old exemplars and models become invalid, well-established patterns of practice disappear, and boundaries between disciplines and specialties are redefined. Scientists then have to adapt to an entirely new way of perceiving their objects of study. Such a gestalt switch is often difficult, and a change of generation of scientists may be needed to complete the transition from the old paradigm to the new one.

If one accepts the definition of a paradigm as a disciplinary matrix, there are obvious affinities between Stent's definition of premature discovery as an event that cannot be connected to canonical knowledge, and Kuhn's claim that scientists work within a disciplinary matrix that includes the knowledge they consider to be proven. The important difference, as I see it, is that Kuhn's perception of science is fundamentally a static one: long periods of quasi inactivity, or rather of painstaking improving and polishing of the existing paradigms, interrupted by short, frenetic periods of radical change. Stent's perception of premature discovery implies a more dynamic vision of science, one linked with progress (or, if one prefers a more neutral wording, with the accumulation of knowledge claims and with improved technical performance). Discoveries that could not fit into the earlier ways of doing science can later become a part of mainstream science, because science has changed in the meantime in many important ways. Change is not the result of violent (and rare) upheavals: it is, in the examples proposed by Stent, the usual way science works. Such a dynamic perception is more akin, I propose, to the vision of science proposed by Ludwik Fleck.

FLECK

In a 1929 article, "On the Crisis of 'Reality,'" Fleck describes science as "an eternal, synthetic rather than analytic, never-ending labor—eternal because it resembles that of a river that is cutting its own bed."[8] For him, science is not only a body

8. Fleck 1986. The comparison between science and a river was first proposed by the Polish philosopher of medicine Zygmunt Kramsztyk (1899). Kramsztyk's writings, I have suggested elsewhere, were in all probability one of the sources of Fleck's ideas. See Löwy 1990a.

of claims about the natural world (the discovery of scientific laws, the elaboration of scientific theories) but also—or rather mainly—a social and cultural endeavor. Science, Fleck explains, is a collective enterprise conducted by distinct "thought collectives" (professional communities), each with its own "thought style" (a concept akin, although not identical, to Kuhn's "disciplinary matrix").[9]

Three elements are important for understanding Fleck's notion of thought style:

1. The role of the socialization of scientists into the thought style of their thought collective. Newcomers to a given scientific specialty acquire the specific style of this specialty. This style becomes, to use a term coined by the sociologist Pierre Bourdieu, their "habitus"—the only possible way to see, think, and act, and one that is so deeply internalized and "naturalized" that it becomes invisible.[10]

2. The incommensurability of thought styles. Observations made with different disciplinary approaches are often incommensurable because different methods are used to measure a given phenomenon. Different thought collectives can therefore produce divergent and not entirely compatible scientific facts.[11]

3. The importance of the circulation of entities produced by scientists. Fleck did not view science as an agglomeration of small, hermetically sealed groups, each producing facts destined for its exclusive use. Just the opposite is true: Fleck was one of the first to argue that science is a collective activity and an institution, and he stressed the importance of interactions among distinct communities of scientists and among scientists and other social groups. Scientific facts, Fleck explained, are seldom limited to the community that produced them: "A set of findings meanders through the community, becomes polished, transformed, reinforced or attenuated, while influencing other findings, concept-formation, opinions and habits of thought."[12] Some things may be lost and some things may be found in this imperfect translation, and the circulation of scientific facts is an important source of innovation in science and in society: "It offers new possibilities of discovery and creates new facts."[13]

9. The term "thought style" is misleading, because it may be understood as referring to concepts and ways of reasoning only, and does not make explicit Fleck's preoccupation with the material practices of science.

10. For example, a bacteriologist loses the "naive" way to looking at a microscopic preparation; a geneticist encounters difficulty thinking about heredity other than as the replication of distinct material entities, the genes.

11. For example, the gene of the classical geneticist is not exactly the same as that of the molecular biologist; the histocompatibility of the transplantation surgeon is not the same as that of the immunogeneticist.

12. Fleck 1976, p. 42. Fleck's book develops in detail an example of such a circulation of a scientific fact, the Wasserman test for the diagnosis of syphilis. The meaning and the uses of this test changed when it circulated among serologists, general practitioners, health administrators, and the lay public. Fleck also shows the mutual influence of the expert and the lay perception of the serology of syphilis.

13. Ibid., p. 110.

Science is at the same time a remarkably stable enterprise (some scientific specialties are built on hundreds if not thousands of years of accumulated knowledge) and one whose aim is to generate novelty, and therefore it is deeply committed to change. Fleck's concept of thought style allows us to account for science's stability *and* its capacity for change. According to Fleck, the thought style of a given scientific community includes the body of knowledge taken for granted in a given area, the questions viewed as legitimate, the accepted ways of answering these questions (the "right" materials, methods, techniques, instruments, experimental systems), and the criteria for evaluation of new knowledge. A thought style ensures stability because it provides a fixed conceptual and material framework for the scientist and a blueprint for the training of newcomers in a given area. It promotes change through its ability to absorb and transform scientific facts originating in other thought styles (or developed at the margins of a given disciplinary style) and to be enriched and modified through this transformation.

A TYPOLOGY OF PREMATURITY STUDIES

Premature discovery cannot be explained through a single mechanism, however important, such as the failure to connect to canonical knowledge or to follow a pattern traced by exemplars. Fleck's term "thought style" captures (imperfectly) the multilayered functioning of science as a dynamic social enterprise.

The definition of a premature discovery as one that cannot be integrated or translated into the thought style of a target scientific community may allow for a more refined analysis of the circumstances in which prematurity may arise, and allow for a rough typology of premature discoveries. The thought style of a given community includes the knowledge taken for granted (probably an approximation of Stent's "canonical knowledge"), as well as the questions viewed as legitimate, the accepted methods for answering these questions, and the ways of evaluating new evidence.

The incompatibility of a given discovery with any one of these four elements makes it unacceptable by a target scientific community: a change in this element occasionally makes possible the integration of a previously discarded discovery. Stent centered his paper on discoveries that, according to him, could not be harmonized with the knowledge taken for granted by a given scientific community. I sketch here some examples of discoveries integrated belatedly into mainstream science because they failed to conform to the three other elements of a scientific thought style.[14]

1. Discoveries that are premature because either the question asked or the methods used to answer this question are viewed as illegitimate. One such instance concerned bacterial variability. In the late nineteenth century, when bacteriologists

---

14. This is but a tentative attempt to provide examples of several types of premature discoveries, not a serious historical study of the adequacy of the classification of these discoveries as premature.

aimed at proving that bacteria form true and stable species, they introduced a homogeneous style of bacteriological investigation. Bacterial cultures too crowded, too old (more than 24 hours after inoculation), or too young (less than 12 hours) were not viewed as serious objects of study, and the observation of bacterial variability was discarded as methodological error.[15] However, in the early twentieth century, when the notion of bacterial species was no longer contested, it became possible to relax the rigid research style in bacteriology and to legitimate curiosity about the possible changes in bacterial morphology under different physiological conditions. This development led to the observation of bacterial variability. Earlier descriptions of the phenomenon were premature because they investigated an "uninteresting" question using "incorrect" techniques: "Species were fixed," notes Fleck, "because a fixed and restricted method was applied to the investigation."[16]

Another example is that of fetal malformations induced by drugs. Teratology, the science of studying fetal malformations, was a flourishing branch of zoology from the eighteenth century on. The investigations in this area rapidly revealed that a variety of chemical substances given to a pregnant female can induce malformation of the fetus. However, teratology was an occupation of the experimentalist busy with artificial production of deformed fetuses. The results were published in zoology and embryology journals and did not reach the general medical public. As late as the 1950s, medical practitioners were mostly unaware of the danger of giving medication to pregnant women or submitting them to x-ray examination, not because evidence on the effect of these treatments on fetuses was lacking (it was abundant) but because it was segregated in specialized journals and, above all, because the question "what are the effects of chemical substances on human pregnancy?" was not perceived by gynecologists and obstetricians as an "interesting" one. Thus the first article on teratogenic effects of thalidomide on fetuses (written by an Australian obstetrician, W. G. MacBride) was rejected by *Lancet* (in June 1961) because it was not judged important enough to be published in a major medical journal.[17]

2. Discoveries that are premature because techniques and research methods that exist at a given time do not allow one to answer the questions being asked, to confirm or disprove a given hypothesis, to develop a given experimental system, or to promote a proposed technological innovation. Premature inventions, which depend on underdeveloped technology—a phenomenon that Hook cites in chapter 1 in this volume—fall into this category.[18]

Another example is "biological individuality." In 1910, Alexis Carrel proposed, on the basis of his transplantation studies in animals, that transplanted organs and

15. Amsterdamska 1987.

16. Fleck 1976, p. 93.

17. Dally 1998.

18. One should add, however, that an invention (unlike a scientific discovery) needs to fulfill additional conditions in order to become successful: possibility of mass production, cost-efficiency, and successful commercialization.

tissues are rejected in mammals because each living organism has a specific "biological individuality"—a unique molecular structure that allows for the recognition and rejection of cells and organs from a different member of the same species. At about the same time, Charles Richet proposed the related notion of "biological personality" on the basis of his studies of anaphylaxis.[19] Both Richet and Carrel were Nobel laureates. Their ideas were widely diffused but were not followed, because in the second decade of the twentieth century immunological and biochemical methods did not allow for the visualization of minute differences between cells and tissues. The concept of "biological individuality" was abandoned, and then rediscovered (in a different context) in the 1950s.

3. Discoveries that are premature because the proof provided by the discoverer does not meet the standards set by a given scientific community. An example of this type of discovery is the role of the mosquito in the transmission of yellow fever, identified by the Cuban physician Carlos Finlay in 1881. Finlay connected the mosquito *Stegomya fasciata* to the transmission of yellow fever on the basis of careful epidemiological observations that linked the sites and the intensity of epidemics of yellow fever with the presence and the density of mosquitoes. He presented his findings first at the International Sanitary Conference in Washington, D.C., in February 1881, then at numerous scientific meetings. He also published a long series of articles defending his hypothesis. However, his "mosquito hypothesis" was politely ignored and did not lead to sanitary action or to investigations by other researchers. Only in 1900 did the investigations of the Reed Commission (of the U.S. Army) in Cuba lead to a widespread acceptance of the "mosquito hypothesis."[20] Experts in tropical medicine, embarrassed by the inattention to the mosquito hypothesis, explained that Finlay's hypothesis was ignored because he was unable to prove his allegations. By contrast, the Reed Commission made carefully controlled (and, incidentally, very dangerous) experiments on human beings, which proved beyond any reasonable doubt that yellow fever is transmitted only by means of the bite of the infected *Stegomya* mosquito.

It is true that Finlay was unable to provide an experimental proof of his mosquito theory. Aware of the importance of such a proof, Finlay tried experiments with mosquitoes and humans, but his experiments were muddled and judged incompetent by the experts. On the other hand, his epidemiological studies (the only ones presented in Finlay's first papers) were not viewed as incompetent. One could imagine a scenario in which the mosquito hypothesis was tentatively accepted on the basis of epidemiological data alone, and then tested through the elimination of mosquitoes from a given area. Such a scenario, plausible today, was not possible, however, in the late nineteenth century. At that time, a successful experimen-

---

19. Carrel 1910; Richet 1964.

20. Finlay 1912; Owen 1911. An additional reason for the interest in the "mosquito hypothesis" in 1900 was the growing evidence at that time that another tropical disease, malaria, was transmitted by mosquitoes.

tation was seen as absolutely indispensable to prove a link between cause and disease. Its absence made Finlay's hypothesis premature.[21]

The investigation of premature discoveries from the perspective of compatibility with specific elements of scientists' thought styles may help to uncover occasional "blind spots" of disciplinary practices in the sciences. Such an investigation may also suggest ways of increasing the "openness" of scientific thought styles without endangering their stability. This may be important, especially when the issue is a practical one. The short delay in the appreciation of Avery's findings, or even the longer delay in the appreciation of Mendel's findings, probably did not seriously affect the development of biological knowledge. By contrast, an earlier awareness of the teratogenic effect of drugs and x rays in humans might have saved many lives and eliminated much suffering.

ON "PROGRESS"

Fleck does not use the word "progress" in his work but views science as a cumulative enterprise in which there is a steady increase in the density of the networks of concepts and "facts." Modern science, he explains, is an increasingly complicated enterprise. There is increasingly more knowledge taken for granted, more questions viewed as legitimate, and more experimental systems, techniques, and tools: "The more developed and detailed a branch of knowledge becomes, the smaller are the differences of opinion. . . . It is as if with the increase of the junction points, according to our image of network, free spaces were reduced. It is as if more resistance were generated, and the free unfolding of ideas were restricted."[22]

The increased density of scientific networks limits the freedom of the individual scientist. The difference between a scientific and an artistic endeavor, Fleck explains, is not one of essence, only of degree. Both art and science produce novelty and become more diversified with time, but artistic endeavors are less restrained by past developments than scientific ones, and the cumulative effect is present more often in science and less in art:

> The artist translates his experience into certain conventional materials by certain conventional methods. His individual freedom is in fact limited; by exceeding these limits, the work of art becomes non-existent. The scientist also translates his experience, but his methods and materials are closer to a specific scientific tradition. The signs (i.e., concepts, words, sentences) and the ways in which he uses the signs are more strictly defined and are more subject to the influence of the collective: they are of more social and traditional character than those used by the artist. If we call the number of interrelations between the members of a collective "social density", then the

---

21. Stepan 1978; Delaporte 1989. Finlay's discovery was delayed, but he cannot be described as an "unknown precursor"—he was in direct contact with the members of the Reed Commission, explained his theory to them, and gave them the mosquito eggs they used in their experiments.

22. Fleck 1976, p. 83–84.

difference between a collective of men of science and a collective of men of art will be simply the difference of their densities: the collective of men of science is much more dense than the collective of art.[23]

The limited freedom of the scientist is not, however, a total absence of freedom. Kuhn depicts science as a static and closed universe, in which incommensurable professional communities bound by a rigid disciplinary matrix resist all change until a true revolution arises. Fleck, we have seen, describes science as a more flexible universe with circulating facts, imperfect translations, and multilevel interactions between "thought collectives." Such a dynamic universe permits continuous progress—that is, a steady increase in the density of scientific networks—but it also leaves room for individual freedom. This freedom is reflected in, among other things, the hesitations and uncertainties experienced by a scientist working at a bench, graphically described by Fleck:

> The first, chaotically styled observation resembles a chaos of feelings: amazement, a searching for similarities, trial by experiment, retraction as well as hope and disappointment. Feeling, will, intellect, all function together as an indivisible unit. The research worker gropes but everything recedes, and nowhere is there a firm support. Everything seems to be an artificial effect inspired by his own personal will. Every formulation melts away at the next text.... The work of research scientists means that in the complex confusion and chaos he faces, he must distinguish that which obeys his will from that which arises spontaneously and opposes it. This is the firm ground which he, as the representative of the thought collective, continuously seeks.[24]

Feelings of uncertainty and confusion are out of place in the self-contained world of a Kuhnian "normal science." Kuhn's vision of science is one of consecutive replacements of incommensurable worldviews. The notion of progress, if present, is grounded in the development of more adequate or more efficient worldviews through dramatic upheavals. Fleck's vision of science as a gradual increase in the density of material and conceptual networks developed by scientists, is more akin to Stent's perception of progress. His perception of the limited freedom of the scientist at the bench (limited, but freedom nonetheless) resonates with Stent's view of the role of individualized approaches in scientific research. Stent's original papers are titled "Prematurity *and* Uniqueness in Scientific Discovery," and I believe that the word "and" is important.[25] It suggests that prematurity and uniqueness are different aspects of the same reality: the material, social, and political organization

---

23. Fleck 1939, 18:10–15; English translation in Löwy 1990a. In the 1930s, artists and scientists were designated as males; Fleck (who perhaps was not familiar with, for example, Marcel Duchamp's "ready-made" objects) seems to believe that the definition of an item as a work of art depends on agreement on minimal technical criteria. The art historian Michael Baxandall has examined the similarities and differences between technological and artistic endeavor and has arrived at a conclusion similar to Fleck's (1985).

24. Fleck 1976, pp. 94–95.

25. Stent 1972a,b.

of scientific work. Such an organization sets limits for the integration of evidence into the collective endeavor of scientists and, at the same time, opens (restricted) spaces for individualized doubt and for individualized creativity, elements that contribute to making science a cumulative, ever-diversifying enterprise.

BIBLIOGRAPHY

Amsterdamska, O. 1987. "Medical and Biological Constraints: Early Research in Variations of Bacteriology." *Social Studies of Science* 17:657–88.

Baxandall, M. 1985. *Patterns of Intention*. New Haven: Yale University Press.

Borges, J. L. 1964. "Kafka and His Precursors." Trans. R. L. C. Simms. In *Other Inquisitions*. Austin: University of Texas Press.

Canguilhem, G. 1974. "L'objet d'histoire des sciences." In *Études d'histoire et de philosophie des sciences*, pp. 11–23. Paris: Vrin.

Carrel, A. 1910. "Remote Results of the Transplantation of the Kidneys and Spleen." *Journal of Experimental Medicine* 12:146–50.

Dally, A. 1998. "Thalidomide: Was the Tragedy Preventable?" *Lancet* 351:1197–99.

Delaporte, F. 1989. *The History of Yellow Fever*. Cambridge: MIT Press.

Finlay, C. 1912. *Trabajos Selectos*. Havana: Secretaria de Sanidad y Beneficiencia.

Fleck, L. 1939. "Rejoinder to the Comment of Tadeusz Bilikiewcz." *Przeglad Wspolczesny* 18:10–15.

———. 1976. *Genesis and Development of a Scientific Fact*. Trans. F. Bradley and T. I. Trenn. Chicago: University of Chicago Press.

———. 1986. "On the Crisis of 'Reality,' " Trans. H. G. Shalit and Y. Elkana. In *Cognition and Fact: Materials on Ludwik Fleck*, ed. R. S. Cohen and T. Schnelle, pp. 47–58. Dordrecht: D. Reidel.

Kramsztyk, Z. 1899. "O znaczeniu wiedzy historycznej." *Krytyka Lekarska* 3, no. 9:253–55.

Kuhn, T. 1970. *The Structure of Scientific Revolutions*. Chicago: University of Chicago Press.

Löwy, I. 1990a. *The Polish School of Philosophy of Medicine from Tytus Chalubinsky (1820–1889) to Ludwik Fleck (1896–1961)*. Dordrecht: Kluwer.

———. 1990b. "Variance of Meaning in Discovery Accounts: The Case of Contemporary Biology." *Historical Studies in the Physical and Biological Sciences* 21, no. 1:87–121.

Owen, M., ed. 1911. *Yellow Fever: A Compilation of Various Publications*. Washington, D.C.: Government Printing Oπce.

Richet, C. 1964. "Anaphylaxis." In *Nobel Lectures: Physiology or Medicine*, pp. 469–90. Vol. 1. Amsterdam: Elsevier.

Schaffer, S. 1986. "Scientific Discoveries and the End of Natural Philosophy." *Social Studies of Science* 16:387–420.

Stent, G. 1972a. "Prematurity and Uniqueness in Scientific Discovery." *Advances in the Biosciences* 8:433–49.

———. 1972b. "Prematurity and Uniqueness in Scientific Discovery." *Scientific American* 227 (December): 84–93.

Stepan, N. 1978. "The Interplay between Socio-Economical Factors and Medical Science: Yellow Fever Research, Cuba, and the United States." *Social Studies of Science* 8:397–423.

CHAPTER TWENTY-ONE

The Concept of Prematurity
and the Philosophy of Science

Martin Jones

In this paper, I shall ask and answer several questions about the precise delineation of Gunther Stent's notion of prematurity as it applies to scientific discoveries. With a specific understanding of the notion thus in hand, I will then make a few points about ways in which thought in the philosophy of science can be seen as having tackled the phenomenon of prematurity, even though the term itself has not entered the standard vocabulary of the philosopher of science; I will also point out that one particular subspecies of premature discovery has not been much discussed.[1] Finally, I will emphasize, or reemphasize, the interest of extending the investigation of prematurity into two areas to which Stent's papers do not themselves draw attention.[2]

THE NOTION ITSELF

Not everyone interprets Stent's original attempt to characterize the notion of prematurity in the same way; moreover, the differences between the various interpretations are the sort of differences which can hamper further inquiry into the topic. Consequently, I begin by trying to pin down a specific understanding of the notion. The understanding I will fix on is one which makes prematurity a more useful notion than it would be on some alternative readings; what is more, it is also more plausibly the notion Stent had in mind.

1. My explicit brief at the conference from which this volume derives was to address the question of how Stent's notion of prematurity might be related, or not, to the history of debates in the philosophy of science. I have taken that as my central aim in this paper, which is essentially an elaboration of those comments.

2. I say "reemphasize" because Ilana Löwy draws some attention to at least one of the areas I have in mind (see chapter 20 in this volume), and Frederic Holmes's chapter (12) in this volume also contains a suggestive discussion of a similar issue.

The following paragraph from the *Scientific American* version of Stent's "Prematurity and Uniqueness in Scientific Discovery" is clearly crucial if we are to understand the notion of "prematurity":

> So why was Avery's discovery [that DNA is the hereditary substance] not appreciated in its day? Because it was 'premature.' But is this really an explanation or is it merely an empty tautology? In other words, is there a way of providing a criterion of the prematurity of a discovery other than its failure to make a full impact? Yes, there is such a criterion: A discovery is premature if its implications cannot be connected by a series of simple logical steps to canonical, or generally accepted, knowledge.[3]

I will refer to the criterion laid out at the end of this passage as the *unconnectability condition*. (As I will point out later, given Stent's application of the criterion, my label may be misleading, but it is at least misleading in precisely the same way as the statement of the condition itself.) Before I go on to raise various questions about this condition, however, it is also useful to quote another passage, from slightly earlier in the paper, so that we can be clear about what Stent means in the above passage when he writes that O. T. Avery's discovery was "not appreciated in its day":

> My *prima facie* reason for saying Avery's discovery was premature is that it was not appreciated in its day. By lack of appreciation I do not mean that Avery's discovery went unnoticed, or even that it was not considered important. What I do mean is that geneticists did not seem able to do much with it or build on it. That is, in its day Avery's discovery had virtually no effect on the general discourse of genetics.[4]

In order to avoid certain other associations that talk of appreciation might have (two of which Stent mentions in this second passage), I will instead call this *lack of integration*. And when the point is that a certain discovery was not integrated at or around the time at which it was made, I will call that *a lack of immediate integration*. The term "immediate," however, should not be taken too narrowly, for as Frederic Holmes points out, in some historical situations the typical rate of integration for a discovery to which the relevant scientific community was entirely receptive would not have been very great by current standards.[5]

There are a number of questions which arise more or less immediately when we ponder the unconnectability condition,[6] one about its intended status, and three about its content; answering them will give us up a better grip on the notion of prematurity.

1. Is the unconnectability condition supposed to spell out a sufficient condition for the presence of some distinct property of prematurity, or is Stent here explaining that what he *means* by the term "prematurity" is unconnectability in the specified sense?

---

3. Stent 1972b, p. 84; cf. Stent 1972a, p. 435. See also chapter 2 in this volume.
4. Stent 1972b, p. 84.
5. See Holmes, chapter 12 in this volume.
6. Perhaps this is especially true if the ponderer is a philosopher by training.

2. Is the term "discovery" being used factively here, as it surely is in at least some contexts? That is, when we say that someone has discovered a certain object or a natural kind in the sense Stent intends here, does that imply that there really is such an object or natural kind? And when we say that someone has discovered *that* such-and-such, does it imply that the proposition in question is true?[7]

3. Is "knowledge" being used factively (as, it is standardly claimed in philosophical circles, it is in its ordinary usage)? That is, can we know only true things, in the sense of the term "knowledge" that Stent has in mind here?

4. Do cases in which a purported discovery *conflicts* with some part of the body of contemporary scientific belief count as cases of unconnectability?

The last of these questions might seem unmotivated at this point—after all, the wording of the unconnectability condition certainly seems to rule out cases of conflict straightforwardly. However, appearances are deceptive.

1. If unconnectability were intended merely as a sufficient condition for the presence of a distinct property of prematurity, we would then have to ask what prematurity is. And canvassing various plausible answers to that question casts considerable doubt on such a reading.

To begin with, the first passage quoted makes it clear that prematurity is supposed to explain lack of immediate integration in at least some cases; and the second makes it clear that a lack of immediate integration can be a reason (and presumably a nontrivial one) for thinking the discovery premature, at least prima facie. It is thus obvious that prematurity is not the same thing as lack of immediate integration.[8]

Next, the ordinary connotations of the term "prematurity" might suggest that prematurity is a matter of lack of immediate integration combined with integration at a later point; but this again would leave us with a notion of prematurity ill-suited to explaining lack of immediate integration. Doubly so, in fact, for we would either be explaining in a circle, so to speak, or explaining the past by reference to the future, or both. What is more, unconnectability would clearly not be a sufficient condition for prematurity in such a sense. That is, there is little reason to suppose that every purported discovery which is unconnectable to currently accepted scientific knowledge at the time of its making goes on (or will eventually go on) to be integrated into ongoing research at some point, and there is some reason to suppose otherwise; and it certainly seems at least possible that the same can be said of real discoveries.[9]

---

7. There are other cases still: for example, someone might be said to have discovered a certain process or phenomenon (such as speciation or continental drift), in which case the question is whether this implies that the process or phenomenon really occurs. And so on.

8. Indeed, the suggestion, in the opening sentence of the second passage, that lack of immediate integration can be *merely* a prima facie reason for thinking a discovery premature makes this especially plain.

9. Perhaps it is also worth noting that unconnectability is presumably not a necessary condition on lack of immediate integration, either. That is, there are reasons other than unconnectability for which

Finally, we might think that to say that a discovery was premature is to say that the scientific community was not receptive to it when it was made (or would not have been, if it had come face-to-face with the discovery). This is not the same thing as saying that the discovery was not immediately integrated, for lack of immediate integration might be due simply to poor advertising, for example. On this reading, then, prematurity is a matter of unreceptivity on the part of the community, and unconnectability is being offered as a sufficient condition for the presence of such unreceptivity (and thus for prematurity). Unreceptivity, in turn, is then supposed to explain some cases of lack of immediate integration.

This reading fares a little better than the others considered thus far, but it is still awkward. The notion of receptivity (and thus the notion of prematurity) is, as things stand, somewhat vague, and Stent does not tell us about any possible sources of unreceptivity other than unconnectability.[10] Accordingly, offering prematurity as the explanation for a lack of immediate integration in various cases, although it is not entirely trivial on this reading, still seems rather unsatisfying: the explanation would be that the scientists in question were unreceptive to the discovery in question. Stent might then go on to explain such unreceptiveness by showing that the discovery was unconnectable to contemporary scientific knowledge; but surely the bulk of the explanatory work here would be done by the talk of unconnectability, leaving the notion of prematurity itself as a somewhat feckless go-between. And yet in the first passage quoted above, it is clear that prematurity itself is supposed to explain the lack of immediate integration of Avery's discovery.[11]

Given all this, it seems to me that the answer to the first of my questions should be that prematurity, in the sense Stent intends, just *is* unconnectability; that is, in stating the unconnectability condition, Stent is telling us what he means by "prematurity." This of course makes it all the more important that we clarify our understanding of the unconnectability condition, and so I turn to the other questions listed above, each of which concerns some important concept involved in the formulation of the condition.

2. I think it is clear that Stent applies the term "discovery" nonfactively; to establish this, however, we should first address another matter.

Stent describes one of his cases, that of research into the idea of macromolecular memory, as an "example of here-and-now prematurity"; and on the same page he refers to "the concept of here-and-now prematurity."[12] This has led some discussants to interpret Stent as introducing a distinction between two different

---

discoveries are not taken up in some cases—not being published in the right place or made by the right people, lack of funding, lack of the requisite technological expertise, and so on. See the section "Further Development" below for some related points.

10. Not that one could not readily think of some.

11. Of course, prematurity as unreceptivity of the relevant community *plus* later integration is a reading afflicted by a combination of the problems mentioned for the readings discussed thus far.

12. Stent 1972b, p. 87; cf. Stent 1972a, p. 438–39.

kinds of prematurity, "retrospective" and "here-and-now." However, one could also read the paper as presenting a single notion of prematurity, but distinguishing between two different ways in which judgments of prematurity can be made: retrospectively and on the spot. Stent's text is perhaps ambiguous between these two readings, but he does introduce the distinction (whichever distinction it is) with the following words: "Does *the* prematurity concept pertain only to retrospective judgments made with the wisdom of hindsight? No, I think *it* can be used also to judge the present."[13] He then goes on to support this claim by arguing that the unconnectability condition is satisfied in the macromolecular memory case.[14] Given this, and the greater neatness involved in thinking of a single concept of prematurity and two vantage points from which it may be applied, I shall opt for that latter reading.[15]

The claim that "discovery" is not being used factively in the statement of the unconnectability condition now follows straightforwardly; for the idea of macromolecular memory, Stent writes, "*whether true or false*, is clearly premature."[16] (He also says that "the results claimed for [the relevant experiments] may not be true at all.")[17] Similarly, Stent's other example of a case in which we can make a judgment of prematurity here and now is that of the purported existence of ESP, and one does not get the feeling from Stent's discussion of this case that he is ready to commit to the idea that there is such a thing.[18]

Of course, this way of reading the unconnectability condition is compatible with being especially interested in those cases of lack of immediate integration involving a genuine discovery. On the other hand, it is far from clear that, as a historian, philosopher, or sociologist of science, one should be any less interested in cases which, by our present lights, do not involve genuine discoveries, for that surely is Whiggishness. No less Whiggish, perhaps, is the idea of explaining the scientific community's reaction to a discovery partly by appeal to the fact that it was (or was not) a genuine discovery.[19] It is worth noting, thus, that reading the unconnectability condition as involving a nonfactive notion of discovery has the advantage that we

13. Ibid., emphasis added.

14. See the quotations from the discussion of macromolecular memory included in the discussion of question 4, below.

15. This also serves to drive home the point that prematurity cannot consist in part of delayed integration, as Stent clearly does not mean to predict that either the idea of macromolecular memory or that of ESP will one day be integrated into scientific research.

16. Stent 1972b, p. 87; emphasis added. Cf. Stent 1972a, p. 438.

17. Stent 1972b, p. 87.

18. It should now be clear why I addressed the distinction between here-and-now and retrospective prematurity first. If one were to read Stent (implausibly, in my view) as distinguishing two *kinds* of prematurity there, then one might worry that the textual evidence I have just adumbrated shows only that here-and-now premature discoveries can fail to be genuine discoveries.

19. This, of course, is at least part of the point of the "symmetry tenet" embraced by proponents of the strong program in the sociology of science. The insistence that we attend in equal measure to the life histories of genuine and spurious discoveries follows from that program's "impartiality tenet." See Bloor 1991, p. 7.

end up with a concept of prematurity which, in one respect at least, we can apply without relying on the putative wisdom of hindsight, and so without opening ourselves to a charge of Whiggishness.[20]

3. A charge of Whiggishness might also arise with the use of the term "knowledge." It is evident, however, that this term, too, is to be read nonfactively, for the simple reason that two of the cases of alleged prematurity Stent discusses involve discoveries he clearly considers genuine, but which he also regards as having been *in conflict* with the "canonical knowledge" of the time. First, concerning the lack of immediate integration of Avery's discovery, Stent claims that "the then current view of the molecular nature of DNA . . . made it well-nigh inconceivable that DNA could be the carrier of hereditary information."[21] And Michael Polanyi's 1914–16 theory of the adsorption of gases on solids is described as having a "basic assumption" which was "irreconcilable" with contemporary thinking about "the role of electrical forces in the architecture of matter."[22] Because Stent regards these discoveries as genuine, at least some of the "knowledge" with which they were in conflict cannot have been knowledge in the factive sense.[23]

4. Answering the third question essentially answers the fourth, too. It is quite clear that Stent means to include in the class of premature discoveries cases in which a putative discovery conflicts with contemporary canonical scientific belief. Not only are two of the three historical cases he discusses at any length cases in which in Stent's view a conflict was present, as we have just seen (the third case being Gregor Mendel's "discovery of the gene"),[24] but so are both the here-and-now cases of prematurity. Concerning the macromolecular theory of memory, Stent writes: "The lack of interest of neurophysiologists in the macromolecular theory of memory can be accounted for by recognizing that the theory, whether true or false, is clearly premature. There is no chain of reasonable inferences by which our present, albeit highly imperfect, view of the functional organization of the brain *can be reconciled* with the possibility of its acquiring, storing and retrieving nervous information by encoding such information in molecules of nucleic acid or protein."[25] The idea of extrasensory perception is similarly described as being "totally irreconcilable with the most elementary physical laws," and one which "cannot be reconciled with what we now know."[26]

---

20. In the interests of brevity, and despite what, to my ear, is a departure from ordinary usage, I will follow Stent in this regard: when I use the term "discovery" without qualification, it should be read nonfactively.

21. Stent 1972b, p. 85.

22. Ibid., p. 87.

23. Or, to put it less neutrally, cannot really have been knowledge at all.

24. Stent 1972b, p. 87.

25. Ibid., emphasis added.

26. Ibid.; 1972a, p. 438. These two quotations are taken from Stent's description of the views of two "future mandarins of molecular biology, Salvador Luria . . . and R. E. Roberts," respectively, views which conflicted with one another in other respects; but Stent goes on to say that it seemed to him that "both

Another reason for taking the class of premature discoveries to include discoveries which conflict with contemporary scientific belief is that, unless we do, we cannot make sense of Stent's conclusion in the final two paragraphs of his discussion of prematurity. There, he takes the phenomenon of prematurity to count as evidence against a "view of the operation of science [which is] commonly held," according to which "the good scientist is seen as an unprejudiced man with an open mind who is ready to embrace any new idea supported by the facts."[27] Instead, prematurity speaks for a view of science described in a passage Stent quotes from Polanyi, according to which "there must be at all times a predominantly accepted scientific view of the nature of things, in the light of which research is jointly conducted by members of the community of scientists. A strong presumption that any evidence which contradicts this view is invalid must prevail."[28] Clearly, the phenomenon of prematurity can provide support for such a view only if at least some premature discoveries count as contradicting canonical scientific belief.

Prematurity can involve conflict, then. So if, as I have maintained, prematurity should be understood as just being unconnectability (in the sense specified in the "unconnectability condition"), this must mean that conflict can count as unconnectability. And apparently it can. After claiming, as we have just seen, that "there is no chain of reasonable inferences by which our present . . . view of the . . . brain can be reconciled with [the macromolecular memory hypothesis]," Stent goes on, in the very next sentence, to describe this as a case of unconnectability: "Accordingly for the community of neurophysiologists there is no point in devoting time to checking on experiments whose results, even if they were true as alleged, *could not be connected* with canonical knowledge."[29]

Given all this, it seems reasonable to complain that the unconnectability condition is misleadingly phrased, for a putative discovery which conflicts with a given body of scientific belief, and is *seen* to conflict with it, surely has been "connected" to the relevant body of contemporary scientific belief by a "series of . . . logical steps,"[30] albeit in what from some points of view might seem a regrettable way. And indeed, the label I have been using to refer to this condition (i.e., the unconnectability condition) thus comes to seem misleading in just the same way. I shall stick with the label nonetheless, both for simplicity's sake and in order to continue to reflect Stent's own formulation of the condition.[31]

Luria and Roberts were right" (1972b, p. 87) and nowhere begs to differ with either of them over the issue of whether the postulation of ESP is in conflict with currently accepted theory.

27. Stent 1972b, p. 88; 1972a, p. 440.

28. Stent 1972b, p. 88.

29. Ibid., p. 87.

30. I have omitted the word "simple" here, in part because I am not sure why the simplicity of the logical steps involved should matter.

31. Interestingly, in the version of the paper which Stent published in *Advances in the Biosciences* (1972a), immediately after first presenting the unconnectability condition as a "criterion" of prematurity, he

Cases of conflict, then, count as cases of unconnectability—but not only cases of conflict. I take it that Stent also means to include cases of what we might call "genuine unconnectability," in which a putative discovery simply fails to make logical contact of any substantive sort with the contemporary body of scientific belief, neither clashing with it nor supporting it, unable to fit naturally into it in any way— cases, we might say, in which the cogs simply spin free of one another. Stent seems to regard Mendel's discovery of the gene as a case of such a failure to mesh, rather than one of outright conflict.[32] He attributes the lack of immediate integration of Mendel's discovery in part to the fact that "the concept of discrete hereditary units could not be connected with canonical knowledge of anatomy and physiology in the middle of the 19th century." Whereas, by the end of the nineteenth century, when the integration of Mendel's discovery began, "Chromosomes and the chromosome-dividing processes of mitosis and meiosis had been discovered and Mendel's results could now be accounted for in terms of structures visible in the microscope." Another factor in the delay, Stent claims, was the fact that "the statistical methodology by means of which Mendel interpreted the results of his pea-breeding experiments was entirely foreign to the way of thinking of contemporary biologists." By the end of the century, however, "the application of statistics to biology had become commonplace."[33]

Whether or not I am interpreting Stent's views about the Mendel case correctly, the passages just quoted suggest some ways in which what I am calling genuine unconnectability might arise: first, because a discovery, while not conflicting with received wisdom, cannot be accounted for by it either; second, because the discovery is formulated in terms which are alien to the conceptual vocabulary of the relevant community; or third, because it can be understood or made plausible only if one understands certain methods which are outside the community's method-

---

adds, "this criterion is not to be confused with that of an *unexpected* discovery, which *can* be connected with the canonical ideas of its day but might overthrow one or more of them," and then goes on to illustrate this notion with reference to "the recent finding of a 'reverse transcriptase' " (p. 435, emphasis in original). This would seem to imply that Stent wishes to rule out cases involving conflict from the category of premature discoveries, but as we have seen, to the extent that it does so, it flies in the face of both his choice of examples and his presentation of those examples. One way to resolve this tension would be to incorporate into the notion of premature discovery the idea that such a discovery does not *actually* lead to an immediate revision in the body of accepted scientific doctrine when there is conflict, whereas an unexpected one does. But this would be tantamount to incorporating lack of immediate integration into the definition of prematurity, at least for cases of conflict, and as we have already seen, this would prevent the notion from being able to do its intended explanatory work.

32. Given that I am discussing Stent's views, I will persist in describing what Mendel did as discovering the gene, even though such a description seems to me questionable in certain respects. (See Holmes, chapter 12, and Löwy, chapter 20, in this volume for some skeptical considerations concerning the individuation and temporal locating of discoveries.)

33. Stent 1972b, p. 86; see also 1972a, p. 437.

ological repertoire.[34] Certainly, if a discovery has one of these features, the failure of the scientific community to integrate it into ongoing work, or to "build on it," will come as no surprise.

One reason for taking the notion of prematurity to include cases of genuine unconnectability is that the formulation of the unconnectability condition strongly suggests doing so. Including genuine unconnectability is also perfectly in keeping with the thought that prematurity might explain some cases of lack of immediate integration, for genuine unconnectability is no less an obstacle to integration than conflict with established theory. Furthermore, prematurity comes to seem a more interesting notion if it encompasses genuine unconnectability, particularly when we consider the question of whether philosophers of science have taken adequate account of prematurity and the issues it raises; and it is that question to which I now turn. In closing this section, however, let me reformulate the unconnectability condition, understood now as giving the content of the notion of prematurity, in such a way as to incorporate the claims for which I have just argued concerning the best way of understanding it:

> To say that a discovery (whether genuine or merely putative) is premature is to say that it cannot be connected to the body of generally accepted contemporary scientific beliefs in such a way as to positively support or further elaborate those beliefs.

PREMATURITY AND THE PHILOSOPHY OF SCIENCE

Citations of Stent's paper are not common in the philosophy of science literature, and certainly the word "prematurity" is not part of the standard technical vocabulary of the well-trained philosopher of science. It is a separate question, however, whether philosophers of science have thought about the phenomenon of prematurity and its implications by another name. And the answer to this question is in part a resounding "yes" (or at least a resounding "in a way") and in part a "not much." The two answers correspond to the two sorts of prematurity we have already distinguished: prematurity which involves conflict, and prematurity which does not, respectively.

Prematurity with Conflict

It will be useful to begin my elaboration of the first of these answers, concerning prematurity with conflict, by briefly rehearsing a cluster of standard issues in the

---

34. One interesting connection here is between the idea that genuine unconnectability might arise in the latter way and a suggestion of Ian Hacking's: "Whether or not a proposition is as it were up for grabs, as a candidate for being true-or-false, depends on whether we have ways to reason about it. The style of thinking that befits the sentence helps fix its sense and determines the way in which it has a positive direction pointing to truth or to falsehood" (1982, p. 48).

philosophy of science. The relevance of these issues to Stent's discussion of prematurity may seem clear enough from the outset, but just a little care is needed in connecting the two, so I will do that explicitly.

The standard issues I have in mind arise when we think about situations in which scientists are faced with recalcitrant data, alleged bits of evidence which conflict with the body of scientific belief currently accepted by those scientists. The fact is that, in many cases, the result is not that preexisting beliefs are overhauled and revised but that the troubling data lose out. This can, of course, happen in any number of ways. The data in question might be carefully examined and then rejected, and reasons for the decision might be laid out; they might be dismissed in a more peremptory fashion ("That just can't be right"); or they may simply be ignored. There is also the possibility that the jury stays out on the data, the clash with received wisdom is regarded as an unsolved puzzle, and the problem is laid aside for another day. In such a case, of course, the worrisome data do not exactly lose out, but neither do they provoke any change in the beliefs of the scientific community.

Such responses to conflict between data and theory have come in for exhaustive discussion in the philosophy of science, especially in the last 30 or 40 years.[35] Indeed, only the most naive form of Popperian falsificationism, according to which a hypothesis is to be cast aside at the first sight of conflict with the data, fails to take these phenomena of scientific practice into account.[36] Karl Popper himself was well aware of the resilience of reigning theory when confronted with apparent counterevidence, even if he was prone to expressing some regret over it.[37] And Thomas Kuhn and Imre Lakatos both placed this feature of scientific practice firmly in the limelight. According to Kuhn, "anomalies" for current theory abound most of the time, and only certain special sorts of anomaly are capable of provoking a crisis.[38] In a similar vein, and with a famous turn of phrase, Lakatos insisted that "in actual history new theories are born refuted."[39]

The fact that the data do not always win out over theory when the two are at odds is also emphasized in a much earlier treatise, for Pierre Duhem famously described at least one sort of rejection of putative observational evidence in *The Aim*

35. I use the term "theory" here for brevity's sake. I mean it to refer to the body of scientific belief accepted at the time in question, and that body of belief typically includes more than just what we would normally call "theories": it may include beliefs about the values of certain constants, beliefs concerning other data which have previously been collected and regarded as good, and so on. (I am also using the convenient term "belief" to refer to the claims which scientists accept at a given time—constructive empiricists should feel free to substitute a different term.)

36. And even that is so only if naive falsificationism is understood as an attempt to describe the actual workings of science rather than as a set of prescriptions for doing science properly.

37. For example, Popper 1970, pp. 52–53 and 55.

38. Kuhn 1970, pp. 65, 81–82, and *passim*.

39. Lakatos, 1970, p. 120, n. 2; see also, e.g., pp. 134–35, 176–77, and 182.

and Structure of Physical Theory.[40] Duhem was especially concerned to draw attention to the ubiquitous necessity of making substantive assumptions about measuring devices, experimental conditions, and the like when arriving at conclusions about the phenomena on the basis of experiment. In the face of a clash between theory and experiment, then, we always have the option of throwing the data into doubt, at least as far as the constraints imposed solely by deductive logic are concerned. Moreover, Duhem was at pains to point out that in many instances in the history of science this was a perfectly rational route to take.

All this is very familiar territory to anyone in a science studies discipline. Indeed, on one way of configuring the issues in the philosophy of science, the implications of the so-called Duhem-Quine thesis are the jumping-off point for a number of well-worn debates: over the theory-ladenness of observation, the underdetermination of theory by the data, the role of pragmatic considerations in theory choice, the significance of the fact that science is a social enterprise, and the role of values in science.[41] In each case, one of the points stimulating the discussion is the observation that, when the data seem to conflict with the beliefs we have already taken on board, the result is very often (although obviously not always) that the data in question are rejected.

How exactly does all this bear on the phenomenon of prematurity? It would be a distortion of the standard philosophical framework to class premature discoveries as *data,* for despite the well-known difficulties involved in delineating a general distinction between data reports and theoretical claims, discoveries of the sort Stent discusses—that DNA is the hereditary substance, that there is ESP, or that the memories of rats are stored in macromolecules, for example—clearly do not count as data by the lights of ordinary philosophical usage.[42] Consequently, we cannot apply the ideas involved in the stock philosophical discussions to the phenomenon of

40. Duhem 1954, chap. 6.

41. Here is one way of drawing out the connections in just a little more detail: First, the fact that so-called auxiliary assumptions concerning experimental conditions and the workings of measuring devices are involved in producing the data leads us to recognize at least one sense in which observation can be theory laden. Next, we notice that it is sometimes (perhaps typically, or always) possible to make more than one theory compatible with the data (or to make more than one theory predict or explain them) by changing our decisions about which of the auxiliary assumptions to embrace and which to reject. This can lead to the conclusion that theory choice is sometimes (or typically, or always) underdetermined by the data. Given *that,* questions will arise as to whether the decision to adopt one theory rather than another actually gets made on pragmatic grounds (such as simplicity, elegance, and, according to some, explanatory power), and whether the idea that theory choices are made in such a way compromises scientific realism. Similarly, we can ask whether various social forces might be at work in influencing or even determining the outcomes of such choices, and if so, whether that is a good or a bad thing, whether it is inimical to the rationality or objectivity of the sciences, and so on. And we can ask similar questions (to the extent that they are separate questions) about the influence of values of various sorts on the process of theory choice.

42. These are difficulties which some would claim are insuperable, on the grounds that there is no clear distinction to be drawn.

prematurity simply by substituting the term "discovery" for the term "data" everywhere.

The problem is that the usual philosophical framework relies on a two-component picture, of data versus theory. One way of solving the problem is to see the phenomenon of prematurity with conflict as involving three components: current theory, the discovery which conflicts with it, and the data presented in support of the claim that the discovery in question has been made.[43] It is then a relatively easy matter to make essentially the same points, and raise essentially the same issues as the ones which arise in the standard philosophical discussions, from a slightly different starting point. We begin with the fact Stent wished to emphasize: namely, that in at least some cases in which a discovery is in conflict with current theory, it is the discovery which loses out. At least in principle, this could happen in a number of ways: the discovery might be ignored altogether; it might be given just the attention needed to reject it peremptorily; it might be carefully studied and rejected for reasons which are given explicitly; or the problem of its conflict with current theory might be put on the shelf for another day. The interesting new twist comes when we consider the third option, reasoned rejection, for now there are two different ways in which this could happen. One possibility is that the discovery claim might be undermined by a rejection of the data which are supposed to support it, whether or not those data conflict with current theory in themselves. Alternatively, and provided that the data do not in themselves conflict with current theory, the inference to the discovery claim might be rejected, while the data themselves are left unchallenged.

In either case, the familiar philosophical lines of thought are off and running. If the data are rejected via the rejection of auxiliary hypotheses concerning measuring devices or experimental conditions, then we get the theory-ladenness of observation, Duhem-Quine, and underdetermination in just the way they arose in the previous framework;[44] and once we have underdetermination, we can start raising questions about pragmatic considerations, social factors, and the role of values in science. If, on the other hand, it is the inference from data to discovery claim which is rejected, then rejection is likely to proceed by way of a denial of some auxiliary hypotheses employed in arguing (however abductively) from data to discovery claim, or by insisting more directly on the underdetermination of the discovery claim by the data; in either case, we have introduced the idea that in the face of conflict, there is often more than one way out, and this idea constitutes an open invitation to return to discussion of the pragmatic, the socially influenced, and the value-laden.

Thus, in essence, the implications of the occurrence of premature discoveries (whether genuine or spurious) of the sort which involve conflict with canonical sci-

43. For a three-component picture which might be adaptable to our needs here, see Bogen and Woodward (1988) and Woodward (1989).

44. The move to underdetermination here is not unproblematic (see, for example, Laudan 1990); my point is only that it can be made with *equal* plausibility here and in the case of a two-cornered fight between data and theory.

entific belief, and the implications of the lack of immediate integration of such discoveries into ongoing scientific research, have been explored extensively in the philosophy of science, albeit not in those terms. What is more, many have drawn, or had already drawn, just the sort of conclusion Stent wished to join Polanyi in drawing.[45]

Incidentally, we are now in a good position to note something interesting about the cases of conflict Stent discusses, at least as they are presented in Stent's article. In each case, in Stent's account, the discovery claim in question met with a relatively flat rejection. The only sense in which reasons were given for the dismissal of the discovery claim is that the people doing the dismissing made some claim about exactly *how* the discovery conflicted with current theory. Avery's claim that DNA is the hereditary substance conflicted with views about its chemical structure; Polanyi's theory of the adsorption of gases could not be reconciled with the contemporary understanding of the nature of the interactions between gas molecules and the solid surface; the macromolecular hypothesis about memory clashed with the then "present . . . view of the functional organization of the brain"; and the idea that there might be ESP was declared by S. E. Luria to be, in Stent's words, "totally irreconcilable with the most elementary physical laws."[46] These may or may not be good reasons for rejecting the discovery claims in question, but they clearly do not involve a close examination of the evidence which purportedly supported those claims, nor a critique of the arguments leading from one to the other. Thus, if Stent's descriptions of the responses with which these discovery claims met are accurate, that would serve to underscore Polanyi's talk of a "*strong* presumption" in favor of the "predominantly accepted scientific view of the nature of things"[47]—a presumption so strong that a detailed response to the experimental evidence cited, and to the subsequent inference to an unsettling discovery claim, is not deemed necessary. I am not in a position to judge the historical accuracy of Stent's accounts of these episodes, but it would certainly be interesting to explore that issue further and, separately, to see whether particular examples can be found of each of the other kinds of possible response to conflict listed above.[48]

Prematurity without Conflict

What now may be said of cases of genuine unconnectability, in which the scientific community fails to integrate a discovery into ongoing research not because it

45. See also the discussion in the preceding section.

46. Stent 1972b, p. 85; 1972b, p. 86; 1972a, p. 438 and 1972b, p. 86; 1972a, p. 438 and 1972b, p. 87, respectively. See also Stent, chap. 2, p. 28, this volume.

47. Stent 1972b, p. 88.

48. In this volume, see Holmes, chapter 12, and Löwy, chapter 20, for criticism of Stent's historical claims, especially concerning the Avery case and (in Holmes's chapter) the Mendel case; and Nye, chapter 11, on the Polanyi case.

conflicts with received doctrine but because there are, so to speak, insufficiently many points of contact between doctrine and discovery?

One might remark at this point that it is not immediately obvious why we should care about such cases. If a discovery simply fails to engage in any way with ongoing scientific work, then it would seem to follow—at least prima facie—that it can have no influence on scientific change and development. And, one might argue, insofar as understanding that process is our central concern, a discovery which is genuinely unconnectable to contemporary scientific belief is thus just irrelevant.

This would be unduly hasty, however. For one thing, a genuinely unconnectable discovery claim *might* be taken seriously in some cases, or at least not dismissed immediately, and it might also seem clear that the discovery in question concerns objects or processes lying within the domain of a currently accepted theory. In that case, it might be disconcerting for members of the relevant scientific community not to be able to encompass the new result within current theory, and yet more so if current theory does not possess even the conceptual resources for addressing the issue. The unease created by this situation might then, in some cases, issue in creative work aimed at elaborating or extending current theory so as to bridge the gap between it and the alleged newly discovered phenomenon. Conflict is not the only possible stimulant of change.

Two connecting threads come to mind here as possibly linking talk of genuine unconnectability with ideas developed in the philosophy of science. The first takes us to debates over the existence, nature, and implications of so-called incommensurability between theories, a set of debates which are at this point just as hoary as, and indeed intertwined with, the debates over theory-ladenness, underdetermination, and the like. The second takes us to a more recent rethinking of the relationship between theory and experiment which has gone on in the philosophy of science; and the defense I have just given of the interest of genuine unconnectability might be seen in essence to extend some points found in Ian Hacking's seminal book *Representing and Intervening* (1983).

Incommensurability first. On introducing the category of the genuinely unconnectable discovery, I listed three possible sources of such unconnectability: that current theory is simply unable to account for the discovered fact or phenomenon, though there is no conflict between discovery claim and theory; that the discovery claim is formulated in terms lying outside the conceptual vocabulary of the relevant community; and that understanding the discovery claim, or finding it credible, requires an understanding of methods which are not to be found on the methodological palette of the community.[49] It is the second and third of these sources, in particular, which call to mind the notion of incommensurability.

49. If either the second or the third factor is present in a given case, then presumably that would also lead to the inability of current theory to account for the discovery. What I have in mind, then, with

Incommensurability is usually thought of as a relation between two relatively overarching theories, such as Newtonian mechanics and relativistic mechanics, or even perhaps between two grander sorts of things called worldviews. There are, in fact, a number of different and independent relations which were initially collected together under the umbrella term "incommensurability," and the conceptual territory has been carved up in more than one way by various authors since Kuhn and Paul Feyerabend first introduced the topic.[50] One component of the idea of incommensurability, however, has always been that competing theories might employ sets of concepts whose differences are sufficiently radical that, perhaps despite appearances to the contrary, the particular claims of one theory cannot be compared to the claims of the other; an important consequence of this is supposed to be that two such theories cannot make conflicting predictions, strictly speaking, so that we are deprived of the option of appealing to "crucial experiments" in deciding between them.[51] Another central component of the idea of incommensurability is that competition is often between whole paradigms, to employ the most notorious Kuhnian term; given, then, that one constituent of a paradigm is a set of accepted methods of inquiry and standards of good problem-solving, this is taken to mean that one difficulty we will inevitably encounter in arbitrating between competing theories is that of finding unbiased standards to employ in the very process of arbitration.[52]

The connections between these two aspects of incommensurability, on the one hand, and the second and third potential sources of genuine unconnectability, on the other, is clear enough: conceptual and methodological gaps, respectively, are at issue in both cases. One interesting difference between incommensurability as it is usually understood and genuine unconnectability, however, is that the latter is a

---

the first item on this list, is an inability to do so arising from less dramatic sources. That is, I am thinking of situations in which the discovery claim and the methods employed in its support are perfectly understandable from the point of view of reigning theory, but in which theory simply lacks the right sort of particular content either to provide the materials needed to explain the discovered phenomenon or to draw out nontrivial connections between that phenomenon and others already addressed by the theory. Incidentally, Löwy, chapter 20 in this volume, provides some nice examples of cases of prematurity arising from methodological differences.

50. See, for example, Doppelt 1978; Newton-Smith 1981, pp. 148–51; and Hacking 1983, pp. 67–74. For the introduction of the notion of incommensurability, see Kuhn 1970 (the first edition of which appeared in 1962) and Feyerabend (1962, 1965). Feyerabend focuses on the first component of the notion listed below, also calling it "meaning variance."

51. Doppelt calls this "incommensurability of scientific meanings" (1978, p. 33); Newton-Smith's label is "incommensurability due to radical meaning variance" (1981, p. 150); and Hacking's term is "meaning-incommensurability" (1983, p. 72).

52. Doppelt covers this under the heading "incommensurability of scientific problems, data, and standards" (1978, 33); Newton-Smith writes of "incommensurability due to radical standard variance" and "incommensurability due to value variance" (1981, pp. 149–50); and this component is also closely related to Hacking's notion of "dissociation," which in part has to do with variation in "styles of reasoning" (1983, pp. 69–72; see also my n. 34 above).

relation between a single discovery and a body of theory rather than between two competing theories of similar scope. Thus the idea that genuine unconnectability can indeed arise from conceptual or methodological divergence serves to draw attention to the possibility that some of the factors which are supposed to give rise to incommensurability between theories might also turn up in the relations between theories and more humble entities.

It should be noted that, in dealing with a case of genuine unconnectability, we are not necessarily trying to make a choice between the unconnectable elements. So, talk of "incommensurability," of lack of a common measure, might mislead. In the face of genuine unconnectability, we may aim at assimilation rather than choice. The implications of genuine unconnectability, then, are not those of incommensurability: it is not that we automatically face an obstacle to making a rational, theory-neutral choice between competing accounts of the phenomena, but that, instead, as Stent describes, we will have difficulty in integrating the discovery into the body of current theory. Nonetheless, questions about the rationality of scientific decision-making may find a foothold in the phenomenon of genuine unconnectability, for if premature discoveries of this variety are ever rejected as dubious, rather than shelved for later consideration, then there is the obvious danger that any argument given to support such a rejection will either fail to make real contact with the discovery claim at all, by virtue of reliance on a set of concepts disjoint from the one the claim invokes, or will appeal to received methodological standards in a way which simply prejudges the issue against the support offered for the discovery claim. And those are essentially the same dangers that proponents of incommensurability theses about theory choice have tended to emphasize.

So much for the connection to incommensurability. The other place at which the idea of genuine unconnectability might be linked to work in the philosophy of science is in the discussions which, beginning in the 1980s, sought to overturn a long-dominant philosophical picture of experiment and observation as practices aimed purely at either testing or providing further support for preexisting theories. As Hacking observed, "Philosophy of science has become so much philosophy of theory that the very existence of pre-theoretical observations or experiments has been denied."[53] Hacking, for one, sought instead to make experimental practice a proper subject of philosophical study in its own right, and issued the much-quoted declaration that "experimentation has a life of its own."[54]

Not only does experimentation have a life of its own, but experiment can drive developments in theory, and not just in the familiar sense that a failed prediction made as part of the process of theory testing can lead to theoretical revision. In two sections of a chapter of *Representing and Intervening* titled "Experiment," Hacking cites a number of examples from the history of optics to make the point that

53. Hacking 1983, p. 150.
54. Ibid.

unexpected observations can provoke new theoretical work, even when the observations were made neither in pursuit of observational support for some theory, nor in an attempt to test one: "The observations," Hacking remarks, "preceded any formulation of theory."[55] Rather, the development of theory "depended on simply noticing some surprising phenomenon."[56]

So far, though, it is not clear that the sort of occurrences Hacking has in mind would count as genuinely unconnectable discoveries, as opposed to being discoveries which actually conflict with current theory, even though they were in fact not made in the pursuit of theory testing.[57] The title of the next section of the chapter, however, is "Meaningless Phenomena." The section is a short one, but remarkably salient in the present context. It begins: "I do not contend that noteworthy observations in themselves do anything. Plenty of phenomena attract great excitement but then have to lie fallow because no one can see what they mean, how they connect with anything else, or how they can be put to some use."[58]

This, of course, sounds exactly like a description of genuinely unconnectable premature discovery, accompanied by a consequent lack of integration. Hacking then goes on to give condensed accounts of two examples intended to make the point: Robert Brown's "painstaking observations," in the early part of the nineteenth century, of what we now know as Brownian motion, and Antoine-César Becquerel's discovery of the photoelectric effect in 1839, a discovery which, Hacking notes, "attracted great interest—for about two years."[59] In neither case was the phenomenon in question integrated into ongoing research until the first decade of the twentieth century, at which point both acquired considerable significance, of course: the first as a decisive piece of evidence in favor of the existence of atoms, and the second as a stimulant to the development of quantum theory.

It would be interesting, then, to investigate closely whether the two cases Hacking cites as examples of "meaningless phenomena" are in fact cases of genuinely unconnectable premature discoveries.[60] There is room for doubt on that score, at

55. See "Noteworthy Observations" and "The Stimulation of Theory" (ibid., pp. 155–58). Hacking actually gives much of the credit for these two sections to the physicist Francis Everitt: see the book's acknowledgements (ibid., p. vii). Quote is on p. 156.

56. Ibid., p. 155.

57. Hacking's claim—that the sort of observations he has in mind in the two sections just discussed are made prior to theory formulation—might seem to suggest that he is focusing on cases of genuine unconnectability. In at least some of the cases he mentions, however, it may be only that the observations in question took place before the development of a theory that could explain them (namely, the wave theory of light), but not before the formulation of a *relevant* theory and, indeed, one with which the observations conflicted (specifically, the corpuscular theory). See in particular his description of the experimental work of David Brewster (Hacking 1983, p. 157).

58. Ibid., 158.

59. Ibid.

60. Incidentally, Hacking reports that Brownian motion had been observed as early as 60 years before Brown's work. I do not know whether Brown discovered it independently or not; if so, we would have at least one case on the books of a premature discovery which was made prematurely more than once.

least in the case of the photoelectric effect, for these days that phenomenon is seen as conflicting straightforwardly with the wave theory of light (as a glance at any introductory physics text will confirm), and the wave theory was achieving ascendancy around the time the discovery was made.[61] So Becquerel's discovery may not have been genuinely unconnectable to current theory;[62] to that extent, however, it also would not perfectly fit Hacking's characterization of the category of "meaningless phenomena," with its talk of lack of connection. But in any case, these seem to be two more good examples of prematurity of some variety or other. And if there are genuinely unconnectable discoveries (perhaps Mendel's is one), then that fact would drive home Hacking's general point that there is more than one way in which experiment can relate, or fail to relate, to theory.

Prematurity and Fruitfulness

I will add one last thought about the notion of prematurity and its relation to the canon in the philosophy of science. The thought is simply that prematurity sounds a lot like the opposite of the theoretical virtue Kuhn calls "fruitfulness," and which he counts as one of the five major criteria of theory choice in the well-known paper "Objectivity, Value Judgment, and Theory Choice."[63] Kuhn characterizes fruitfulness thus: "A theory should be fruitful of new research findings: it should, that is, disclose new phenomena or previously unnoted relationships among those already known."[64] In a footnote to that characterization, he adds: "The last criterion, fruitfulness, deserves more emphasis than it has yet received. A scientist choosing between two theories ordinarily knows that his decision will have a bearing on his subsequent research career. Of course he is especially attracted by a theory that promises the concrete successes for which scientists are ordinarily rewarded."[65]

So characterized, the notion applies only to theories and is relevant to situations in which a scientist or a community of scientists must choose between competing theories. Nonetheless, it is easy to see how we might extend the concept of fruitfulness so that it can apply to discoveries; and when a discovery is announced, there are often other scientists who face a decision about whether to devote their time,

---

61. See the account of this period by Jed Buchwald (1989, chap. 12). "By, at the latest, the early 1840s," Buchwald observes, "there are scarcely any physicists or mathematicians who dispute the wave theory's fundamental principles" (p. 308). Of course, it is a more delicate question whether the theory had yet been developed to the point at which it could manage to conflict with the photoelectric effect, or whether any conflict that did exist was recognized as such. I mean only to suggest that the modern understanding of the relation between the photoelectric effect and the wave theory at least introduces the possibility that the two were in conflict at the time of the discovery, a possibility which must be ruled out for the case to count as an instance of genuine unconnectability.

62. Thanks to Gonzalo Munévar for emphasizing this point during discussion at the conference.

63. Kuhn 1977, pp. 320–39.

64. Ibid., p. 322.

65. Ibid.

energy, and funding to pursuing the implications of that discovery. Thus, a discovery which is premature, perhaps especially if it is of the genuinely unconnectable variety, might equally well be called an unfruitful one. It is thus interesting to note that fruitfulness is still an underexplored criterion in the philosophy of science.[66]

FURTHER DEVELOPMENT

In closing, I have two brief suggestions as to how the idea of prematurity might be further developed. One, echoing remarks by Ilana Löwy and by Frederic Holmes, is to look at place as well as time, and the other is to look at theory as well as observation and experiment.

First, the emphasis in Stent's discussion of prematurity, as the very choice of label obviously suggests, is on bad timing: he shows particular interest in discoveries which could not be integrated into ongoing research at the time they were made, due to the nature of the contemporary body of accepted scientific belief, but which were seized upon and assimilated later, sometimes in very important ways, once theory had changed.[67] But, we might ask, might it not be the case that sometimes a discovery is badly placed rather than badly timed?

The idea here is that sometimes a discovery might simply be made in the wrong scientific community for it to be integrated into ongoing work. Of course, in one sense any premature discovery takes place in a scientific community which is wrong for it. It is possible, however, that there might be *at the time* another scientific community which *would* have been able to integrate the discovery into its ongoing research, had that community been presented with the discovery in question. Because of its unlucky place of birth, however, the discovery goes unappreciated everywhere.[68] The communities in question may be in different countries—a circumstance more likely to have proven an obstacle once upon a time than it would now, of course—or they may simply be divided by disciplinary, or even subdisciplinary, barriers.[69]

---

66. David Hull makes some points very similar to the points I make in this subsection, although he uses the term "promise" to connote an opposite of prematurity, and makes a connection to discussions in Lakatos and in Laudan, rather than Kuhn (see Hull in this volume).

67. Of course, I have argued that it is no part of the notion of prematurity itself that later integration takes place. Also, and on (I have argued) a related note, the cases Stent discusses in which judgments of prematurity are made here-and-now may or may not turn out to be of this sort sub specie aeternitatis.

68. In scientific discovery, perhaps, location is (sometimes) everything.

69. Löwy, who distinguishes and illustrates the various ways in which integration of a discovery can be delayed, seems to have this possibility in mind at more than one place in her chapter (see Löwy, chapter 20 in this volume). Holmes also draws attention to the role of subdisciplinary boundaries in his critical remarks concerning Stent's account of the Avery case (see Holmes, chapter 12 in this volume). The Avery case is, in Holmes's account, not an instance of the sort of thing I have in mind, however, for it is not a case in which a discovery goes unappreciated everywhere; and neither author seems to me to place clear emphasis on precisely the possibility I am trying to pinpoint. In any case, it cannot hurt to add another voice to the call for further work on this sort of prematurity.

Second, it is an interesting question whether there are ever premature discoveries at the level of "pure theory," as opposed to premature discoveries which take place in the process of investigating the phenomena more directly. The distinction I have in mind here, however, is not the philosopher's standard distinction between theory and observation, a distinction of mixed reputation. The discovery that DNA is the hereditary substance is not a discovery of an observable entity, process, or state of affairs; nor is the sentence "DNA is the hereditary substance" an "observation report" by anyone's lights. I am also not drawing attention to a distinction between single hypotheses (such as that DNA is the hereditary substance) and the more complex sorts of structures which tend to be called theories, for Polanyi's theory of the adsorption of gases on solids is presumably already one of those. The idea, rather, is that sometimes, perhaps especially in some branches of contemporary physics, discoveries are made not so much by observing, experimenting, and trying to think through the significance of one's (or someone else's) observations and experimental results but, rather, by playing with mathematical structures of various sorts, perhaps in a highly abstract way.[70] And if discoveries are indeed made that way on occasion, then there seems to be no obvious reason why premature discoveries might not be. In any case, this seems like an interesting avenue to explore if we wish to develop the notion of prematurity further.

BIBLIOGRAPHY

Bloor, D. 1991. *Knowledge and Social Imagery.* 2d ed. Chicago: University of Chicago Press.

Bogen, J., and J. Woodward. 1988. "Saving the Phenomena." *Philosophical Review* 97:303–52.

Buchwald, J. Z. 1989. *The Rise of the Wave Theory of Light: Optical Theory and Experiment in the Early Nineteenth Century.* Chicago: University of Chicago Press.

Doppelt, G. 1978. "Kuhn's Epistemological Relativism: An Interpretation and Defense." *Inquiry* 21:33–86.

Duhem, P. 1954. *The Aim and Structure of Physical Theory.* Trans. P. P. Wiener. Princeton: Princeton University Press. Translated from the second French edition, 1914.

Feyerabend, P. K. 1962. "Explanation, Reduction, and Empiricism." In *Minnesota Studies in the Philosophy of Science: Scientific Explanation, Space, and Time,* ed. H. Feigl and G. Maxwell, pp. 28–97. Vol. 3. Minneapolis: University of Minnesota Press. Reprinted in *Realism, Rationalism, and Scientific Method: Philosophical Papers,* ed. Feyerabend, vol. 1 (Cambridge: Cambridge University Press, 1981), pp. 44–96.

———. 1965. "On the 'Meaning' of Scientific Terms." *Journal of Philosophy* 62:266–74. Reprinted in *Realism, Rationalism, and Scientific Method: Philosophical Papers,* ed. Feyerabend, vol. 1 (Cambridge: Cambridge University Press, 1981), pp. 97–103.

Hacking, I. 1982. "Language, Truth, and Reason." In *Rationality and Relativism,* ed. M. Hollis and S. Lukes, pp. 48–66. Oxford: Basil Blackwell.

70. Indeed, it is a criticism sometimes voiced against current work in quantum gravity and in string theory that it proceeds in such a way to far too great an extent.

————. 1983. *Representing and Intervening.* Cambridge: Cambridge University Press.

Kuhn, T. S. 1970. *The Structure of Scientific Revolutions.* 2d ed. Chicago: University of Chicago Press.

————. 1977. *The Essential Tension.* Chicago: University of Chicago Press.

Lakatos, I. 1970. "Falsification and the Methodology of Scientific Research Programmes." In *Criticism and the Growth of Knowledge,* ed. I. Lakatos and A. Musgrave, pp. 91–196. Cambridge: Cambridge University Press.

Laudan, L. 1990. "Demystifying Underdetermination." In *Minnesota Studies in the Philosophy of Science,* vol. 14 *Scientific Theories,* ed. C. W. Savage, pp. 267–97. Minneapolis: University of Minnesota Press.

Newton-Smith, W. H. 1981. *The Rationality of Science.* London: Routledge and Kegan Paul.

Popper, K. 1970. "Normal Science and Its Dangers." In *Criticism and the Growth of Knowledge,* ed. I. Lakatos and A. Musgrave, pp. 51–58. Cambridge: Cambridge University Press.

Stent, G. 1972a. "Prematurity and Uniqueness in Scientific Discovery." *Advances in the Biosciences* 8:433–49.

————. 1972b. "Prematurity and Uniqueness in Scientific Discovery." *Scientific American* 227 (December): 84–93.

Woodward, J. 1989. "Data and Phenomena." *Synthese* 79:393–472.

PART SEVEN

Closing Considerations

CHAPTER TWENTY-TWO

Prematurity and Promise

Why Was Stent's Notion of Prematurity Itself So Premature?

David L. Hull

This essay is divided into three parts. Initially, I try to clarify such concepts as retrospective prematurity, here-and-now prematurity, postmaturity and so on. "Getting clearer" is not something anyone can do in advance. Until one learns what readers think one has written, one cannot decide what needs clarifying and what can be left untouched. With respect to prematurity, this process is only beginning. Prematurity was largely ignored when Gunther Stent introduced it, and thus far few people seem to think that it needs rehabilitation. Ernest Hook has attempted to change that. He hopes to make prematurity less premature by inviting many scholars to apply this notion to the areas of science they know best. The putative examples of prematurity mentioned in this volume and others proposed in the literature are a motley group. Many turn out not to be premature at all. However, making decisions about which instances are premature requires that we get much clearer about precisely what makes a discovery premature.

Although no one seems to have been all that taken with the idea of prematurity at the time, sociologists and scientists exhibited more interest than either historians or philosophers of science, so I ask why. I also introduce a concept that seems related to the preceding cluster of ideas: promise. The term "prematurity" can be usefully contrasted with "postmaturity," but I see an equally instructive contrast between these two terms and "promise." Why do some ideas engender little or no interest among scientists at the time of their introduction, while others are judged as showing great promise? Is the notion of promise any less slippery than the notion of prematurity? Thus far, it has been largely ignored. Is it destined to remain so?

PREMATURITY

As Holmes shows, both Stent's notion of prematurity and the concept of postmaturity introduced by Harriet Zuckerman and Joshua Lederberg assume a normal rate for scientific change.[1] Premature and postmature discoveries are deviations from this norm. However, as Holmes argues, this notion of a normal rate of scientific change is highly questionable. As in the case of biological evolution, rates of change differ from one lineage to the next at any one time, just as radical changes may occur in any one lineage through time. Sometimes scientific change occurs at a stately pace, sometimes quite rapidly. In periods of rapid change, premature discoveries do not stay premature for long. Hence, one should expect to find the clearest instances of prematurity in periods of relatively slow, steady change.

Löwy notes that close attention is required to each of the substantive terms in Stent's condition for prematurity. As defined by Stent, "A discovery is premature if its implications cannot be connected by a series of simple logical steps to contemporary canonical, or generally accepted, knowledge."[2] "Discovery" cannot be limited only to particulate instances in which a scientist puts the last piece in a puzzle. Discovery must also include theories, sets of observations, and experimental studies (see also Hetherington). And as Holmes argues, discovery is a process, sometimes a protracted and complex process. Oxygen was not discovered on 15 June 1775 or any other particular date. The most that can be said is that it was discovered sometime between 1774 and 1777. Any narrower date makes its discovery more particulate than it actually is.

As a result of all this complexity, determining which of the discoveries turn out to be the "same" is highly problematic. As Holmes asks, "But how can we be sure that the earlier discovery was, in its own time, the same discovery that it later appeared to be?"[3] To use a currently fashionable term, scientific knowledge is situated. Because it is situated, extracting an item from its historical context necessarily distorts it. One can understand Mendel's discoveries only if they are viewed in the context of the period in which he made them. They were "rediscovered" in a quite different day and place. Each of the rediscoverers at the turn of the century had his own agenda. So did William Bateson. Quite unconsciously, the founders of what came to be known as "genetics" read into Mendel what they needed. If we are to discover any instances of prematurity, we cannot be all that persnickety. Similar enough has to be good enough. Discoveries, when defined too precisely, become unique. If we make our standards stringent enough, one and the same discovery can never recur; on more reasonable construals, the same discovery can be made more than once.

Canonical knowledge turns out to be an even more problematic idea. In this connection, Stent quotes Polanyi to the effect that there must be "at all times a predominantly accepted scientific view of the nature of things, in the light of which

1. Zuckerman and Lederberg 1986; Holmes, chapter 12 in this volume.

2. Stent 1972b, p. 84.

3. Holmes, chapter 12 in this volume; see also Ruse, chapter 15 in this volume. For various senses in which a discovery is or is not a "true" discovery, see Zinder, chapter 5, and Hook, chapter 1, in this volume.

research is jointly conducted by members of the community of scientists."[4] Thus, what counts as canonical knowledge is relative to particular communities. Each research community has its own canon. When Stent observes that Avery's discovery was premature, it was premature at the very least with respect to the people who were part of Max Delbrück's bacterial virus group at the California Institute of Technology in 1948,[5] and possibly with respect to molecular geneticists in general. Scientists as a whole may share a canon, but if they do, it concerns only very general beliefs about reason, argument, and evidence.

Ghiselin points out that within a scientific community "not everybody agrees on the definition of canonical knowledge." If a "minority of scientists accept a discovery, or even pay serious attention to it, then the discovery is not altogether premature in the Stentian sense."[6] Contrary to a strict and uncharitable reading of Thomas Kuhn, agreement on canonical knowledge within a community need not be unanimous. Some disagreement can and does exist.[7] If so, then why do such communities of scientists seem so homogenous in their beliefs? First, such communities are usually recognized only in retrospect after lots of the internal disagreements have been eliminated and reconciled. In retrospect, canonical knowledge looks much more homogenous than it actually is. Second, members of such communities tend to play down their differences in the presence of outsiders: "We all agree on everything. Well, at least on basics." Finally, each scientist in a community thinks that his or her views capture the essence of the canonical knowledge of the community to which he or she belongs:[8] "What is the canon of my group? My canon."

The system of beliefs within a particular scientific community turns out to be more heterogeneous than one might expect. In addition, communication between these groups is more frequent and successful than a strict and uncharitable reading of Kuhn would imply. According to early Kuhn, paradigms are "incommensurable." They cannot be brought into conflict. Perhaps they cannot be brought into absolutely sharp conflict, but nevertheless, rough and crude conflicts are possible and may suffice. Scientific communities are not as closed-minded and insular as some commentators would have us believe. Contrary to a strict reading of Kuhn, cross talk does occur among disciplines and scientific communities. Even so, part of the explanation of prematurity is that members of certain communities may not be all that aware of what is going on in other communities. If you do not read the relevant journals, you just might miss out.[9]

---

4. Stent 1972a, p. 439.

5. See Holmes, chapter 12 in this volume.

6. Ghiselin, chapter 16 in this volume.

7. See Löwy's discussion of similarities in the notions of paradigm, disciplinary matrix, and thought style, chapter 20 in this volume.

8. Hull 1988. Editor's note: See also Zinder, chapter 5 in this volume.

9. See Holmes, chapter 12, and Löwy, chapter 20, in this volume.

Löwy reads the phrase "series of simple logical steps" in Stent's definition as referring to the "target community's ways of asking relevant questions, of producing experimental results, and of examining new evidence." Anyone reared as a philosopher is sure to read this phrase in a much more rigorous way, but, with respect to Stent, Löwy may well be right. However, the connections that Löwy lists are anything but simple. The net effect of all the preceding is that what appeared to be quite straightforward uses of the terms "discovery," "canonical knowledge," and "simple logical steps" turn out to be extremely complicated.

Hetherington distinguishes between a discovery being unknown and unappreciated, saying that if a discovery is unknown, then it cannot be appreciated or unappreciated, and that to be known but not appreciated is a precondition for any analysis of prematurity.[10] Hook distinguishes between discoveries that are overlooked, deemed irrelevant to current work, or outright rejected.[11] Stent distinguishes between two sorts of prematurity: retrospective prematurity and here-and-now prematurity. The contrast is between how scientists view an idea in retrospect and how they viewed it at the time. When we study the history of science, certain instances catch our eye. One is the unappreciated precursor.[12] An unappreciated precursor is a scientist who published a view that we now take to be important, but whose fellow scientists did not know about it or at best did not appreciate it. Why was it unknown or unappreciated?

Then there are all those ideas that were unappreciated when they were first published, and that have remained so to the present. Some of these ideas may be false, but many are merely unconnected. If one dips back into old journals, the vast majority of papers seem to be of this latter sort, and sociologists have reaffirmed this impression more quantitatively in studying citation patterns. The vast majority of citations are to a small percentage of papers. Many papers never get cited at the time—or later. In general, publishing a scientific paper is roughly equivalent to throwing it away.[13]

The problem is where to fit in Stent's here-and-now prematurity. Stent asks, "Does the prematurity concept pertain only to retrospective judgments made with the wisdom of hindsight?"[14] He answers, "No, I think it can be used also to judge the present. Some recent discoveries are still premature at this very time." He gives two examples: information being stored by an animal in nucleic acids or other macromolecules, and ESP. The results of the experiments that gave rise to these putative phenomena could not be connected to any canon in 1972. As far as I can tell, they cannot be so connected to this day. It seems that the gustatorial trans-

10. Hetherington, chapter 9 in this volume.

11. Hook, introductory remarks circulated to conference participants, 22 August 1997. See also Hook, chapter 1 in this volume.

12. Sandler 1979.

13. See comments by Auden 1973, p. 8; see also Stern, chapter 18 in this volume.

14. Stent 1972b, p. 87.

mission of knowledge was an artifact of the experimental design being used.[15] In part because ESP still cannot be fitted into any accepted canon, it remains questionable.

As far as I can make out, all it takes for a discovery to be an instance of here-and-now prematurity is that, at the time it is made public, it is noticed but cannot be fitted into (i.e., connected to) any canon—regardless of later estimations of its truth or significance.[16] A discovery can fail to fit into a generally accepted canon in two ways. First, it can fail to fit into a canon yet not threaten it. Philosophers of science term such putative discoveries "curiosities" (see Comfort for an example). Second, a phenomenon can fail to fit into a canon because it conflicts with it. Philosophers of science term such phenomena "apparent falsifiers."

Stent terms this second class of phenomena "unexpected" discoveries and views them as being connected to the canonical ideas of their day because they contradict one or more elements of this canon.[17] This implies that here-and-now prematurity, in Stent's view, applies only to curiosities. They are noticed, they do not contradict the relevant canons of knowledge at the time, but they do not fit in either, and scientists at the time can make these estimations.

My only criticism of this notion of prematurity is that it has such a huge extension. Instances of retrospective prematurity are relatively rare and interesting, while instances of here-and-now prematurity are commonplace and not very interesting. Elsewhere I have argued that scientists do not read the scientific literature to discover the truth but to find results that bear on their own research.[18] If a result does not bear on their work, they ignore it. If a result confirms their developing canon, they accept it, usually without testing it. Testing is reserved for Stent's unexpected discoveries, that is, apparent falsifiers.

As I mentioned previously, one problem with attempting to apply Stent's notion of prematurity is his definition in terms of "simple logical steps." According to Löwy's gloss, this relation is anything but "simple" and "logical." In the ongoing process of science, propositions are transformed in all sorts of subtle and not-so-subtle ways as the theory develops.[19] Hence, the distinction between connected and unconnected implications is not very sharp. For example, Motoo Kimura in his early publications argued that the discovery that most mutations are adaptively neutral required the rejection of Darwinian (or neo-Darwinian) theory and its replacement with his own Neutral Theory of Evolution.[20] Hence, using Stent's terminology, the discovery that most mutations are neutral is "unexpected." However, other evolutionary biologists thought otherwise. They simply integrated Kimura's views into their own theory and kept the name the same.

15. Collins and Pinch 1993.
16. See Hook, chapter 1 in this volume.
17. Stent 1972a, p. 435.
18. Hull 1988.
19. See Holmes, chapter 12 in this volume.
20. Kimura 1983.

Because Stent's notion of "simple logical steps" is so complicated, I prefer his more implicit operational (or sociological) criterion of what scientists are able to do with a discovery. The fault that geneticists found with Avery's discovery was that they "did not seem to be able to do much with it or build on it."[21] One can recognize that scientists are using a particular discovery and building on it even if one is unable to decide whether a discovery can be connected by a series of simple logical steps to a particular canon.

STENT'S PREMATURE NOTION

Self-reference or reflexivity, as it is often called, is usually employed as a club to annihilate one's opponents. Certainly that was how it was used in the early days of what has come to be known as "Science Wars."[22] How can advocates of the Sociology of Scientific Knowledge spend so much time gathering evidence to show how irrelevant evidence actually is in the decisions that people make? However, I think that reflexivity can play a more positive role in the study of science. Humbug and self-deception are relatively easy to see in others, much harder to uncover in oneself. Reflexivity is one way to puncture one's own pretensions and delusions. In general, scientists are unduly committed to their own ideas. OK, how about me? Am I unduly committed to my own ideas or am I an exception to this general principle? In this essay I utilize reflexivity but in a positive fashion.

When Stent set out his notion of prematurity in 1972, it was largely ignored. Why? Because it was premature! Others were unable to connect it to canonical knowledge. More operationally, others working on the nature of science were not able to do much with it, to build on it. The trouble is that no single canon of knowledge relevant to prematurity existed at the time. Several different groups of scholars were intent on explaining science—historians, philosophers, sociologists, and scientists themselves. These four groups of workers had very different responses to Stent's notion of prematurity.

Why was Stent's notion of prematurity ignored by students of science in 1972 and after? The quick answer is that the vast majority of ideas were, and are, ignored. If 99 percent of publications are ignored, no one should be surprised if a particular paper is ignored. Being noticed is what demands an explanation.

Hook suggests that Stent's paper was ignored because it was published in *Scientific American,* which students of science at the time did not read. Perhaps Stent would have been wiser to publish in such journals as *Isis* and *Philosophy of Science,* assuming that they would publish such a paper (see below). But of all the journals published at the time, *Scientific American,* which caters to a wide audience, had to be reasonably high on the list of journals read by students of science.

21. Stent 1972b, p. 84.
22. See Laudan 1981, 1982.

Right from the start, Stent anticipated the hostility of historians of science because prematurity sounds Whiggish, and Whiggism at the time was the bête noire of history. Historians must write histories in the context of the time, in its own terms, not evaluate earlier periods in terms of what happened later. One individual who disagreed with Hook's contention that the notion of prematurity is useful and has heuristic value, comments: "The concept seems heuristic only with hindsight (because you have to know what came later to label something premature with respect to it). But hindsight is exactly what historians label 'Whiggish.' Although we might, from our own points of view, be interested in our precursors, these points of view cannot be used to interpret how the past unfolded (on pain of appeal to pernicious teleology). The same is true of evolutionary explanations."[23]

Retrospective prematurity certainly looks Whiggish. In reading an old journal, one happens across a discovery that was ignored or unappreciated in its own day, but which looks both important and right to us today. The strong tendency is then to elevate the author of this unappreciated achievement to the status of an unappreciated precursor.[24]

I think historians are right to be highly suspicious of precursoritis. In most cases, such claims are simply false. The putative precursor was not presenting anything like the later ideas that had such success. For example, as Comfort argues so cogently, McClintock was not unappreciated or a precursor: McClintock's evidence for transposition was accepted immediately. Her interpretation of its wider significance, however, was and remains doubtful in most scientists' minds. The prematurity argument about McClintock founders on two counts: she was neither ignored at the time, nor proven right later.

But, and of equal importance, truly unappreciated precursors do not count. They did not have much of an effect on the course of science in their day, and it is too late now. Mendel's is a case in point. It is unclear whether Mendel understood his own work in the way that later commentators claim, but assuming that Mendel was a genuine precursor to Bateson and later geneticists, the only use that these workers made of Mendel's paper was to deflect a priority dispute among its rediscoverers. If an obscure Moravian monk actually discovered the basic laws governing hereditary, then there is no point in the rediscoverers engaging in an unseemly priority dispute.

I suspect that Stent introduced the notion of here-and-now prematurity in an attempt to head off the charge of Whiggism. For me, his maneuver does not work, but then I do not find Whiggism (or presentism, as it is sometimes called) the unalloyed evil that so many historians do.[25] In any case, whether justified or not, one reason why historians of science did not take to Stent's category of prematurity is

23. Cited by Hook, chapter 1 in this volume.
24. See Holmes, chapter 12, and Löwy, chapter 20, in this volume.
25. See Hull 1979.

that retrospective prematurity appeared too Whiggish for their tastes, while here-and-now prematurity seemed too broad a category to be of much use.[26]

Among most philosophers of science in 1972, the reigning influence was still some version of logical empiricism. Even though the literature of the day was filled with criticisms of logical empiricism, the hope remained for many philosophers of science that this way of doing philosophy of science could be salvaged.[27] Stent's reference to "logical steps" is certainly compatible with logical empiricism because of the heavy emphasis on inference in this view of science. What is scientific method? It is the formulation of hypotheses and the testing of observation statements derivable from these hypotheses. What is theory reduction? The derivation of a higher level theory from a lower level theory. What is explanation? The derivation of the explanandum from the explanans.

I agree with Gonzalo Munévar that taking the phrase "simple logical steps" at face value is a mistake. Stent probably meant something much closer to Löwy's gloss.[28] But any philosopher of science reading Stent's paper at the time would have taken this phrase at its face value and responded positively as a result of this misreading. If inference played such a big role in logical empiricism, and philosophers at the time misread Stent's reference to "simple logical steps" to mean simple logical steps as they understood this phrase, then why did philosophers of science ignore Stent's notion of prematurity? It fit into our canon and, as Carpenter observes, we are all happier to accept an idea that fits into our own canon than one that does not.[29]

Several reasons can be given for the failure of logical-empiricist philosophers of science to incorporate prematurity into their own canon. In fact, this failure is overdetermined. First, Stent took his paper to be a contribution to scientific discovery, when one of the cornerstones of logical empiricism is that the subject matter of philosophy of science is justification, not discovery.[30] Second, although here-and-now prematurity has no temporal dimension, retrospective prematurity does. In cases of retrospective prematurity, scientists at the time judge a contribution unconnectable. Later scientists conclude that the contribution was not only connectable but important.

However, the sorts of inference that logical-empiricist philosophers use to analyze science are atemporal. When an event occurred, relative to other events, is ir-

---

26. On Mendel as a precursor, and on the role of credit for contributions in atomizing scientific discoveries, see Holmes, chapter 12 in this volume.

27. This claim about the status of the logical empiricist analysis of science can be tested by reading the papers published at the time in the premier journal of the field—*Philosophy of Science*. Most authors who published in this journal merely puzzled over this or that aspect of logical empiricism, some criticized it, but no one provided anything like an alternative to logical empiricism.

28. Gonzalo Munévar's made these comments at the symposium "Prematurity and Scientific Discovery," held at the University of California, Berkeley, 2–4 December 1997. See also Munévar, chapter 23, and Löwy, chapter 20, in this volume.

29. See Carpenter, chapter 7 in this volume.

30. Brannigan 1981.

relevant. For example, a common distinction between explanation and prediction of particular events is that in explanation the event has already occurred and in prediction it has yet to occur. Logical empiricists respond that no matter, explanation and prediction are symmetrical. They exhibit precisely the same logical form, temporal differences be hanged.

Logical empiricists are also not interested in the psychology of scientists. A recurrent dispute in the philosophy of science is whether added weight should be given to the derivation of as-yet-unknown phenomena in the confirmation of an hypothesis. If you already know that a particular sort of phenomenon occurs, then you can build it into your hypothesis. However, if a particular sort of phenomenon is as yet unknown to you, then you cannot. Hence, the derivation of as-yet-unknown phenomena should be given greater weight than the derivation of known phenomena. However, this difference is purely psychological. Inference is inference, regardless of when the phenomenon referred to actually occurs.

If one turns from logical steps to Stent's sociological criterion, the distance between prematurity and logical empiricism only increases. Logical empiricists are not interested in psychology. They are even less interested in sociology—or "mob-psychology," as one unsympathetic critic termed it.[31] However, scientists deciding which contributions they can use in their own research hardly seems to manifest mob psychology. In any case, the view of science produced by logical-empiricist philosophers of science is not all that concerned with the behavior of scientists but instead is interested in the logical relations between propositions. As a result of all the preceding, the failure of philosophers of science to take notice of Stent's prematurity is understandable.[32]

As Stent notes, of those scholars who study science itself, Mertonian sociologists took the greatest interest in prematurity.[33] The notion of prematurity fits nicely into the Mertonian panoply of classificatory terms such as "universalism," "communalism," "disinterestedness," "organized skepticism," "humility," "originality," and "priority." Prematurity is one form of discontinuity in science. Postmaturity is another. According to Zuckerman and Lederberg, premature discoveries are those that were made but neglected, whereas postmature discoveries were not made but could have been.[34] For a discovery to qualify as postmature, it must have three attributes: it must have been technically achievable prior to the time at which it was

31. Lakatos 1970, p. 140.

32. Although Stent's term "prematurity" has not played a role in the philosophical literature, classifications similar to his were published both before and after the appearance of his paper in 1972. For example, Lakatos (1970, p. 116) distinguishes between progressive and degenerating problem shifts, while Laudan (1977, p. 17) distinguishes between conceptual and empirical problems. Laudan divides the latter sort of problems into those that are unsolved, solved, and anomalous. In the case of Lakatos and Laudan, at least, their metalevel concepts have proved not to be premature. They have had an impact on the relevant canon.

33. Stent 1972a, p. 448. Editor's note: See also Stern, chapter 18 in this volume, written subsequently.

34. Zuckerman and Lederberg 1986, p. 629.

actually achieved, it must have been comprehensible to working scientists at the time, and scientists at the time must have been able to appreciate the implications of the discovery. The example that Zuckerman and Lederberg give is the discovery of sex in bacteria. The techniques for discovering sexual recombination in bacteria were available in 1908, and if anyone had run the relevant experiments, geneticists, and possibly even bacteriologists, would have appreciated the results.[35]

However, in an admittedly haphazard and sketchy search of the Mertonian literature, I did not come up with many references to Stent or his notion of prematurity. As Zuckerman's advocacy indicates, prematurity could be connected to Mertonian sociology of science, but apparently other Mertonian sociologists did not join in her enthusiasm. They did not see how they could do anything with prematurity. Thus, the Mertonians present a problem case. According to Stent, those authors who ignore or reject a new discovery are right to do so if they cannot connect it to the canon of the day. But at least one Mertonian did pay attention and attempted to integrate prematurity into Mertonian sociology of science, while others did not. What happened?

Degeneration of research programs is the likely cause. Stent came along with a concept that fitted nicely into Mertonian sociology of science right when that school was about to be eclipsed by advocates of the Sociology of Scientific Knowledge.[36] Stent hitched his wagon to a falling star.

The final category of workers who might have used Stent's notion of prematurity is scientists themselves. Zuckerman is a scientist—a sociologist, not a geneticist or a bacteriologist. In the Zuckerman-Lederberg study mentioned earlier, Zuckerman played the role of a "sociologist-observer." Lederberg was one of the founders of what came to be known as microbial genetics, but in this paper he was playing the role of a "scientist-participant." As such, both belonged to a larger group that is most neutrally described as "students of science." They are studying science by whatever means available. More recently, Philip Tobias, a paleoanthropologist, has taken up the cause of prematurity.[37] Although in a 1996 paper he discusses ten examples of prematurity, he concentrates on one in his own field—anthropology. However, in this paper Tobias is functioning not as an anthropologist but as a student of science.

I do not mean to define out of existence the category of scientists who found Stent's notion of prematurity useful. But when they make such evaluations, they are functioning not simply as scientists but also as students of science. If a reasonably large number of scientists were led to step out of their usual roles and take on a second in connection with prematurity, then I would conclude that these scientists found Stent's idea promising. Thus far, however, they have not.

35. See also Hook, chapter 1 in this volume.
36. Hull 1993.
37. Tobias 1996.

PROMISE

To such categories as prematurity and postmaturity, I add a third: promise. Most scientific publications engender little or no response. Other scientists cannot see how the data, hypotheses, or theories presented in a paper can help them in their own research. Hence, these ideas lie buried in the literature, rarely bursting forth later as an important contribution to science. But the opposite phenomenon also occurs. Sometimes a scientist publishes a paper, and everyone jumps on it, launching yet another scientific bandwagon. Stent discusses James Watson and Francis Crick's early papers on DNA in the context of uniqueness. If Watson and Crick had not published, how long would it have taken for someone else to do so? (Historians do not like questions like this one either.) I would like to use Watson and Crick's early publications as an example of promise. When many of Watson and Crick's contemporaries read these papers, they concluded that everything fit together so elegantly that this model of the structure of DNA just had to be right. But more than being right, this model exhibited promise. If it were correct, hundreds of papers could be published, and even a few Nobel Prizes lay in wait down the road. Watson and Crick's contemporaries were able to anticipate the use that they could make of this model. They could build on it.[38]

The notion of promise is not new in the literature on science, but it has proven to be almost as premature as Stent's notion of prematurity. Nevertheless, I find it to be of extreme importance. Why has not the idea of promise seemed more promising to other students of science? Historians can have no objections to it because it is not in the least Whiggish. It is not retrospective but prospective. Certainly, promise does not fit all that well into the old-fashioned logical-empiricist literature, but alternatives to logical empiricism have been generated in which such temporal notions fit quite naturally. For example, Imre Lakatos defines his notion of progressive and degenerating problem shifts in terms of a series of theories related by descent, not single, static theories.[39] Larry Laudan also views science as a temporal process.[40] In this book, Laudan introduces a distinction between the context of acceptance and the context of pursuit.[41] In the context of acceptance, Laudan advises scientists to "choose the theory (or research tradition) with the highest problem-solving adequacy" (italics omitted). However, in the context of pursuit, Laudan argues that it is "always rational to pursue any research tradition which has a

---

38. Crick (1988, p. 73–74) sees the introduction of the Watson-Crick model of DNA quite differently: "It took over twenty-five years for our model of DNA to go from being only rather plausible, to being very plausible." Our differences may lie in the distinction between the context of acceptance and the concept of pursuit. Scientists who did not fully accept the Watson-Crick model were willing, nevertheless, to risk their careers on it—not a small commitment.

39. Lakatos 1970, p. 119. See also Hull 1988.

40. Laudan 1977.

41. Ibid., p. 109.

higher rate of progress than its rivals."[42] Furthermore, "scientists can have good reasons for working on theories that they would not accept" (italics omitted).

Quite obviously, Laudan is talking about promise, but both of the preceding references appear in the 1970s. What has happened since? Once again, on the basis of an admittedly sketchy examination of the literature, I am forced to conclude that nothing like promise has played all that large of a role in the study of science. What is wrong? Are students of science unable to connect promise to their canon? Are they unable to find anything to build on in the notion of promise? Even so, I remain stubbornly wedded to this notion. With respect to prematurity, postmaturity, and promise, I take promise to be the most promising research topic. But for now, it remains another example of prematurity.

CONCLUSION

When Stent introduced the notion of prematurity in 1972, he received numerous letters on the subject, but it was ignored by nearly all students of science. To historians of science, it seemed Whiggish. To philosophers of science, it seemed to go beyond inference to make reference to psychological and sociological factors. Mertonian sociologists of science found it relevant to their way of studying science, but Mertonian sociology of science was already being eclipsed by relativist advocates of the Sociology of Scientific Knowledge. Many scientists may well have found the notion of prematurity interesting and important, but as scientists they could not incorporate this notion into their own research. Prematurity is not something a scientist working on phages could incorporate into a research program. It is a metascientific concept. We will have to wait to see how successful Hook turns out to be in encouraging students of science to build on this idea.

BIBLIOGRAPHY

Auden, W. H. 1973. "Letters." *Scientific American* 228 (January): 8.

Brannigan, A. 1981. *The Social Basis of Scientific Discoveries.* Cambridge: Cambridge University Press.

Collins, H., and T. Pinch. 1993. *The Golem: What Everyone Should Know about Science.* Cambridge: Cambridge University Press.

Crick, F. 1988. *What Mad Pursuit: A Personal View of Scientific Discovery.* New York: Basic Books.

Hull, D. L. 1979. "In Defense of Presentism." *History and Theory* 18:1–15.

———. 1988. *Science as a Process: An Evolutionary Account of the Social and Conceptual Development of Science.* Chicago: University of Chicago Press.

———. 1993. Review of *Making Science: Between Nature and Society,* by Stephen Cole. *American Journal of Sociology* 99:839–40.

Kimura, M. 1983. *The Neutral Theory of Molecular Evolution.* Cambridge: Cambridge University Press.

42. Ibid., pp. 111, 110.

Lakatos, I. 1970. "Falsification and the Methodology of Scientific Research Programmes." In *Criticism and the Growth of Knowledge,* ed. I. Lakatos and A. Musgrave, pp. 91–195. Cambridge: Cambridge University Press.

Laudan, L. 1977. *Progress and Its Problems: Towards a Theory of Scientific Growth.* Berkeley and Los Angeles: University of California Press.

———. 1981. "The Pseudo-Science of Science?" *Philosophy of the Social Sciences* 11:173–96.

———. 1982. "A Note on Collins's Blend of Relativism and Empiricism." *Social Studies of Science* 12:131–32.

Sandler, I. 1979. "Some Reflections on the Protean Nature of the Scientific Precursor." *History of Science* 17:170–90.

Stent, G. 1972a. "Prematurity and Uniqueness in Scientific Discovery." *Advances in the Biosciences* 8:433–49.

———. 1972b. "Prematurity and Uniqueness in Scientific Discovery." *Scientific American* 227 (December): 84–93.

Tobias, P. V. 1996. "Premature Discoveries in Science, with Special Reference to *Australopithecus* and *Homo habilus.*" *Proceedings of the American Philosophical Society* 140:49–64.

Zuckerman, H., and J. Lederberg. 1986. "Postmature Scientific Discovery?" *Nature* 324:629–31.

CHAPTER TWENTY-THREE

Reflections on Hull's Remarks

Gonzalo Munévar

I comment here on two aspects of David Hull's main themes: Gunther Stent's notion of prematurity, and the notion of promise.

PREMATURITY

Hull's first order of business is to try to clarify Stent's notion of prematurity. He does so by listing a series of concerns about the Stentian idea that sometimes a scientific discovery is not accepted at the time because "its implications cannot be connected by a series of simple logical steps to contemporary canonical, or generally accepted, knowledge."[1] By "logical steps" Stent did not mean formal logical inferences, as analytic philosophers may expect, but, as he explains, a "chain of reasonable inferences." These inferences should be taken, furthermore, to be reasonable to the practitioners in the field. As I read Stent, his insight fits well with a similar one made by Hull: scientists tend to accept results that bear on their own research and ignore others. Surely, if a putative discovery cannot be connected to a field's canons of research, it is likely to go unappreciated and other scientists will not pursue it.

Most of Hull's concerns have to do with the complexities we encounter once we try to apply the notion of prematurity. One such complexity is that the notion of canonical knowledge is problematic. The problem arises not from lack of unanimity but rather from our expectation that canonical knowledge should be determined by necessary and sufficient conditions or in some other essentialist manner. But here as in so many other instances, it pays to take a populational approach, as Hull and I often advocate: Since we are dealing with human beings, of course there is going to be some variation in individuals' beliefs about exactly what the canon is.

---

1. Stent 1972.

But ideas about the canon of research (in a well-formed discipline at any rate) will cluster around some central requirements. And even if the versions of the canon in any given discipline are far more loosely distributed than the conjecture suggests, if the recognized practitioners see no way of connecting the new discovery to their own version of the canon—whatever it is—they are likely to ignore said "discovery." Expanding the canon, therefore, does not affect the point that work which cannot be connected to it will go unappreciated and thus be premature.

One of Stent's main examples is Gregor Mendel's discovery of the gene in 1865. Hull brings up Ilana Löwy's and Frederic Holmes's remarks on the complexity of discoveries to question whether what Mendel really discovered in his own day was the same as what the "rediscoverers" of Mendelian genetics attributed to him some 35 years later. Hull decides to cut prematurity some slack: "Similar enough has to be good enough." It seems to me that we can go a little further. The context in which any discovery is interpreted is bound to change, and I conjecture that the more important the discovery, the greater the change. The reason for this is that important discoveries affect profoundly the way science is done, and new ways of doing science transform the significance that we attribute to a variety of important results. In the new science, some of the old problems and concerns of the initial discoverers are left behind. Think, for example, of how quaint and arcane Johannes Kepler's astronomical investigations using ideal solids seem today, even though we accept his three laws. Thus the discoveries that produce the new science are reinterpreted (at least as to what is important and what is not) in light of the new science. This reinterpretation is bound to be much more radical in the case of premature discoveries, for there are likely to be even greater differences between the aims, procedures, and style of the initial investigator and those of his or her future admirers. If Mendel's investigations were to be connected to the canon of early-twentieth-century experimental biology, *of course* they had to be interpreted in the light of such canons, and *of course* such interpretations would have seemed rather foreign to him. But then, as Hull indicates, similar enough is good enough. We thus have, in Mendel's work, another case of a discovery unappreciated in its day, and unappreciated because there was no canon to which it could be fruitfully connected. This is clearly a case of prematurity.

Hull worries a good deal about what he calls "Stent's 'here-and-now' prematurity." He criticizes that notion because it presumably has a "huge extension." As he says, "Instances of retrospective prematurity are relatively rare and interesting, while instances of here-and-now prematurity are commonplace and not very interesting." His reason for this claim appears to be that the vast majority of scientific papers are unappreciated when they are first published and they remain so, presumably forever. As he points out, most papers are never cited at the time of publication (or later). Hull believes that the ideas in these papers are not false, just unconnected to the canon. And since Stent argues that a discovery that is unconnected to the canonical knowledge today can be said to be premature "at this very time," Hull concludes that most work published today is premature at this very time (that is, it is here-and-now premature). But I think that Hull is wrong on this point. Most scientific work published today is not unconnected to the canon. On the contrary, it fits into the canon all too well: most papers published today are pedestrian technical applications of the present scientific knowledge. The same point could be made about most of the work done in times past, and I am sure the same will be said about work done in the future. Much of it is ignored not because it is unconnected but because it is considered trivial or mediocre.

What Stent intended, I believe, was to use his insight about prematurity to explain the rather perplexing attitude of many scientists towards certain unusual "discoveries." For example, if the claims about ESP and macromolecular memory were true, they would have astonishing consequences. Why then do bright scientists fail to consider these questions worth their time? The reason is that they have no way to connect them to their canonical knowledge: there is nothing they can do with such claims given the tools and procedures at their disposal. If *some day* such claims turn out to be indeed true, then we will be able to say that they were premature.

PROMISE

I find Hull's interest in the promise of scientific ideas quite congenial. He has a precursor in Thomas Kuhn, who said that "the success of a paradigm . . . is at the start largely a promise of success. . . . Normal science consists in the actualization of that promise."[2] Now, why are some scientific investigations found promising and others not? I offer the following suggestions, all connected in some way with Stent's notion of prematurity.

There are two main ways in which we can understand scientific promise. The first deals with the realization that an alleged discovery can be connected to canonical knowledge. The second deals with the realization that an alleged discovery makes an inviting case for the acceptance of revised, or even of new, canonical knowledge.

When an obstacle blocks the way to a connection between the discovery and canonical knowledge, as the understanding of DNA prior to 1950 blocked work by geneticists on O. T. Avery's results, then we see no promise in the discovery. Conversely, a discovery is seen to have much promise when such connections come readily to the minds of several practitioners in the field. Sometimes, however, grasping the promise of a particular development requires a great deal of perceptiveness, imagination, or even genius. Take, for example, the invention of the laser by C. H. Townes.[3] Developments in the theory of light might have suggested the possibility of lasers, but it took someone with a strong background not only in science but also in engineering to realize that possibility, let alone make it a reality. He saw promise where others were completely impervious to the possibility.

2. Kuhn 1970, pp. 23–24.
3. See Townes, chapter 4 in this volume.

Then there are cases in which a scientist realizes that there *should* be a connection between a particular development and canonical knowledge, even though none comes readily to mind. Promise lies, again, in such a realization. A good example is the discovery of x rays by Wilhelm Röntgen. One evening Röntgen noticed a strange glow in his laboratory, a glow that he traced to his cathode-ray tube. This was actually at odds with canonical knowledge, for such a glow signified energy, and Röntgen's balanced equations did not have room for this unaccounted-for energy. Nevertheless, Röntgen thought, physics (his canon) should account for it. Three months later it was clear that he had found a new form of electromagnetic radiation. In the end Röntgen extended canonical knowledge in order to assimilate his observations of the glow.

And finally we have the second type of promise, in which we see promise in an idea only when we also see that the idea will usher in a new way of doing science, a new canon. A good example is Antoine-Laurent Lavoisier's discovery of oxygen, which did usher in a new scientific canon. I think this result justifies Norton Zinder's claim that truly important discoveries cannot be connected to the canon. The reason is that, if they are truly important ("revolutionary"), they are likely to bring about a largely new way of doing science. In any event, if the idea, or investigation, cannot be connected to the scientific canon of the day, it can still be seen as promising if it suggests how it can fit into an enticing new canon.

This result in turn suggests that Stent's notion of prematurity needs to be modified. A discovery is premature if it is not appreciated in its day and if it cannot be connected to the prevailing scientific canon or provide a new scientific canon. Here I have been influenced by some Kuhnian ideas (on the assimilation of x rays by Röntgen and on revolutionary changes), as well as by Stent's notion of prematurity, in my investigation of the notion of scientific promise.

BIBLIOGRAPHY

Kuhn, T. S. 1970. *The Structure of Scientific Revolutions.* 2d ed. Chicago: University of Chicago Press.
Stent, G. 1972. "Prematurity and Uniqueness in Scientific Discovery." *Scientific American* 227 (December): 84–93.

CHAPTER TWENTY-FOUR

Comments

Gunther S. Stent

I thank the authors of the preceding essays for the effort they expended on the re-examination of an essay I published 30 years ago that addressed, in part, prematurity in scientific discovery. In midcareer and less experienced at that time, I thought that prematurity was a fairly obvious and straightforward historical concept. I shared this evidently mistaken idea with Martin Jones, who expresses the opinion in his contribution to this volume that "all this is very familiar territory to anyone in a science studies discipline."

In writing my essay I had overlooked some very cogent examples of premature discovery brought forward by some of the contributors to this volume. They include continental drift, global warming, catastrophic mass extinction (which, as it turns out, would not have been premature had it been discovered a century earlier), and finally (what is probably the most consequence-laden case of the prematurity of a discovery in world history), nuclear fission.

SOME GENERAL PRINCIPLES

My essay was meant to address the history and sociology of science, and not (or hardly) its philosophy. Thomas Kuhn's *Structure of Scientific Revolutions* had appeared only recently, and I had not yet fully appreciated what I would later come to realize had been one of Kuhn's main messages (which he, in turn, had derived from the Polish immunologist Ludwik Fleck's *Genesis and Development of a Scientific Fact*). I refer to the insight that the history, the sociology, and the philosophy of science are actually a single, indivisible subject. This lesson is implicit in the analysis of the acceptance of Darwin's theory of natural selection presented in Michael Ruse's chapter. Ruse first sets forth two of what he perceives as the main reasons for Darwin's success—the convincing scientific argumentation and the ideological context in which the theory was introduced. And then he continues: "The third reason Darwin succeeded with natural selection is more sociological. . . . [It is Robert Merton's] 'Matthew Principle': major scientists get lots of credit for the work they have done—more so than minor scientists would for the same work."

So I was less concerned with the metaphysical truth of discoveries than with the psychology of their acceptance as a truth by (in Fleck's parlance) "thought collectives." (The affinity of my concept of prematurity to the ideas published by Fleck in the 1930s—of which I became aware only after I wrote my paper—is treated in Ilana Löwy's excellent contribution to this volume.) Thus in the case of extrasensory perception (ESP) as an example of prematurity in the here and now, I referred to ESP as a discovery because the claims for its existence were supported by more empirical statistical data than the results of most other psychological experiments. Yet even if ESP had been (metaphysically) true, its truth could not have been accepted by the neurobiological thought collective because there was no theoretical explanation for it: that is, its members could not have connected ESP with the collective's canonical knowledge.

In his essay in this volume, Michael Ghiselin finds that "Stent . . . tried to make a categorical distinction between that which is premature and that which is not. . . . Stent's model suffers from a lack of realism, partly because of its typological and nonquantitative formulation." It is not the case that I tried to make a "categorical" distinction between that which is premature and that which is not, in the Aristotelian sense of an absolute difference or unbridgeable, either-or conceptual gap between the members of two sets.

Aware of Ludwig Wittgenstein's argument that natural objects can be classified only in terms of "fuzzy sets" (as my Berkeley colleague Lotfi Zadeh called them), I would never have tried to make an Aristotelian categorical distinction between what is and is not premature.

I am surprised that Ghiselin, of all people, would misinterpret my classificatory scheme. Did he not himself reject the Aristotelian categorical concept of biological species and refer to species as "individuals"? I readily agree that some discoveries may be more premature than others may, but the introduction of a metric of prematurity suggested by Ghiselin strikes me as—may I say it?—premature.

WHAT IS A DISCOVERY?

In ordinary speech, as well as in philosophical discourse, the word "discovery" refers to the act of making known something that was previously unknown. And this is also the meaning I had in mind when I wrote my essay. The term implies tacitly that the novel knowledge made known by the discovery is true. And if it should turn out at some later time that it is not true, the appellation "discovery" is withdrawn. There can be no false discoveries.

At the outset of their essays both Zinder and Hetherington trivialize the concept of prematurity. Hetherington has it that "all great discoveries *virtually by definition* may suffer initial underappreciation, because such discoveries, as a condition of subsequently recognized greatness, must counter prevailing belief and, ultimately, change the scientific canon" (emphasis added). And Zinder declares that "all true discoveries are premature; all other 'discoveries' are at best just clever, logical extrapolations, although occasionally they also entail brilliant technical innovation." (Zinder's choice of the predicate "true" is not felicitous, since, bearing in mind that the word "discovery" implies truth, his phrase "true discovery" is a pleonasm. So let us give him the benefit of doubt by supposing that, like Hetherington, he really means "great" rather than "true.") Thus both Hetherington and Zinder claim in essence that the statement "This great discovery was premature" is what Immanuel Kant classified as an analytic proposition, one whose truth is entailed by the meaning of its words (e.g., "no bachelor is married"). But their claim is falsified by what Kant referred to as a synthetic proposition (e.g., "no bachelor is happy"): namely, by the empirically true proposition that not all discoveries to which the predicate "great" is applied do suffer an initial underappreciation. For instance, James Watson and Francis Crick's discovery in 1953 of the DNA double helix, which is widely regarded as one of the greatest discoveries of the twentieth century, suffered no initial underappreciation. It was generally appreciated many months before the crystallographic data supporting it had even been published and years before experiments were carried out that would prove that Watson and Crick's ingenious proposal of the mechanism of self-replication suggested by it was correct. The reason the molecular-biological thought collective immediately accepted the DNA double helix on mere hearsay (and relegated its few doubters to the ranks of cranks) was that the DNA double helix was so perfectly in harmony with the molecular-biological canon.

Hetherington seems to believe that theories do not belong to the set of things that can be discovered, as implied by his remark that "Stent applies his concept of prematurity only to discoveries, not to theories." That theories, contrary to Hetherington's opinion, are among the things that can be discovered is set forth in Ruse's contribution. According to the examples from organic evolution chosen by Ruse, there are two kinds of knowledge that can be discovered, namely, facticity (e.g., evolution did occur) and causal theory (e.g., both genetic drift and natural selection brought about the origin of species). (Ruse's third kind, namely, phylogeny, is logically included in facticity and thus lower in the epistemic hierarchy than the other two.)

In any case, Hetherington's assertion that I did not apply my concept of prematurity to theories is counterfactual. The excellent essay by Mary Jo Nye in this volume disagrees with me about Polanyi's adsorption theory, one of my three main examples of a premature discovery. Nye and I agree, of course, that Polanyi's theory *was* a discovery, and she disagrees with Polanyi and myself only in regard to the factual question of whether his theory was or was not appreciated without undue delay.

DELAY IN APPRECIATION: A NECESSARY
BUT NOT SUFFICIENT CRITERION FOR PREMATURITY

The first question that has to be addressed in deciding whether any given discovery was premature is whether there was a significant delay in its appreciation by the particular scientific community for which it would eventually be of the greatest importance. For if there was no such delay, then the question of prematurity is moot.

As I stated in my essay, by "lack of appreciation" of a discovery I did not mean that it "went unnoticed, or even that it was not considered important." Referring to my paradigmatic case of Avery, I said that what I did mean by a lack of appreciation of his discovery was that no one seemed to be able to do much with it, or build upon it, except for the students of the transformation phenomenon per se. That is to say, I maintained that, for many years, Avery's discovery had virtually no effect on general genetic discourse.

Most of the criticism of my categorization of Avery's discovery in 1944 as premature did, in fact, allege that there was no delay in its appreciation. Some of my critics pointed out that the failure of Max Delbrück and his American Phage Group (which I joined in 1948) to catch on to the importance of DNA for many years does not prove that the less narrow-minded folks who really counted did not appreciate it right away. Norton Zinder's essay provides a typical example of this argument in his account of a meeting held at Rockefeller University in 1994 to celebrate the fiftieth anniversary of the publication of the report of the Avery experiment. He reports that several of the early pioneers of proto–molecular biology present at this celebration provided anecdotal evidence of how their work had been influenced immediately by Avery's discovery. Yet, at least as told by Zinder, none of them mentioned that it led them to do any experiments prior to the early 1950s (the date of the publication of the Hershey-Chase experiment) that involved the concept of DNA as the carrier of genetic information.

That there really was a delay in appreciation of what eventually became the main conceptual lesson to be drawn from the Avery experiment is shown by its virtual lack of citation in the literature of general genetics in the decade after 1944, including absence of its mention in even highly speculative essays addressing the problem of the nature of the gene. On this crucial point Holmes was misled by Olby, whom he quotes as having made the statement (easily proven as counterfactual by an examination of the literature) that the "significance [of Avery's discovery] was quickly grasped by such leading figures as Theodosius Dobzhansky, Herman Muller, Sir Henry Dale, and Macfarlane Burnet." In their papers cited by Olby, only Burnet (in my opinion, the most intelligent and creative of the four "leading figures") said that DNA might be acting as a gene. Dobzhansky and Dale interpreted the action of DNA as mutagenesis rather than intercellular transfer of genetic information, while Muller, who did not believe in the first place that Avery's DNA extract was protein-free, suggested that the transforming principle consisted of bacterial chromosomal nucleoproteins that engaged in ordinary crossing over with the chromosomes of the recipient bacterium.

DOES THE PREMATURITY CONCEPT ASSUME A NORMAL RATE FOR THE PROGRESS OF SCIENCE?

Nye believes that it is "inaccurate" for me to describe Polanyi's work on the adsorption theory as premature, because Polanyi lost the debate only "in the short run." Evidently, the decade or more of lack of appreciation was long enough for Polanyi to write a paper about it, in which he said that, in view of the need for orthodoxy in science, this delay was justified. Nye's implication that only appreciations delayed "in the long run" qualify for prematurity would render my designation of the Avery case as paradigmatic even more inaccurate than my designation of the Polanyi case.

In view of her "short-" and "long-" range distinction, Nye can be justly criticized for "presuming"—as Frederic Holmes puts it in his contribution—that there is "some 'normal' rate of scientific advance and a normal lapse of time between the first presentation of a discovery and its assimilation into a field." Precisely because her argument is based on that false presumption, I do not accept her statement that Polanyi lost the debate only "in the short run" as a valid justification of her claim that his case was not really an instance of prematurity. For as Holmes justly states, "Whether the acceptance of a discovery appears to be rapid or slow, accelerated or delayed . . . depends not only on the interval measured in months, years, or scientific generations but also on the subjective perspectives of those involved and of those who interpret such events historically."

From this true proposition of Holmes's, it does not follow at all, however, that the notion of a delay in appreciation of a discovery presumes that there is "some 'normal' rate of scientific advance." (In his chapter David Hull accepts this claim as having been "shown" by Holmes.) On the contrary, Holmes's proposition shows that exactly the opposite is the case. There is no such presumption, and the judgment of delayed appreciation is obviously understood as a subjective call based on a comparison of the differential time lags in the appreciation of various discoveries within in the scientist's personal experience or historical ken.

When I made the entirely commonplace, not to say platitudinous, assertion that there was a long delay in appreciation of Avery's experiment as having proved that DNA is the carrier of genetic information, my use of the predicate "long" did not refer to some absolute lag in sidereal time on a clock driven by the metric of a universal rate of normal scientific advance. It was based on my perception of the great diachronic difference between the decade-long lag in appreciation Avery's result and the virtually instantaneous appreciation of the DNA double helix. Thus, on this comparative scale of lags in appreciation, Zinder's account of his discovery of the virus-mediated transfer of genetic information represents a case of a moderate lag, one intermediate between Avery's and Watson and Crick's.

WHAT DID MENDEL ACTUALLY DISCOVER?

In the 1972 version of my paper reprinted here I wrote that "probably the most famous case of prematurity in the history of biology is that of Gregor Mendel, whose discovery of the particulate nature of heredity in 1865 had to wait 35 years before it was 'rediscovered' at the turn of the century." In his discussion of my interpretation of Mendel's case, Holmes states correctly that, in his 1865 paper, Mendel did not mention the particulate nature of heredity, and that "it requires considerable hindsight to infer these conceptions from discussions in Mendel's paper." Both of these propositions, albeit true, strike me as unwarranted criticism (inspired by Robert Olby as well, as Holmes acknowledges).

As for Holmes's second proposition, its obviously disapproving intent implies his surprising denial of the epistemological commonplace that looking back on the past by means of hindsight is part of the historian's solution, not of his problem. And while Holmes's first proposition is literally true, in that Mendel did not mention the particulate nature of heredity explicitly, he did so implicitly. As an actual reading of Mendel's paper readily shows, he mentioned two future central concepts of genetics: "traits" (also known as "phenotypes") and "formative elements" (also known as "genes"). It seems most unlikely that Mendel, who was trained both in the physical sciences and in the litany of the Roman Church, had any meaning of "element" in mind other than that derived etymologically from the Latin word *elementum*, namely, one of the irreducible parts of which all matter is composed.

Here are some of the things Mendel said about "elements." My quotations are taken from the English translation of Mendel's papers edited by Curt Stern and Eva Sherwood.

> It is presumably beyond doubt that in *Pisum* a complete union of formative elements from both fertilizing cells has to take place for the formation of a new embryo. How else could one explain that both parental [traits] recur in equal numbers and with all their characteristics in the offspring of hybrids?
>
> This development [of hybrids] proceeds in accord with a constant law based on the material composition and arrangement of the elements that attained a viable union in the cell.
>
> In those hybrids whose offspring are variable [in their traits] a compromise takes place between the differing elements of the germinal and the pollen cell great enough to permit formation of a cell that becomes the basis for the hybrid. However, this balance between the antagonistic elements is only temporary and does not extend beyond the life of the hybrid plant. . . . In this manner the production of as many kinds of germinal and pollen cells would be possible as there are combina-

tions of potentially formative elements." [In his paper, Mendel supports this proposition with statistical calculations whose validity depends critically on his assumption that the formative elements are particulate and distributed at random over the daughter cells.]

It requires only minimal (terminological) hindsight for a contemporary geneticist to infer that Mendel's conception of "formative elements" corresponds fairly closely to the particulate hereditary determinants that geneticists would later call "genes" and which implement the expression of the distinct traits that geneticists would later call "phenotypes."

PRAGMATIC UTILITY CAN REPLACE CONNECTABILITY TO CANONICAL KNOWLEDGE

I regret that I did not mention in my paper this very important proposition set forth by Ernest Hook in his contribution. I suppose that at the time I did not know about the medical cases adduced by him here, but I certainly did know about Wilhelm Röntgen's x rays. Moreover, I learned about the ultimate theoretical validation of the long-ridiculed therapeutic use of leeches when I started to use them as my experimental material the very year in which my article appeared. Yet there will always remain the epistemological problem of how, absent a theoretical nexus to canonical knowledge, one can go about deciding whether a medical discovery actually does work. I suppose that is why acupuncture is still not accepted as a genuine therapy in the Western medical community, despite massive statistical data gathered in China in its support.

IS THE CONCEPT OF PREMATURITY WHIGGISM?

In his "Coda on Prematurity," Nathaniel Comfort designates the prematurity concept as Whiggism. According to Herbert Butterfield, who in 1931 gave the term its now commonly accepted meaning, "Whiggism" refers to making the present the absolute judge of past controversies and the sole criterion for the selection of episodes of historical importance. With Joseph Agassi's lambasting of Whiggism in the 1960s, it became a term of facile abuse, such as "Fascism" or "reductionism." And so in polite society, people came to speak of "presentism" rather than of the odious "Whiggism."

As Hetherington points out in his chapter, "the charge of Whiggism . . . is a potent but indiscriminate club, one too quickly raised by contextualists and prigs against those who would use their own experience in science to help understand and empathize with the intellectual state of past researchers." In fact, if there were any connection between Whiggism and the prematurity concept, it would be that prematurity happens to be a case of "reverse Whiggism." For the prematurity concept makes the past (i.e., my own past experience in science) the absolute judge of present controversies (e.g., the reasons for the long-delayed appreciation of Avery's discovery).

CODA

Finally, I want to make clear that I prepared the preceding critical remarks only because the editor pressed me to do so.[1] They must not be taken as representative of my overall view, since I enthusiastically endorse by far the larger part of the matter presented in the essays included in this volume.[2]

1. This is not to imply, of course, that he agrees with them.

2. I express my gratitude to my Berkeley colleague, Ernest B. Hook, for organizing and securing financial support for a conference devoted to a reconsideration and clarification (not to say exhumation) of a concept I developed such a long time ago. Fortunately, Professor Hook ignored my advice—that he forget about holding this conference—when he first proposed the project to me. I warned him that he would have trouble getting anyone to cross the San Francisco Bay Bridge to attend the event he was planning in Berkeley. And the possibility that he would be able to persuade a bevy of distinguished historians, sociologists, and philosophers of science to fly in from distant points in Canada and the United States to present major papers on the prematurity concept seemed to me highly implausible. My glum prognosis was completely mistaken.

CHAPTER TWENTY-FIVE

Extensions and Complexities

In Defense of Prematurity in Scientific Discovery

Ernest B. Hook

Gunther Stent's essays on prematurity in scientific discovery have stimulated such a range of comments and viewpoints that one might regard his papers as analogous to a Rorschach test for those working in the natural sciences or their metastudies. Part of the interest of these responses lies in their self-reflective quality. They indicate how individuals highly trained in at least one discipline, be it scientific or metascientific, react to the "stimulus." But, in light of the responses, rather than develop this theme, I think it more important to attempt some further defense of Stent's core notion and its utility.

TERMINOLOGICAL ISSUES

"Premature" in Stent's Sense Does Not Mean "Ahead of Its Time"

Clearly, use of the term "prematurity" brings with it some associated freight. This has led to some discussion in this volume implicitly or explicitly at cross purposes with the specific technical definition Stent suggested. Michael Ruse, for instance, in discussing the "prematurity of Darwin's theory of natural selection," prefers to use the term "premature" in close accord with its ordinary-language sense, "ahead of its time."[1] He proposes four reasons why he believes that "in major respects natural selection, even in Darwin's *Origin,* was a premature idea." These are: (1) conceptually biology was "not ready" for natural selection, in that biologists then were more interested in describing homologies than investigating or interpreting adaptive complexity; (2) the quality and adequacy of the concept of natural selection (as then perceived) appeared inadequate to explain evolution because the mechanisms of transmission were not well understood; (3) there was no vigorous

---

1. See Ruse, chapter 15 in this volume.

champion to push the concept; and (4) the concept lacked "cash value," that is, there were no professional niches for work in natural selection.

Ruse argues persuasively that each factor "delayed" or impeded acceptance of natural selection as at least an important evolutionary force by a significant number of professional biologists: that is, that for each of these reasons, when Darwin published the *Origin,* the concept's time had not yet come, to paraphrase Ruse. But only one factor he mentions—the second, which implies difficulty in understanding how any trait selected "naturally" in a single individual could spread into future generations without being diluted in each transmission—appears to be even a candidate for classifying natural selection as premature in Stent's sense. Ruse does not propose explicitly any of the reasons he mentions as such a candidate. But his discussion usefully illustrates the distinctiveness of Stent's formulation, which applies at most to one among the many meanings included in the common understanding of the term "premature."

Terminology: Implications for Charges of Whiggism and Concerns about Hindsight

I had hoped that what struck me as modest, primarily terminological, alterations I proposed in Stent's original formulation of prematurity would help to deflect charges of Whiggism and concerns with hindsight.[2] Nevertheless, some still perceive these as major difficulties for that formulation. Perhaps I insufficiently emphasized my proposed alterations. In any event, consider Elihu Gerson's statement that the "notion" of prematurity does not "stand up": "In Stent's view, a discovery is here-and-now premature if there is no way of carrying out the inference steps that will connect it to canonical knowledge. But we cannot decide if a discovery is premature, wrongly posed, or simply irrelevant without the after-the-fact knowledge of whether or not a suitable interpretive apparatus has been developed and applied successfully."

If in passages where Stent addresses prematurity one replaces the term "discovery" with "proposal," "claim," "suggestion," "hypothesis," "interpretation," or a related term that implies no knowledge or judgment as to the subsequent fate of whatever is proposed, then, I maintain, this problem disappears. Stent's paper—in particular his discussion of here-and-now prematurity and examples of extrasensory perception and macromolecular transfer of memory—implies that he *intended* the term "here-and-now prematurity" to apply to a claim or hypothesis or proposal *whatever* its subsequent career, that is, whether it is eventually discarded or accepted.

"Then-and-There Prematurity"

These considerations suggest it is useful, in defending Stent's notion, to have also a term that clearly and unambiguously denotes a past claim premature at the time

2. See Hook, chap. 1 in this volume.

proposed, whatever its subsequent fate. To extend his usage, I suggest "then-and-there prematurity." A claim or hypothesis then-and-there premature at some time and place in the past may be, at present, either integrated into part of a currently recognized discovery—endorsed by some but still considered disconnected by the rest of the community and thus (for them) still here-and-now premature—or dropped, that is, embraced by no one. I find this notion implicit within Stent's discussion of the prematurity of past episodes and think making it explicit may help avoid present misunderstandings. The term emphasizes the contexts of prematurity in some presumed time, place, and/or scientific community. It also enables a broadening of the concept as discussed below. It facilitates viewing a claim at some time as (then-and-there) premature to one individual (or group of individuals) but not to another.

AN EXPANSION OF PREMATURITY

Individuals within any scientific community may disagree about what is connectable to canonical knowledge. And the preponderance of individuals in one community may disagree with those in another. Such variation justifies an approach that may classify a claim as appearing then-and-there premature to some individuals or communities but not others. From this perspective, for instance, the hypothesis that DNA was the biochemical substrate of heredity could have been (then-and-there) premature to almost all members of the community of geneticists in the 1940s, as Stent suggests, but not premature, as others have emphasized, to some individual biochemists and other investigators, including even some geneticists at the periphery of their field.

Indeed, this accounts for the fact that some, such as Joshua Lederberg, have contested Stent's classification of the work by Avery and colleagues as premature. These and other reasons discussed below appear to confirm the utility of discussing prematurity and canonicity from the perspective of individuals as well as that of a scientific community. Gonzalo Munévar endorses this view.[3] I suspect that here and above we deviate from Stent's intention in proposing that one entertain for some purposes also a more variable, relativistic psychological construction of prematurity, separate from a social one (defined by connection to generally accepted knowledge) that gives less weight to psychological variability.

Certainly Stent's intended conception of prematurity appears more homogeneous and monolithic than this proposed extension. But I think the key aspect of Stent's concept remains in this reformulation: the failure of any individual member of a scientific community to follow up some claim or hypothesis may be explained, under some circumstances, by his or her failure to connect it by "simple logical steps" to a body of knowledge that she or he regards as canonical. That is, to him or her it is (or was) premature. And it may (or may not) also have been re-

3. See Munévar, chapter 23 in this volume.

garded as premature by all or almost all members of a scientific community, as in Stent's original formulation.

STRUCTURALISM AND PREMATURITY

In the first part of his essay, Stent discusses canonical knowledge as a purely social phenomenon, that is, as "generally accepted" knowledge. At the very end of his paper, after a lengthy excursion into the "uniqueness" of discovery—a section not reprinted in this volume—Stent returns to discuss canonical knowledge from a psychological perspective, invoking a structuralist approach and citing Jean Piaget and the work of neurophysiologists of vision.

Canonical knowledge is, from Stent's later perspective, simply the set of preexisting "strong" mental structures with which primary scientific data are made congruent. Data (and presumably claims, hypotheses, or proposals) that cannot be transformed into a structure congruent with this knowledge constitute a "dead end" and are "meaningless"—implicitly "here-and-now" premature—until a way to so transform them has been shown. "To make congruent with strong structures" then becomes the equivalent of, or at least does the work of, "to connect by a series of logical steps to canonical knowledge."

One may or may not regard this altered perspective as an advance in our understanding of the original concept. But whatever the psychological (or neurophysiological) underpinnings of canonical knowledge, and however great the interest of proposed mechanisms, I do not see these as directly relevant to the meaning and utility of prematurity as a category of historical explanation.

Elihu Gerson, a sociologist, apparently does see Stent's discussion of structuralism as pertinent in this regard, albeit flawed.[4] He criticizes Stent's structuralist comments as "highly idealistic and highly reductionist." They do not accord with experience, he states, because "the discovery process is a matter of social organization over time, of reliable results that can be incorporated in new lines of effort. The issue," he writes, "is not one of changing minds but of changing or inventing conventional practices."

But before one changes practices, one must first change one's mind! And one changes one's mind not solely because of social influences and organizational factors defining conventional practices. The vagaries of personal experience, knowledge, and awareness of logic and the scientific method—aside from immediately acting social forces—affect whether one may try a new practice and/or adopt it. Idealistic and reductionist strong structures need not be invoked to defend this latter view.

I find the thrust of this point so strong that I suspect I may have misunderstood or misread Gerson's objections, for I infer, especially from his comment about

---

4. See Gerson, chapter 19 in this volume.

"changing minds," that he regards psychological factors as irrelevant to discovery processes and, by implication, to a discussion of prematurity. In any event, by no means do I—or Stent, as I understand him—deny the importance of social factors to discovery, or deny the possibility that these may affect individuals' cognitive factors and the nature and interpretation of personal experience and practices. But clearly, there must be significant variation in cognitive and other psychological factors among individuals within any scientific community, variation relevant to understanding why one individual embraced a claim, hypothesis, or proposal later recognized as a discovery, but another rejected it.[5]

OTHER EXTENSIONS OF PREMATURITY

Ilana Löwy in chapter 20 offers a broader notion of prematurity than Stent does, albeit one not as broad as the notion "ahead of its time," the ordinary-language sense implied by Michael Ruse's discussion. She is concerned not only with connectedness to what is approximated by "generally accepted" knowledge but also with incompatibilities between a scientific claim, hypothesis, or proposal and three elements of the "thought style" of a scientific community, as conceived by Ludwig Fleck. That term, she states, "captures (imperfectly) the multilayered functioning of science as a dynamic social enterprise."

Happily, Löwy offers a typology of what she has in mind and some criteria, which I reorder here to facilitate discussion. A proposal is premature, in her sense, because (1) it, in essence, meets Stent's criterion, or (2) the ways of evaluating new evidence are unacceptable to the reigning thought style, or (3) the questions posed or methods used may not be viewed as legitimate within the accepted thought style, or (4) within the existing thought style there is no apparent way to use or extend available methods to follow up on the proposal.

The second criterion, I believe, is subsumed within the first. Part of existing knowledge taken for granted is knowledge about knowledge, and pertains to methods and evaluation of evidence offered to extend knowledge.

Regarding the third criterion, Löwy's examples indicate that she intends "not legitimate" to mean, simply, "uninteresting" or "unimportant." For reasons I will develop, it is misleading to designate uninteresting or unimportant claims, hypotheses, or proposals as in any sense illegitimate.

Certainly, lack of interest in, or the perceived unimportance of, a scientific claim, hypothesis, or proposal later recognized as a discovery has long been acknowledged as grounds for delay in its acceptance. But citing this as grounds for terming a scientific proposal "not legitimate" within the accepted thought style is a separate matter. "Illegitimate" implies a far stronger barrier to acceptance than does "uninter-

---

5. Differences in personal "psychic capital" invested in some parts of the canon are likely the most significant source of this variation.

esting" or "unimportant." It implies that something is *wrong* with the work, for either logical-methodologic or social-ethical reasons.

Employing a term in a new technical sense, beyond or more constrained than its ordinary-language sense, may serve a useful purpose, as does Stent's use of "premature." But by long-sanctioned usage, "illegitimate" denotes objection on ethical or methodological grounds. I see no utility in employing the word in a markedly new sense (to mean "uninteresting") and then making it yet another referent of "premature."

Löwy's remaining proposed criterion for designating a discovery as premature is investigators' inability to "do" anything with a claim, hypothesis, or proposal offered by another. This is analogous to but separate from viewing the work as unimportant. This criterion might be termed "unfruitfulness." It comes closest to the category "no immediate relevance," one of the grounds for rejection mentioned in chapter 1, in that one type of irrelevance arises from perceived likely unfruitfulness. Löwy cites as an example the notion of biological individuality developed around 1910. Even though the notion lay fallow, presumably because contemporaries saw no obvious means to follow up on it, at the time one could nevertheless connect the idea conceptually to generally accepted knowledge. It was, therefore, not premature in the Stentian sense.

Certainly Stent implies, in his introductory remarks about the work of Avery and colleagues on DNA, that he had something like Löwy's last criterion in mind when he developed his thinking about prematurity. As he says, "My *prima facie* reason for considering Avery's discovery premature is that it was not appreciated in its day. . . . By [that] I do not mean . . . [it] went unnoticed, or . . . was not considered important, [but] that . . . no one seemed to be able to do much with it." But whereas David Hull, as I understand him, reads this passage as indicating Stent here offers an implicit operational or sociological criterion for lack of connectability to canonical knowledge, I read the passage simply as an exegesis of how Stent's idea evolved—how he got to the final notion—not as offering an operational equivalent of his formal definition of prematurity.[6]

Is unfruitfulness equivalent to Stent's definition of prematurity? Martin Jones implies it is. He contends, in essence, that when an announced claim, hypothesis, or proposal is "genuinely unconnectable"—and, therefore, premature in Stent's sense—then it "might equally well be called an unfruitful one."[7] Löwy's example of biological individuality, however, illustrates how a proposal may be unfruitful yet not premature in the Stentian sense. But is the opposite possible? Can there be a premature claim, hypothesis, or proposal that is nevertheless fruitful?

6. See Stent, chapter 2, and Hull, chapter 22, this volume. After writing this chapter, I asked Stent precisely what he meant by "my *prima facie* reason for considering [some] discovery premature is that it was not appreciated in its day." All he meant, he said, was that lack of appreciation made such a discovery a *candidate* for consideration as premature.

7. See Jones, chapter 21 in this volume.

I argue yes, for two reasons. One is trivial. An investigator trying to disprove the results offered in support of a claim viewed as premature in the Stentian sense may find an unanticipated, productive new direction, or may even find evidence confirming the heretical notion. More substantively, consider the hypothesis nested in the implications of the paper by Avery and colleagues: that DNA was an informational molecule, perhaps even the informational molecule of genetics. If premature, this hypothesis was nevertheless fruitful for Erwin Chargaff.[8] And it was more fruitful still for James Watson and Francis Crick![9] These investigators may have previously accepted the general canonical view that protein was the informational molecule, but this was not sufficient to deter them from their "mad pursuit," as Crick titled it.

A premature claim, hypothesis, or proposal in Stent's sense can therefore be fruitful. But I suspect such cases are very rare and only occur when the investigator's pertinent canonical knowledge is not highly structured or deeply ingrained. In such a case, she or he will be less shackled by conventional beliefs and more willing to take the risks inherent in pursuing a claim, hypothesis, or proposal that may provide an important new connection to, or even significantly change, the canon. And such a case is more likely to occur among gifted investigators who focus more intensely on a problem than the rest of the scientific community does, as illustrated by Glenn Seaborg's and Charles Townes's accounts of their own discoveries, in chapters 3 and 4 of this volume, respectively.

David Hull prefers what I have termed "unfruitfulness" as a criterion for prematurity, instead of the criterion of connectability by "simple logical steps," because the latter is so complicated. But if these ideas are different, as I maintain, why abandon Stent's usefully distinctive notion because it is more complex? If we term some past scientific proposal "unfruitful," that simply tells us it went nowhere because no one could do anything with it. It does not tell us why. If we call it (thenand-there) premature, we advance an interesting hypothesis as to why. Evaluation of that hypothesis will expand our understanding of the historical reception of the proposal.

DELAY IN DISCOVERY

In this volume Frederic Holmes implies that the ideas of delay, prematurity and postmaturity imply—inappropriately and unhistorically, as I understand him—a normal rate of scientific advance and a "normal lapse of time between the first presentation of a discovery and its assimilation into a field."[10] Yet there are useful and important questions one can ask about historical events that clearly and un-

8. See Chargaff 1978, pp. 82–89.

9. See Watson 1980, pp. 12, 18, and Crick 1988, pp. 36–38, for comments on the influence of the report by Avery and colleagues on their own work.

10. Holmes, chapter 12 in this volume.

problematically imply some sort of delay without invoking any "normal" rate of scientific advance. For instance, why was nitrous oxide not taken up for inhalation anesthesia in 1800 when Humphrey Davy first published the idea? Or why was nuclear fission, as it was later termed, not pursued when Ida Noddack first proposed the concept in 1934? Similar kinds of queries are implied by historians' references to findings or hypotheses integrated "belatedly" into mainstream science, as Ilana Löwy terms it in discussing her own examples.

Suppose some interval of significant magnitude exists between the time a hypothesis or claim is first proposed to a scientific community and the time it is generally accepted by that community. It appears reasonable to believe some sort of delay has occurred and to view the extent of the delay as the length of that interval. Merely to talk about such an interval for any particular episode, one must be able to specify at least roughly its endpoints, and that what was first suggested was in fact what was later accepted. The interval between the recognition of nitrous oxide's utility as an inhalation anesthetic and its use for that purpose began with Davy's proposal in 1800. The end of this interval lies sometime between its first use by Horace Wells, around December 1844, and widespread recognition of inhalation anesthesia about two years later.[11] In the case of nuclear fission, the interval—which lies between Ida Noddack's second paper in 1934 and the Hahn-Strassmann-Meitner-Frisch work in December 1938 and January 1939—lasted roughly five years.[12] Many of the questions one may ask about such a temporal interval imply at least a perception of some type of delay, and one that begs for a close look at the episode in a search for explanation.

Certainly, on closer examination, one may determine that a discrepancy exists between what was actually first proposed at the time, that is, at the apparent beginning of the interval, and what was later retroactively thrust upon it conceptually, so to speak.[13] Moreover, one may not be able to provide any beginning date

11. See, for example, Bergman 1998, pp. 272–82. Actually, one might claim the delay was even longer, because, although the use of nitrous oxide led to the widespread use of ether in 1846, nitrous oxide itself was not widely adopted as an anesthetic outside Hartford, Connecticut, until many years later. But there is a more important sense in which this example may have to be qualified. Davy wrote, "As nitrous oxide in its extensive operation appears capable of destroying physical pain, it may probably be used with advantage during surgical operations in which no great effusion of blood takes place." To our modern eyes, this appears to be a suggestion of our current understanding of inhalation anesthesia. But the qualification "in which no great effusion of blood takes place" has never been satisfactorily explained. Bergman in fact suggests that Davy was not concerned about mitigating pain per se, that is, alleviating the suffering of the patient, but only, in accord with the then-popular Brunonian principles of medicine (formulated by John Brown [1735–88]), about mitigating whatever adverse consequences pain might have on recovery (pp. 279–82). Davy's comment may well be an example in which our current perspective thrusts upon some past proposal an implication not intended by the proponent or inferred by his or her contemporaries. In that case, of course, it was not (then-and-there) premature.

12. See Hook, chapter 10 in this volume.

13. For example, the question of whether Mendel proposed discrete separable inherited units. See Holmes, chapter 12 in this volume, for elaboration and references.

for the process. Under these or other circumstances, the extent of and even the existence of delay becomes problematic. But if, with the proposed criterion, one judges that acceptance of a claim, hypothesis, or proposal has been delayed, then it appears reasonable to ask why, from a current perspective, such was integrated belatedly into mainstream science. And prematurity in Stent's sense, from this viewpoint, is simply one of many potential causes of delay. In the case of nuclear fission, prematurity was one contributor. But prematurity was not responsible for the delay in the introduction of nitrous oxide for anesthesia.

PREMATURITY AS A NONSTARTER IN THE SOCIAL SCIENCES

George Von der Muhll's discussion of potential prematurity in the field designated as political science appears strongly influenced by Thomas Kuhn's formulation of paradigms, although he does not acknowledge this explicitly. No one generally accepted paradigm exists in political science, but there have been and are contenders. Von der Muhll implies that, in the case of the entire community of political scientists—unlike the natural sciences community—the field has not reached the stage at which claims, hypotheses, proposals, and so on can be premature in Stent's sense. Kuhn, he implies, would label it "pre-paradigmatic."[14]

But there is a sense in which Stent's formulation can still be useful here, whereas Kuhn's remains barren. Many political scientists, Von der Muhll informs us, have strongly embraced one theory, that of rational action. Indeed, in one case a whole department has adopted it as an organizing focus for research. Therefore, there are presumably doctrines, claims, hypotheses, or proposals currently rejected by the many contemporary adherents of that school because they cannot connect them by a series of logical steps to their rational-action canon. Admittedly, I have not found any concrete examples of this claim, although I have yet to search extensively. And certainly the term "here-and-now premature" may not be useful to a rational-action political scientist to indicate this lack of connectedness. But the concept appears to be identical to at least the extension of Stent's concept that I have sketched above regarding the natural sciences.

Moreover, to say that no widely accepted organizing theory (or theories) exists in political science—that is, that there is not, in Kuhn's sense, some unifying theoretical paradigm—does not exclude some collection of empirical evidence or descriptive material, at least of a recent historical nature, on which there is broad agreement among those in the field and which constitutes canonical knowledge for almost all, however unsophisticated and undeveloped such knowledge may appear to be to those seeking grand theories. There must be some knowledge on which political scientists in any university department generally agree, if only so they can

14. See Von der Muhll, chapter 17 in this volume.

teach an introductory course and adopt a textbook for it. I suspect, moreover, that most of my impressions and generalizations about political science could be extended to the other social sciences. In any event, these considerations indicate that, even in a field labeled "pre-paradigmatic" by some of its practitioners, the notion of prematurity may find application.

THE UTILITY OF STENT'S FORMULATION
The Example of Extrasensory Perception

Stent's discussion of alleged extrasensory perception (ESP) I found useful in thinking about this controversy. An explanation of why may provide a concrete example of one way his formulation is useful whereas Thomas Kuhn's is not. I rejected (and reject) the evidence for ESP out of hand, but I felt unease about doing so. I regarded myself as an open-minded individual with an appropriate intuitive understanding of something termed the "scientific method." I recognized implicitly that my attitude to ESP appeared inconsistent with this. Moreover, I knew of apparently highly reputable psychologists, statisticians, and other scientists, as well as philosophers of science, who, having examined more of the claimed evidence for extrasensory perception than I, endorsed, or at least took some aspects of, the phenomena seriously or otherwise believed in them. This reinforced my unease.[15]

For these reasons, I was not able to confront the perplexity and disquiet I felt about rejecting the evidence offered for ESP. Thinking about ESP as a paradigm, or at least thinking that it implied one different from or incommensurable with that which I held on matters in neuroscience, provided no useful insight or fruitful implication. Stent's characterization of the claims as here-and-now premature, however, provided me with a helpful heuristic framework.[16] And his approach has admirable social utility in implying mechanisms for mediating disagreements about ESP and similar issues involving controversy. Where individuals differ about whether a proposal may be connected to a canon, Stent provides a neutral bridge to a useful dialogue between skeptics and believers.

His formulation implies that supporters of a claim that appears premature to a scientific community should address explicitly (and indeed focus their efforts on) those aspects of the skeptics' canon that proponents must alter to get the critics to take the alleged phenomenon seriously. This implication is precisely where I find Stent's notion preferable to the implications of Kuhn's on different paradigms. Stent

---

15. For discussion by a philosopher of science that implies one should take the evidence for the claimed phenomena seriously, see, for example, Scriven 1964. Jessica Utts (1991) offers statistical support. Brian Josephson, a Nobel Prize–winning physicist, believes the known laws of physics are compatible with some of the claims (Josephson and Pallikari-Viras 1991; Josephson 1992).

16. I do *not* intend to imply by this that the social and science-policy value of the term "here-and-now premature" applied to current claims is as a terminological or psychological euphemism for nonsense!

implies a basis for dialogue between those on different sides and a useful target for the research program of a beleaguered minority whose claims are outside the canon. If Kuhn's formulation of incommensurable paradigms implies something as useful, I do not see it.

As Stent implies with regard to extrasensory perception, those who continue examining the wrinkles of the same experiments, and their sympathetic statistician colleagues who report on the infinitesimal probabilities that the data result from chance, will not convince the skeptics. But one of three types of findings might lead us to alter our skepticism, the first two of which are clearly also implied by Stent: (1) the finding of some hitherto unsuspected brain structure apparently capable of signaling or receiving the necessary information, (2) the discovery of some new, pertinent principle of physics, or (3) the demonstration of some clear economical and/or practical utility of the concept—that is, that it "works" in some significant sense, even if one cannot explain why.[17] I exclude from the latter the cash value of an entertainment performance because I would suspect the latter to be fraudulent.

The Utility of Prematurity to History

A New Social Role for Historians

One virtue of Stent's formulation is that it creates a conceptual bridge from past controversies and disagreements to present ones. It reinforces the value of, indeed the need for, history. By understanding the resolution of past examples of then-and-there prematurity, we are in a better position to deal with episodes of here-and-now prematurity, some of which may involve great controversy. And this provides an additional social role for historians of science and historically oriented philosophers of science. Who else but those familiar with the fine grain of past episodes of then-and-there prematurity and their resolution could more usefully moderate and facilitate such interactive discussion about present controversies? The slender list of possible "applied" activities of the historian of science[18] might well be expanded to include such a contribution.

A Distinctive Approach to Historical Analysis of Scientific Developments

Can one defend as well the utility of the notion of (then-and-there) prematurity in historical analysis—that is, for the usual professional work of historians of science? My comments above in the section on delay sketch the outlines of such a defense. Nathaniel Comfort, however, suggests its main utility is rather as a foil and target, a red flag that, when unfurled, challenges historians to reject the hypothesis of prematurity and "place a discovery in the context of its time, place, and contemporary ideas."[19] Certainly, the initial impression of prematurity should serve as a stim-

17. By "works" I mean it is employed consistently to achieve a non-trivial social or economic outcome.
18. Heilbron 1987.
19. Comfort, chapter 13 in this volume.

ulus to a closer examination of the episode. But *any* historical explanation of why a scientific proposal appears, from a later vantage point, to have been belatedly integrated into mainstream scientific knowledge must be placed in the same context. At the end of that process, in any particular case, I maintain, the concept of prematurity as formulated by Gunther Stent will provide one useful and distinctive category of historical explanation.

BIBLIOGRAPHY

Bergman, N. A. 1998. *The Genesis of Surgical Anesthesia.* Park Ridge, Ill.: Wood Library–Museum of Anesthesiology.

Chargaff, E. 1978. *Heraclitian Fire: Sketches from a Life before Nature.* New York: Rockefeller University Press.

Crick, F. 1988. *What Mad Pursuit: A Personal View of Scientific Discovery.* New York: Basic Books.

Heilbron, J. 1987. "Applied History of Science." *ISIS* 78:559–63.

Josephson, B. D. 1992. Letter. *Physics Today* 45:15.

Josephson, B. D., and F. Pallikari-Viras. 1991. "Biological Utilisation of Quantum Nonlocality." *Foundations of Physics* 21:197–207.

Scriven, M. 1964. "The Frontiers of Psychology: Psychoanalysis and Parapsychology." In *Frontiers of Science and Philosophy,* ed. R. G. Colodny, pp. 95–106. London: George Allen and Unwin.

Utts, J. 1991. "Replication and Meta-Analysis in Parapsychology." *Statistical Science* 6:363–403.

Watson, J. D. [1968] 1980. *The Double Helix: A Personal Account of the Discovery of the Structure of DNA.* Ed. G. S. Stent. New York: W. W. Norton.

Compositor:	Impressions Book and Journal Services, Inc.
Text:	10/12 Monotype Baskerville
Display:	Baskerville
Printer and binder:	Edwards Brothers, Inc.